THE PLANNING MOMENT

The Planning Moment

COLONIAL AND POSTCOLONIAL HISTORIES

Sarah Blacker
Emily Brownell
Anindita Nag
Martina Schlünder
Sarah Van Beurden
Helen Verran
EDITORS

FORDHAM UNIVERSITY PRESS   NEW YORK   2024

Copyright © 2024 Fordham University Press

All rights reserved. No part of this publication may be reproduced, stored in a retrieval system, or transmitted in any form or by any means—electronic, mechanical, photocopy, recording, or any other—except for brief quotations in printed reviews, without the prior permission of the publisher.

Fordham University Press has no responsibility for the persistence or accuracy of URLs for external or third-party Internet websites referred to in this publication and does not guarantee that any content on such websites is, or will remain, accurate or appropriate.

Fordham University Press also publishes its books in a variety of electronic formats. Some content that appears in print may not be available in electronic books.

Visit us online at www.fordhampress.com.

Library of Congress Cataloging-in-Publication Data available online at https://catalog.loc.gov.

Printed in the United States of America

26 25 24   5 4 3 2 1

First edition

Contents

FOREWORD
by Dagmar Schäfer   ix

Entanglements of Colonial and Postcolonial Planning:
An Introduction   1

Census: New Hebrides/Vanuatu, 1967
Alexandra Widmer   20

Charcoal: Dar es Salaam, Tanzania, 1973
Emily Brownell   29

COBOL: The Pentagon, United States of America, 1959
Benjamin Allen   37

Computing: United States of America, 1949
Benjamin Peters   46

Constitution: India, 1950
Monika Kirloskar-Steinbach   56

Dam: South Korea, 1961
Aaron S. Moore   64

Dodecahedral Silo: Spain, 1953
Lino Camprubí   76

EMES Sonochron: Federal Republic of Germany, 1986
Martina Schlünder   84

Famine: India, 1877
Anindita Nag 96

Fertility Survey Workforce: Puerto Rico, 1949
Raúl Necochea López 104

Fertilizer: South Korea, 1952
John DiMoia 113

Grid: New York, United States of America, 1972
Robert J. Kett 124

Hackathon: India, 2012
Lilly Irani 133

Kishikishi: Belgian Congo, 1956
Sarah Van Beurden 144

Land Parcel: Lebanon, 1990
Mona Fawaz and Nada Moumtaz 152

National Budget: Sudan, 1946
Alden Young 160

Orangutans: Borneo, 1962
Juno Salazar Parreñas 168

Parasite: Liberia, 1926
Gregg Mitman 176

Riverbed: South Korea, 2008
Chihyung Jeon 186

Seeds: German East Africa, 1892
Tahani Nadim 195

Steel Plant: Orissa State, India, 1955
Itty Abraham 204

Surnames: Brazil, 1979
Ana Carolina Vimieiro Gomes 212

Taxonomer: United States of America, 1923
Laura J. Mitchell 220

Treasures: Palestine/Israel, 1979
Tamar Novick 230

Water Samples: Treaty 8 Territory, Canada, 2012
Sarah Blacker 235

Weeds: Laos, 2006
Karen McAllister 245

Zoomorphic Wickerwork Figure: Australian Administered British New Guinea, 1908
Helen Verran 254

The Planning Moment: Avenues for Analysis 265

ACKNOWLEDGMENTS 275

ARCHIVAL SOURCES 277

BIBLIOGRAPHY 279

LIST OF CONTRIBUTORS 311

INDEX 315

Foreword
Dagmar Schäfer

"[...] a bee puts to shame many an architect in the construction of her cells. But what distinguishes the worst architect from the best of bees is this, that the architect raises his structure in imagination before he erects it in reality. At the end of every labour-process, we get a result that already existed in the imagination of the labourer at its commencement"
(KARL MARX, 1887, *CAPITAL: A CRITIQUE OF POLITICAL ECONOMY*, VOLUME 1, BOOK ONE: THE PROCESS OF PRODUCTION OF CAPITAL, PAGE 127).

Comparing humans to bees, Karl Marx observed that humans act upon their thinking. The ways in which we act upon our thoughts is apparent in everyday life—we organize our days, leisure, and work activities considering which socks to wear and which pen to use, which laboratory experiment is pursued. We make plans on a daily basis. We are gathering, using, and generating knowledge through actions, with the aim of making things work. While planning we rely on previously made plans and the physical and intellectual orders that they created – especially technologies such as washing machines and dryers, ink and paper, housing, infrastructures of roads or thinking, such as chemical taxonomies. In short, planning is not only a ubiquitous human activity; it is also inherently technical in nature, as Marx thought. He was wrong about two major things, though: plans rarely play out as envisioned by other humans, and also, other-than-humans have plans, too.

The working group "Decolonial Planning" was a formative project in the early years of "Artifacts, Action, Knowledge," Department III at the Max Planck Institute for the History of Science, Berlin, and this book offers a panoramic view of the methodological opportunities that can arise from such research. The contributions elucidate the multiple and diverse histories of colonial and postcolonial worlds and reflect on planning as a technology of collective action and knowledge making.

The motivation for forming a group focusing on "scale and scope" within the histories of planning was to shift focus from concepts to practice and processes. This research turns away from the (over-researched) eureka moments of science and technology; instead drawing attention to the everyday, material, and often quite ambiguous character of producing, using, and implementing knowledge orders. As a ubiquitous human activity, planning allowed us to address the many biases that inform our understanding of knowledge production. On the most basic level, nowadays knowledge production is closely associated with creativity and chance, whereas it is in fact more often than not the outcome of human efforts of careful planning, such as organizing family life to be able to work; of setting up the laboratory to pursue one's experiments; of educating the next generation over a decade to follow suit. Planning means to investigate the past and resource the present. It means to imagine a future and anticipate it. This is why planning is also substantially concerned with producing and using knowledge.

Historically, we have an extraordinarily continuous documentation of the many ways people planned: vessels excavated from early times were used as storage containers to preserve food over winter times, star maps were used to navigate the lands or sea, or the documents of past bureaucracies that reflect the messy realities of life and death pressed into the grand visions of contemporary elites. Planning produces the empiricism through which we learn, live, and survive. Planning relates knowledge (or its objects) to power along a temporal trajectory. A bag of rice, given away for free, has to be produced as surplus and transported to places of need as people discuss who and how one should plan and which knowledge, or information needs to be documented, conveyed, or systematized. Is thinking on a larger scale better than tending to details? Should all be centrally organized? How could flexibility be achieved and creativity promoted and control still be maintained?

Objects are subjected to planning powers and textual accounts unfold debates over planning kinds. In the historical Chinese world people hotly debated over types of planning. In eleventh-century Song (960–1279) China, the renowned Chinese philosopher Zhu Xi (1130–1200), for instance, assumed that the key to the success of large schemes was to bring order to the small things—that is, everyday needs. For him, the proper placing of the ancestral shrine in each individual's home was a first step towards organizing society and state. The principle of big planning was to understand the major effects that could result from small details. His *Guidelines of Family Rituals*, published ca. 1169, thus instructs on the ideal location of an ancestral shrine in each individual's home. A century earlier men like Wang Anshi (1021–1086) had propagated a different approach in which grand setups mattered more than the small details. This was a period in which the Song state was gradually losing political control over the northern plains. The state lacked access to the traditional source of horse and cattle central to transportation and warfare. Taking a different approach to Zhu Xi, the state officials in charge opted to institutionalize offices and publication of pharmaceutical literature to promote state-run large-scale livestock holdings. Such planning was visionary, imagining grand schemes of self-subsistence and efforts over a long-durée.

A comparison of different approaches to planning in Chinese history shows that each approach to planning brought forth distinct formats and fields of knowledge and know-how. Song scholars created a field called "methods to prevent diseases or malfunctions," which, besides veterinary care and medicine, included hydraulic engineering, crop selection, and moral training, as well as the study of philology and philosophy. Studying planning hence provides an opportunity to look at the formation of disciplines, professionalism, and expertise as a fluid dynamic rather than a definitive act that produced particular sciences or technologies.

In the past as much as in the present world, planning meant juggling complex situations but also deciding whether long-term vision requires long-view hindsight, or taking a risk. Accordingly, people gathered empirical data, performed divination, or calculated measurements. And often, we can see how the shadows of yesterday's plans turn into iconic templates for the future. A good example is a bronze plate, excavated in the 1970s in Hebei that has gold and silver inlays depicting the contours of the fourth-century BC tomb of King Cuo where it was found. Engraved measurements suggest that the plate was used in construction. Inscribed along the rim, an official decree identifies the plate as part of a complex imperial administrative apparatus. It once functioned as an actual instrument of construction, yet its placement into the tomb achieved a new function as ritual device accompanying its owner into the afterlife.

Studying the histories of planning is about how goals are set, and skills and materials promoted or identified. We study when, how, and why guidelines, models, recipes, and blueprints are generated to coordinate and organize — but this research also raises questions about the logic of materialities and thought, and how these logics are stabilized by creating precedents, algorithms, or contingencies. This suggests taking a closer look at how the ontologies of objects operate in relation to the writing down of plans—theoretical considerations on knowledge making—and epistemic approaches. Thus a key component of research on planning is to develop an enhanced view of the historical mediality of knowledge production; which objects come to play; how and what kind of knowledge is written down or situated/located in artifacts, people, ritual performance or the surrounding landscape.

During the fifteenth century, the Ming state owned and operated a complex system for the manufacture of silk. One of its products from 1470 was a golden shirt excavated from a tomb in 2006. By 1470, political control lay at the capital and court in Beijing while silk was produced across the territory, mainly in Jiangnan and Sichuan. Terminological specification lay at the heart of organizational schemes. Thus the décor and colors were spelled out in the names of silks. Fifty years lie between the second garment which was excavated a couple of years later that carries the same name. But whereas the earlier version was made of silk in yellow dye, this one is made of a silk thread wrapped in gold foil. The texture changed from a single to a double twill weft. Bolts from the later tomb have a banderol which gives the names of the officials and craftsmen who produced the bolt. By that time more and more weavers were recruited on demand to satisfy the people's demand for excellent silk. Information that had previously been collected in accounting books was now attached to the product in order to ensure that the chains of tasks and responsibility remained clear. Shifts in techniques reveal changing modes of trust and responsibilities and indicate new forms of labor and production techniques.

Among the many questions that the study of planning raises, shifts in response to changes in the scale or scope of knowledge-making are crucial. Especially since capitalism and globalization, scale has mattered as a way to tell the history of science and technology: models that worked at a small scale failed once they were scaled up. Large-scale plans had to shrink; long-term processes were sped up. Sciences and technologies changed along with the scope of a theory and the scale of operations. Disciplines were formed to provide a formal structure for the management and manipulation of scale. Scale and scope play out in historical knowledge dynamics in many ways: on territorial and temporal scales, ideas of synergy, extrapolation, performance and scalability. Scholars were invited to explore contingencies, informalities, political practices, and interstices in the history of planning. While historical research may not have ignored practices, procedure, and processes, it has tended to restrict the study of planning to the implementation of planning with systems of education (formal or informal), institutional history, or human habitat construction (architecture, etc.). Planning itself was often perceived as an act of imagination that curbed creativity and spontaneity in an attempt to avert risks and make life safe.

Especially in the twentieth century, the planning of science, technology, and medicine was ostensibly divorced from social or political values. Although even then, the story was more complicated; it was through this divorce that the ideologies informing planning practices became consequential and inflicted violence, particularly as colonizing technologies. In *The Planning Moment: Colonial and Postcolonial Histories*, both the diversity and the enormous power of planning are given recognition in scholarly research.

With a range of viewpoints that are both global in scope and procedural in nature, the contributions in this book offer substantial insights into the dynamic entanglements of knowledge forms, the relationship between management and methodological varieties, and the role of systemic choices and procedural improvisations in the identification of systematic knowledge.

Dagmar Schäfer

THE PLANNING MOMENT

Entanglements of Colonial and Postcolonial Planning: An Introduction

Sarah Blacker, Emily Brownell, Anindita Nag, Martina Schlünder, Sarah Van Beurden, and Helen Verran

What is planning? Epistemically focused answers tend to be short: planning is "a link between knowledge and action,"[1] where planning is understood as the epitome of "rational choice,"[2] as an attempt at "controlling the future,"[3] or as a form of "storytelling about the future."[4] Whereas all of these short definitions might be right, or at least partially right, this book aims to complicate the answer by historicizing the question. In particular, we ask if forms of planning that emerged first in colonial and then postcolonial contexts have a specific epistemic or knowledge structure, and if so, what kind of knowledge has been involved in planning, and what kind of planning has been involved in historical practices of knowledge production?

Beginning with the planning practices of imperial expansion in the nineteenth and twentieth centuries, the plans in this book originate in colonizing technologies; yet, as an epistemic infrastructure, they are subsequently used in myriad new and sometimes oppositional ways. There is no foundational link between planning and modernity; planning as a practice of states, institutions, and individuals can be found in every epoch of history: planting a seed is, we believe, a form of planning. However, with the rise of Western ideals of science and of truth making through fact making, planning has become involved in scientific practices. Through the planning of experiments, observations, and education, planning has become entangled with science's ideal of rationality, understood as absence of affect. It is in this historical moment, particularly in colonial and postcolonial settings, that this book is situated.

The essays in this collection might be mapped onto an imaginary line drawn from the bright new futures of the post-Enlightenment to the crumbled horizons of the twenty-first century's world order. They are populated by the afterlives of colonialism and its tendency to plan as the means to organize and define populations, reshape labor and production, solidify and demarcate cultures and social groups, and remake space.[5] Yet, colonial plans—devised on paper and in theory—only become part of history through what we define as planning moments. Each of the following essays engages with a unique episode of planning and offers a temporal snapshot of a planning moment that catches plans unfolding in intended and unintended ways. These episodes highlight what Ernest Alexander noted in a recent article in *Planning Theory*: "There is no planning—only planning practices."[6] The planning moment reveals a particular historical juncture and the manner in which a plan made its appearance; its temporal particularity. A plan's contingent trajectory expresses the on-the-ground tension between two idealized extremes. On the one hand, the hubris of the assumption that the plan will be fully realized and, on the other, the fears (fully justified in the light of experience) that, subject to the contingencies of the real world, the plan will fall apart, and all impetus for change will be lost. Drawing on these planning moments helps us trace how colonial plans re-emerge as the decolonizing technologies of newly independent nation states. There, they are often refracted into counter- and sub-plans, and, in time, reworked as the epistemological infrastructures of neoliberal late capitalism.

Through focusing on materiality, power, and inequality, these essays argue for a new conception of planning which recognizes the crucial importance of artifacts and non-human bodies to unmask the so-called rationality of planning, while also indicating the unevenness and messiness planning produces in practice. In focusing on the colonial and postcolonial world, our book calls for a careful reconsideration of the dominant yet unnoticed Eurocentric epistemic frameworks that shape planning and the writing of planning histories. Such an approach helps us to show how coloniality manifests itself beyond the historical institution of colonization. Thus, this book also argues that we—as scholars trained in Western, modern academic styles of thinking and writing—need to look into forms as well as contents if we want to create new infrastructures of knowing.

This volume first emerged from a series of conversations between visiting scholars at the Max Planck Institute for the History of Science in Berlin about what exactly would constitute a history of planning. Was there a set of terms we could agree upon as fundamental to such a history? Our early conversations laid bare the fact that a common conceptual vocabulary around planning is limited to what is deployed by planning practitioners and urbanists, often contained in glossaries. In these texts, we found highly regulated and specific definitions. Indeed, the history of planning by the mid-twentieth century reveals the centrality of efforts to universalize and standardize a planning language by practitioners. For example, in 1937, the Dutch planner J. M. de Casseres published his "Principles of Planology," in the *Town Planning Review*, with a footnote stating that, although the article dealt primarily with spatial planning in Holland, "it is applicable to most countries."[7] Indeed, this was the point of "Planology." De Casseres attempted to cement planning as a "science of the spatial organization of the community" that would allow for "a place for everything and everything in its place."[8] The need for planning in recent years, according to de Casseres, had expanded rapidly from concerns about villages or roads in cities to the need for regional, national, and now global plans. "Let us hope," writes de Casseres, for the "eventual elimination of material and physical disorder and imperfection which hinder the highest human evolution."[9]

De Casseres was not the only scholar in the 1930s to propose both a new name for the science of planning and to attempt to standardize and universalize its nomenclature. The planning scholar M. R. G. Conzen, a year later in the same journal, suggested the name "Geoproscopy."[10] While in step with de Casseres' universalizing desires, Conzen also noted that the practice of planning was becoming "more and more complex" due to the "growing specialization of work and a corresponding need for co-ordination." The tensions between complexity and universalization clearly sat at the heart of the quickly expanding profession. Without a good name, Conzen worried that there would be "confusion" which would "retard the clarifying of systematic problems." Indeed, he was vexed that what was known as "town planning" in England was "city planning" in America and "Städtebau" in Germany and that these were not "correct literal translations" of each other, not to mention the French term of "urbanisme." These terms "betray their origin from practice," he wrote.[11] A new name would eschew this provincialism of planning as an art and invite "an even wider frame" because "the science of planning in its scope as well as in its name must, systematically speaking, comprise the whole of the earth's surface." In returning to our own desires to consider a common conceptual language of planning, we realized nothing could possibly sound more imperial than the ambition of planners themselves to gloss their practice in universalizing forms until it covers "the whole of the earth's surface."[12]

We realized that our own work to define a history of planning risked reproducing these problematic ambitions. And yet we also find ourselves in the midst of a small-scale publishing boom in academia of "keywords," "lexicons," "glossaries," and "abecedaries."[13] This reflects a renewed interest in concepts and a desire to find shared analytics with which to work across disciplines and to realign knowledge. These projects also capture the urgency of finding new humanist perspectives for the Anthropocene by reworking and destabilizing a form of universalist knowledge that has produced untold harm. Nevertheless, this urge towards consilience and shared conversation still carries the fraught history of standardization and canonization.

Instead, we decided to seek out histories of planning emerging from the diverse historical settings in which planning occurs and plans are made. From these initial conversations about terms and concepts, we convened anthropologists, STS scholars, historians, planning practitioners, museum curators, and philosophers for two week-long workshops. Each participant produced focused and relatively short texts rich in empirical detail. These sparked a broader conversation about distinguishing and connecting the many extant histories of planning.

Since most of our participants did not readily see themselves as scholars of planning, we first asked them to think broadly about where in their work they encountered colonial and postcolonial planning and what it might look like to decolonize the history of planning. To facilitate this, we invited workshop participants to choose a story of planning to recover and share from their own research. We asked them to anchor their stories of planning practices around an artifact—a thing—that emerged in the process of planning, rather than focusing on the planners or on state histories. We hoped that these stories would reveal new perspectives on what has many times over been neatly packed together and glossed as a rational entity—the plan. While processes of planning function through abstraction and flattening, we wanted to place the messy emergent aspects of planning back in view.

The result of those workshops is this book's twenty-seven case studies of planning, each a separate microcosm of plans and planning practices that attend to how plans function as epistemic infrastructures. Their temporal focus is on colonialism and its aftermaths, including the possibilities and limitations of decolonization. Their combined content allows us, in the book's conclusion, to consider how (post)colonial planning creates culture, infrastructure, difference, space, and time. Our aim is not to synthesize or to conclude but rather to open up further discussions, not only about the persistent knowledge infrastructures of planning but also about the ways that we, as scholars, think and rely on similar knowledge infrastructures.[14]

Alphabetical glossaries and keyword manuals are central to the practice of planning and academic knowledge production at the moment; so let us start by first considering the example of the Roman alphabet as a colonial plan. This example also demonstrates that the scope of the history of planning must be widened beyond its frequent focus on spatial and economic planning.

The Alphabet as a Colonial Plan?

In the mid-nineteenth century, a group of English colonial officials, missionaries, and scholars set about making alphabetical history by planning a grand reordering of language use in India. They sought to institute what they saw as rational language use, deploying the Roman alphabet as their tool. The plan did not become an official policy of the British, and evidently failed to achieve its aim of replacing the Perso-Arabic alphabet in Indian life. A London firm published a collection of texts promoting the plan to rationalize the alphabet of Indian languages, along with an opening interpretive essay, in book form in 1859. Edited by "Monier-Williams, M.A., of the University of Oxford: Late Professor of Sanskrit at the East India College, Haileybury," the book offers thirty-six "documents" arranged, more or less, in chronological order of publication.[15] We have chosen it here as an example of a colonial plan in part because the text uncomfortably echoes our own initial attempt to gloss colonial planning. The book presents an evangelist's efforts to capture twenty-five years of activism and lobbying by a wide variety of professionals arguing in support of their plan to alter writing of the languages of India. What, then, is the plan here? The alphabet? The efforts to implement the alphabet? Or, the work to gather such arguments together, perhaps in hopes of resurrecting efforts in another time and place?

Born in Bombay and a product of the British Empire, Monier-Williams no doubt spent much of his life and career thinking about translation in various forms. But he and his peers did not aim to merely supplant one alphabet with another one; it would be more apt to call it conversion (he also hoped to Christianize India). The plan was to intervene—to change understandings as well as the conduct of the British Empire and to aid colonial administration. As one entry in the volume, first published in the "Watchman and Wesleyan Advertiser," in 1858, notes: "The method of Romanising the written and printed languages of India is one so convenient to the Government, so useful to the student, so propitious to the native mind, so conducive to the spread of the science and ideas of the West, and so likely to be subservient to Christianity, that, though no one wishes to obliterate all the native alphabets, we hope it will, in the course of another century, gradually supersede them in popular use."[16]

The British imperial project was full of such grand plans to translate and convert—plans that variously failed, succeeded, or were transformed, co-opted, and recalibrated on the ground. This alphabet plan's heady ambitions—its focus on replacing a subcontinent's worth of scripts with one universal alphabet to aid in movement between contexts and populations — mark it as a poignant example of colonial planning. And yet, despite its grandiosity, the Monier-Williams' alphabet project rested on the pedestrian work associated with perennial contingencies of planning: translating, transliterating, printing, and dutifully distributing copies of new Christian Bibles into "Indian languages." Indeed, these more quotidian plans on which the larger aspirations of Monier-Williams and his cohort relied become the real effects of this failed plan, with their own afterlives in communities and in history. They are also the kind of tasks (plans) that we hope to bring into view in this volume, placing the utopian visions of master planners as background to the work on the ground. From our point of view, the assembled texts of Monier-Williams' book are an artifact that offer a glimpse of these colonial plans in the making.

Plans forecast certain futures, or are consolidated in retrospect, but they are also multiple other things in the present in which they unfold. In this way, we suggest that plans propose themselves, like this book and the alphabet itself, as knowledge infrastructures. We routinely delude ourselves that a plan for a future is a rational object, and further, that rationality can be done a priori, before the collective action it imagines itself as prescribing. And certainly, a volume, such as Monier-Williams' book, that gathers efforts retrospectively aims to prove the rationality of such undertakings.

Not originally designed to produce the proliferation of hierarchies that its orders have ushered into being, the Roman alphabet itself points to the ways in which plans persist, doing different kinds of work long after tasks they were designed to attend to are completed.[17] As an infrastructure of writing, alphabets offer a judicious starting point for interrogating the epistemic assumptions inscribed in the conceptual frameworks employed in studies on planning. Alphabets signify the segmentation of the world into small units and form the basis for a common foundation of knowledge with far-reaching social and political consequences. They attempt to subsume other linguistic infrastructures, or any other forms of vernacular/marginalized ways of knowing the world. And yet letters and words can also be co-opted and reappropriated, constructed into counter-planning missives. Alphabets are thus not simply meaning-making devices that provide a common foundation of knowledge; rather, they involve a complex structure of power configurations and exchanges.

The Planning Moment: Three Warp Threads

While we present here manifold histories of planning, this volume also argues that there is reason to read and narrate together certain planning histories.

The essays here are peripatetic, covering an ambitious swath of time and space, but are woven around three warp threads. Together, these threads render the "planning moment": interventions in a particular situation which work to embed particular epistemic commitments.

The first thread is that of coloniality and colonialism. Each essay is either rooted in the colonial moment or its aftermath, exploring how the academic study of planning—and the concept of planning itself—has helped to perpetuate a global imperial matrix of power, frequently by seeking to naturalize and obscure planning practices as apolitical "development."[18] The second thread used to render the planning moment is the effort to make visible the sorts of knowledge produced in and through planned intervention. The final thread running through each essay is the focus on planning at the smallest scale, through everyday objects, which we hope brings into the foreground practices that show the diverse, context-dependent, and situated nature of planning.

While plans can determine how people are placed within and outside of societies, cultures, polities, and economies, the practices by which plans are implemented, and the subjectivities they elicit, are not always within the control of those implementing plans. In fact, contrary to the hegemonic nature of the plan in James Scott's description of high modernism, there is often a decided difference between the prescriptive ethos of plans and their material and subjective effects.[19] Nor are plans solely the prerogative of states and other top-down institutions. Rarely fully hegemonic, plans tend to have multiple and at times unintentional expressions and effects on subjectivities. Beyond top-down classification and identification, planning is enacted through spatial relations, material affinities, economic conditions, systems of identifications, political conflicts, and so on. They can also be shaped in opposition to the epistemic infrastructures of (post)colonial plans, or as by-products, infra-plans, or even counter-plans. Or sometimes, (post)colonial plans simply go nowhere. Such processes, variously internally contradictory and paradoxical, are richly illustrated in this volume.

The "moment" in "planning moment" as we use it here, has both particular historical situation and particular political situation. Each of the essays fleshes out these aspects, often revealing how and where the historical and the political are entangled. The "planning moment" recognizes that these entanglements are subject to what we might think of as cultural drift. Situations, as expressions of historical and political forces, shift in particular directions at particular rates. Such shifts are partially intervened in as participants'—both colonized and colonizers—plan to exert influence. As interventions, plans or designs for interventional collective action, are informed by particular epistemic traditions, not just modern epistemics. As practiced in particular times and places, plans enacted express particular types of inferentiality and temporality.

The analytic concept of the planning moment sees each of the histories collected here as to some extent pointing to a set of tensions: the temporal particularity of a plan, and the tension between a sense that plans are always about to prevail and to fall apart. On the one hand, planners cleave to that sweet spot where success seems assured, and on the other, in covering their backs, so to say, they readily point to precarity and crisis, where failure seems inevitable. Despite failures, unexpected consequences, and hubris, plans often beget plans, both small and large. As such, the plans in this volume attest to the infectious hubris of a particular moment of colonial modernity, one that lives on, despite processes of political decolonization.

This collection of essays has its own planning moment. It is an expression of a particular historical and political situation. The book emerged from the workings of an elite European academic institution in the closing years of the second decade of the twenty-first century, a moment when many historians sought to promote decolonization in the wake of European imperial ambitions. This was in line with a focus on demonstrating and analyzing past epistemic injustice, which was increasingly evident in the academy in general.

Thread 1: Planning and Colonialism

Empires are massive planning institutions. At least in part, it was in seeking to understand, reshape, and exploit colonial landscapes and labor that the natural and social sciences of the twentieth century emerged explicitly as modes of knowledge production. As empires consolidated colonial rule, backing administrative legal structures with coercive policing and military force, they also found that legitimacy called for legibility. The paring down of entangled socio-cultural environments into a governing legibility requires what Theodore Porter calls "mechanical objectivity": the repetition of standardized procedures of measurement, demarcation, quantification, and reportage.[20] Colonial governments found in regimes of planning a reliable mode through which to reduce messy complexities to linear narratives, flattening out layers into smooth formulations and visions that could support colonial illusions of "progress." It was referred to in colonial planning literature with terms like "development," "betterment," and "efficiency," reflecting the apparent scientific and rational facets of the colonial enterprise.[21] These plans were not simply imperial because they were imported from the metropole and then executed in the margins, but because planning emerged as a product of colonial relations and evolved as the main practice of rule.[22] Colonies were the "experimental fields," or "laboratories of modernity" for testing these new forms of knowledge.[23]

The centrality of planning in the colonial world is also an apt illustration of planning's iterative nature: colonial administrators planned institutions to generate knowledge about colonial subjects and environments in order to create policy to inform plans for the future. In turn, these plans generated data that forecast future plans. For example, this took the form of research institutes that still exist and shape postcolonial knowledge production while obscuring alternate methodologies and their epistemologies: the outlines of urban planning schemes written into the landscapes of cities; the lines on maps that determined tribal "homelands"; the entrenchment of academic disciplines or cash crops with commodity prices that bend at the will of the global economy.

James Scott reminds us that the epoch of high modernism and its obsession with master plans cuts through all political spectrums, so what, then, is so special about colonial planning?[24] First, we know there is much that is not special about it but can instead be lumped more accurately into historical epochs than set aside as "colonial." Across a broad range of political landscapes, high modernism prioritized the bird's-eye view of the state and sought to erase local, idiosyncratic systems of planning. While not necessarily intended by Scott, this has become a somewhat generalized definition of the legacy of state planning in the twentieth century: the clumsy and violent ways that states assume they know better than local populations how to order time and space. What, then, makes planning unique in colonial situations, when states also seek to standardize space and time within their borders, frequently with little input from those they are planning for?

One could argue that Scott's examples are all of colonial endeavors that just happen to take place within nation states. But there are other aspects that make colonial planning important to think with. First, colonial planning is planning subjects rather than citizens; colonial populations have little political or legal recourse to escape planners and their plans. Planning in colonial settings is also predicated on difference. While difference pervades planning within nation states and frequently justifies its need, difference is the incontrovertible "truth" that buttresses planning in colonial settings. This is not to say that "difference" is always real, but that it is insisted upon by those who plan. This is first and foremost racial difference, which is accompanied by the insistence on cultural, social, and environmental difference. These are the sorts of difference that can seemingly only be surmounted by opening research institutes to study "tribe" and soil, climate and "customary" law. This is difference that can be identified within colonial borders, but that primarily exists between the colony and the metropole. Underlying all of this is also the predication of cognitive difference: that the "native," among other things, is incapable of the foresight needed to enact planning. The future, as something distinct and better than the present, is the domain of the colonizer, never the colonized, who is dragged down by the tugs and pulls of tradition, into the void of history.[25]

Colonial planning is also distinct from other forms of planning in its temporal orientation. It aims to cement particular visions of the past, the present, and the future simultaneously. Pre-empting the planning of future uprisings was a dimension of all colonial planning. In the contexts of settler-colonialism, practices of colonial planning are dedicated to reframing the history of the place as that of *terra nullius*—the assertion of an absence, effacing the histories of the originary inhabitants of a now-colonized place—in order not only to justify the presence of colonial governance but also to produce a common sense understanding of colonial governance as the past, present, and future source of resources to sustain life.

The ubiquity of planning in colonial contexts has meant that, while frequently not labeling their books as such, many historians of colonialism are also writing histories of planning. Across a broad range of topics—labor, hygiene, agriculture, education, economics and urban life—these books illustrate the gaps, tensions, and negotiations between the plans of colonial administrators and the lives of communities subjected to the state's imaginary of what a colony should resemble, what it should produce, and how it should plan for the future. In *Unlearning the Colonial Cultures of Planning*, Libby Porter defines planning as a mechanism through which the interests of a colonial regime are "continually mediated and reconstituted."[26] Plans cannot be proposed and implemented only once, but must be continually reinforced and reintroduced in different forms and through different epistemic practices. As Porter shows, the colonial regime presents its plans as produced by rationality, but plans take hold through an affective register. Colonial planning furnishes its plans with an aura of necessity and inevitability, continually impressing on colonial subjects that there can be no alternative to the colonial present. The essays collected here aim to reveal the mechanisms and effects of colonial planning by considering such examples as what was erased in planning craft economies in Belgian Congo; how "common sense" functioned in assigning surnames to former enslaved people in Brazil; or how both humans and plants resisted capture in Laos (Van Beurden, Vimiero, McAllister, this volume).

The colonial archive is first and foremost a planner's archive. Colonial planning regimes order the archives in which we work and thus also shape the histories we write. Frequently we have at our disposal only the sources that were preserved in colonial archives, punctuated by traces that document forms of resistance, alternative forms of planning, or simply other forms of life. This volume points to the spaces where what we call "counter-plans" were hatched, even if sources remain elusive. In this way, the volume gives form to what Ann Laura Stoler calls "the disjuncture between prescription and practice, between state mandates and the maneuvers people made in response to them, between normative rules and how people actually lived their lives."[27]

Some of the entries in this volume reflect the shape of this colonial/planning archive, while others seek out the occasional intrusion of incongruent logics of planning. It is in these moments that we can glimpse the enormity of what has been left out of the archive. Many authors in this volume explicitly address the problem of a lack of sources and how they have navigated these absences and written histories attentive to what has been left out. Grounding entries in objects and materials that bear the imprint of colonial planning practices help make visible what sits outside the archive.

From origins to endpoints, essays here also examine how historically colonizing plans become decolonizing tools, in the shape of the varied projects of the nation state. This repackaging of planning knowledge infrastructures draws attention to the long-term impact of colonizing practices and how hard it can be to escape them. They have, as both Stoler and Timothy Mitchell might call it, "durability," whether that is the enduring "grid" serving as the organizing principle of spatial planning or the centrality of certain cash crops to national economies (Kett and Young, this volume).[28] This is not to suggest a reductive view on postcolonial societies or to negate historical decolonization but to consider what makes attempts at undoing colonial planning so fraught and sometimes seemingly impossible. We discuss some of these lingering durabilities in our closing interpretive essay.

Thread 2: Planning and Knowing

The second thread running through the essays addresses how planning both creates and undermines certain types of knowledge. What kind of knowledge is used for planning for the future in writing India's constitution? What kind of knowledge is created through surveys of family planning in Puerto Rico? How does a census redefine work and productivity in the New Hebrides (Kirloskar-Steinbach, López, Widmer, this volume). Following from these questions, what kinds of gaps and interstices are produced in planning knowledge and how do these gaps include and exclude? How do intermediaries who are tasked with collecting planning knowledge sometimes also undermine flows of power and information? We extend Simin Davoudi's conceptualization of planning as a "practice of knowing," illustrating how knowledge is not merely something that is possessed by planners, but rather knowledge is a set of practices "that planners do."[29] Planning is best understood as a form of sequenced, structured, repetitive, procedural practice that enables a succession of actions and materialities. As such, it resembles the character of infrastructures with a similar function: when planning remains transparent, actions flow from it seamlessly; by the same token, planning is most visible when plans fail. The similarities in epistemic practice between planning and infrastructure are striking: both require sequential collective action and knowledge-making; remain suspended in a state of never being complete; require continual updating; and always fall into disrepair once maintenance activities are stopped, generating future ruins. It is in order to reflect these similarities that we suggest planning is best understood as an epistemic infrastructure.

To further understand the entanglements between planning and knowledge, we can turn to the experiences of Bouvard and Pécuchet, the two protagonists in Gustav Flaubert's unfinished novel *Bouvard et Pécuchet*, published in 1881. Flaubert's novel describes the experience of two middle-class Frenchmen employed as copy clerks, who become fast friends following an accidental meeting at a park bench. A large inheritance by Bouvard allows them to leave Paris for the Norman countryside, where they plunge into an obsessive pursuit of knowledge on a variety of subjects—geology, biology, medicine, chemistry, agriculture, politics, and archaeology. Subsequently, they apply their theoretical knowledge to the reality of nature. The "two idiots," as Flaubert calls them, acquire a considerable amount through reading books but their accumulated knowledge is never properly put into practice. They fail miserably in all their grand schemes to improve the human lot, until finally, they return to their old profession as copy clerks.

One of the defining features of modern planning is the elimination of contradictions and paradoxes in order to become a universal system of knowledge. At a time when new studies are providing a critical reading of the modernist, mainstream histories of planning, what might we learn from a fictional work? Especially so when the author's indictment of a culture obsessed with individual knowing and institutional knowledge resonates with our own understanding of planning. Similar to Flaubert's representation of knowledge as contingent, disorderly, and resistant to systematization, planning knowledge fails to fit within the narrow boundaries of a classification system.

Recent studies on planning have foregrounded the role of "story telling" as a method to demonstrate how planning practices can be made more inclusive by vernacular narratives and the lived experiences of its users. Yet how can stories be used to understand planning histories and to reveal the unexpected and unorthodox spaces of planning? Evidently, in the histories explored here, stories are not mere narrative structures but are a "rendering of events which are being connected with each other."[30]

Our volume also brings into critical focus the concept of the "counter-plan" that fundamentally questions the predominant power hierarchies and the place of modern knowledge systems in theorizing planning. In using the term "counter-plan," we invoke its double meaning. First, it suggests the possibility of being counter or against master plans or master planning practices. In contrast to the historical and conceptual literature on planning, this book does not focus only on plans in the written form. Plans and planning can be found on paper, but they can also be embodied in artifacts as well as in non-human bodies. Non-human and more-than-human entities have also become recognized as actors in planning theory.[31] The essays in this book that deal explicitly with non-human entities like bees and museum ethnographic collection items (Novick and Verran, this volume) clearly support such an avenue of thought.

Second, our use of counter-plans unearths the cracks in practices of planning while simultaneously tracking down how spaces of exclusion and injustice are reconfigured through everyday planning practices. Counter-plans illuminate power asymmetries and allow us to write what Leonie Sandercock calls "insurgent planning histories"— inclusive histories that acknowledge that the plan is constantly shifting.[32]

Alternatives to state-controlled planning have circulated in planning literature since the 1980s, and postmodern planning theory has distanced itself from its identity as a modernist tool, attempting to work with multiple epistemic traditions.[33] Yet it has been difficult for planning theory and practice to escape from its top-down view, a dynamic we hope to avoid through our inclusion of counter-planning practices.

Furthermore, the counter-plan reveals the contradictions in the classificatory logics underpinning planning. The point here is not simply to question the practice of taxonomy, but also to highlight the hegemonic structures of power that determine what is worthy of classification. Counter-plans thus become a fundamental tool of possibility that enables us to detach planning from the universalizing logics of modernity/coloniality, opening up an alternative space from where ethical and political alternatives can be imagined.

To what extent does the knowledge produced through planning practices reflect the epistemic structures of each particular planning paradigm? The heterogeneity of knowledge produced is informed by what scholars have characterized as the interdisciplinarity of planning practices.[34] Pinson calls planning "the undisciplined discipline" in reference to the ways in which planning practices borrow from many different fields but lack a core set of practices.[35] However, despite its interdisciplinary origins, dominant Western forms of planning have been mobilized as machineries of colonialism and are now intertwined with local planning practices in complex ways. There are often-overlooked forms of dialogue between local planning practices and colonial master plans. Gautham Bhan argues that while Western "planning theory" remains dominant in much of the Global South, "Southern" forms of planning have long held these plans at bay.[36] Shubhra Gururani shows how everyday forms of Southern (counter-)planning can function insurgently against top-down plans.[37]

Thread 3: Planning and Things

The third and final thread woven through each essay examines what kind of material relations are triggered by planning. What is the range and durability of these relations and how are previous material relations reassembled through the practices of postcolonial planning? The conceptual framework of the book is informed by the approach of material semiotics in science and technology studies, which has its roots in a feminist appropriation of techno-scientific politics as well as in the use of semiotics in actor-network theory.[38] It does not understand material as given and stable but rather wants to examine how matter materializes differently in a set of diverse practices. Material semiotics sees humans and non-humans entangled in chains of socio-material practices in which agency is distributed among the different (human/non-human) participants. Whereas in a traditional Western cosmology, non-humans lack the intention to plan, you will find here plants, animals, and artifacts participating in planning activities.

Oddly, concern with material semiotics and socio-materiality has only recently developed in planning theory, despite the fact that socialities and materialities are central to planning's practices. In 2015 Robert Beauregard, a pre-eminent planning academic in the United States, began his book *Planning Matter: Acting with Things* with a series of lists that he named as "ontographies" created with "the intent of maximizing variety and minimizing obvious affiliations between adjacent [named] items." He later tells us that he learned the trick of creating disruptive ontographies from reading Bruno Latour. Each of Beauregard's lists names a group of things that "happen" together in planning. The contents list of our book might also be read as an ontography, although the things of these titles are very different sorts of things. While we might wish for a more nuanced and skeptical reading of what he calls object-oriented philosophy, there is no doubting Beauregard's careful and informed skepticism towards planning's modernist instrumentalism. *Planning Matter* can be read as an exemplar of an STS-informed technoscience practitioner intervention.[39] Our book, in turn, picks up on this practitioner enthusiasm.[40] Unlike Latour's utopian speculative texts which often have practitioners "getting hold of the wrong end of the stick," the detail-rich essays offered here contribute to these aims in an on-the-ground way.

What kinds of narratives emerge when we rethink the relation between materiality and concepts? Does the antagonism between storytelling and analyzing disappear? Here, you will find histories of planning told through things: for example, through weeds as a thing of counter-planning, through bags of fertilizer tasked with nation-building in Cold War Korea, and through alarm clocks that "proved" some German women guilty of terrorism in 1980s Cologne (McAllister, DiMoia, Schlünder, this volume). For the most part, the things in these essays are exemplary modern objects — materialized spatiotemporal entities, but some essays highlight activities that have been turned into "a thing," like computing, or dredging (Peters, Jeon, this volume). Or they are a proprietary product: COBOL, for example (Allen, this volume). Some things symbolize, like a statue of a gorilla in a library (Mitchell, this volume). Some things are events — the running of a hackathon — as things to plan with and through, events embed and express a particular logic of planning (Irani, this volume). We hope this volume works to recognize that stories are things too, and they are active participants in contemporary practices of planning.

Planning Matters: A User's Guide

In his cartography of Western academic writing, Sean Sturm differentiates between expository and exploratory essays.[41] Most of the texts we produce and consume as scholars are of the former nature: they display knowledge and work like an epistemic round trip, since their trajectory is completely tautological: they always end where they have begun. First, there are (kind of rhetorical) questions to be asked, then the means for answering are introduced, an argument is crafted and tools (concepts, theories, literature) are applied to material; in the concluding section, the answers to the posed questions are summed up.

Like cruise ships that are protected through their powerful size and technical equipment, formats keep academic writing on track, on well-known epistemic paths and narrative itineraries, especially in encounters with "the other" (disciplines, languages, cosmologies, styles, and gender). Expository formats and genres secure academic writing against the dangers of the unknown, the unchartered territories of thinking and writing, against drifting, or straying off course. This security comes with a price: this kind of epistemic thinking does not get in touch with anything beyond itself. It shields itself from being challenged, questioned, affected by others' ways of thinking; scholars usually take formats, styles of writing and arguing for granted, leaving their historical burden, their epistemic legacies, their in-built power mechanisms unquestioned. Can we find "generative ways" of telling planning stories differently?[42]

Despite the reversed relations between material and concept, this book does not refuse conceptualization or expository writing per se—it rather strives to reflect our order and forms of thinking, writing, and arguing in "generative" ways. Several contributions are concerned with particular concepts in postcolonial planning activities like development, population, economic growth, resistance, or epistemic concepts like representation. But they pay attention to the construction or deconstruction of these terms rather than assuming their meanings.

The stories in this book do not build on each other in the same way as we know it from expository writing: they do not aim at an interior, in-built hierarchy that pushes one singular argument. In other words, the plan that inhabits this book invites users to create their own trajectories. Readers should feel free to draw different things or threads together, searching for patterns, or even adding more stories to see how they change the fabric and modify the patterns.

Notes

1. John Friedmann, *Planning in the Public Domain* (Princeton, NJ: Princeton University Press, 1987).

2. Paul Davidoff and Thomas A. Reiner, "A Choice Theory of Planning," *Journal of the American Institute of Planning* 28, no. 2 (1962): 103–15.

3. Aaron Wildavsky, "If Planning is Everything, Maybe It's Nothing," *Policy Sciences* 4, no. 2 (June 1973): 127–53, on 129.

4. James A. Throgmorton, "Planning as Persuasive Storytelling About the Future: Negotiating an Electric Power Rate Settlement in Illinois," *Journal of Planning Education and Research* 11, no. 3 (October 1992): 17–31; and Merlijn van Hulst, "Storytelling, a Model *of* and a Model *for* Planning," *Planning Theory* 11, no. 3 (August 2012): 299–318.

5. The historical "colonial" at the heart of this book is the colonial enterprise of the nineteenth and twentieth centuries, an endeavor that was largely (although not exclusively) driven by the Global North. We do not mean to imply, however, that planning did not form part of the precolonial context of this particular colonial episode, or that other (and non-Western) colonizing eras of the past did not rely on planning as a technology of expansion and rule.

6. Ernest R. Alexander, "There is No Planning—Only Planning Practices," *Planning Theory* 15, no. 1 (February 2016): 91–103.

7. J. M. de Casseres, "Principles of Planology: A Contribution to the Scientific Foundation of Town and Country Planning," *Town Planning Review* 17, no. 2 (February 1937): 103.

8. de Casseres, "Principles of Planology," 104.

9. de Casseres, "Principles of Planology," 111.

10. M. R. G. Conzen, "Towards a Systematic Approach in Planning Science: Geoproscopy," *Town Planning Review* 18, no. 1 (July 1938): 1–26.

11. Conzen, "Geoproscopy," 1–2.

12. Conzen, "Geoproscopy," 3.

13. Abecedaries enjoy a growing popularity in academic writing while pursuing quite different agendas with it. Some examples are: J. M. Bernstein, Adi Ophir, and Ann Laura Stoler, eds., *Political Concepts: A Critical Lexicon* (New York: Fordham University Press, 2018); Susan Squier, *Poultry Science, Chicken Culture: A Partial Alphabet* (New Brunswick, NJ: Rutgers University Press, 2011); Vinciane Despret, *What Would Animals Say If We Asked the Right Questions?*, trans. Brett Buchanan (Minneapolis: University of Minnesota Press, 2016); A. K. Thompson, Kelly Fritsch, and Clare O'Connor, eds., *Keywords for Radicals: The Contested Vocabulary of Late-Capitalist Struggle* (Chico, CA: AK Press, 2016); Imre Szeman, Patricia Yaeger, and Jennifer Wenzel, eds., *Fueling Culture: 101 Words for Energy and Environment* (New York: Fordham University Press, 2017).

14. See Geoffrey Bowker, "Sustainable Knowledge Infrastructures," in *The Promise of Infrastructure*, ed. Nikhil Anand, Akhil Gupta, and Hannah Appel (Durham, NC: Duke University Press 2018), 203–22.

15. Monier Monier-Williams, ed., *Original Papers Illustrating the History of the Application of the Roman Alphabet to the Languages of India* (London: Longman, Brown, Green, Longmans, and Roberts, 1859).

16. Monier-Williams, *Original Papers*, 234.

17. For more on this, see Judith Flanders, *A Place for Everything: The Curious History of Alphabetical Order* (New York: Basic Books, 2020).

18. For more on the politics of development, see Arturo Escobar, *Encountering Development* (Princeton, NJ: Princeton University Press, 1994).

19. James C. Scott, *Seeing Like a State: How Certain Schemes to Improve the Human Condition Have Failed* (New Haven, CT: Yale University Press, 1998).

20. Theodore M. Porter, *Trust in Numbers: The Pursuit of Objectivity in Science and Public Life* (Princeton, NJ: Princeton University Press, 1995); and see Lorraine Daston and Peter Galison, "The Image of Objectivity," *Representations* 40 (October 1992): 81–128; Lorraine Daston, "Objectivity and the Escape from Perspective," *Social Studies of Science* 22, no. 4 (November 1992): 579–618.

21. Frederick Cooper and Randall M. Packard, "Introduction," in *International Development and the Social Sciences: Essays on the History and Politics of*

Knowledge, eds. Frederick Cooper and Randall M. Packard (Berkeley: University of California Press, 1997), 1–42.

22. Libby Porter, *Unlearning the Colonial Cultures of Planning* (Farnham: Ashgate, 2010), 3.

23. Helen Tilley, *Africa as a Living Laboratory: Empire, Development, and the Problem of Scientific Knowledge, 1870–1950* (Chicago: University of Chicago Press, 2011); Vinh-Kim Nguyen, "Government-by-Exception: Enrolment and Experimentality in HIV Treatment Programmes in Africa," *Social Theory & Health* 7, no. 3 (2009): 196–217; and, on "experimental exuberance," see Michelle Murphy, *The Economization of Life* (Durham, NC: Duke University Press, 2017), 78–94. See also Kenny Cupers, "Editorial: Coloniality of Infrastructure," *E-Flux architecture*, September 2021, https://www.e-flux.com/architecture/coloniality-infrastructure/412386/editorial/.

24. Scott, *Seeing Like a State*, 4.

25. For the most infamous example of the claim that Africa had no history, see Hugh Trevor-Roper, *The Rise of Christian Europe* (London: Thames and Hudson, 1965), 9–11; and reiterated in Hugh Trevor-Roper, "The Past and Present: History and Sociology," *Past and Present* 42, no. 1 (February 1969): 3–17.

26. Porter, *Unlearning the Colonial*, 16.

27. Ann Laura Stoler, *Along the Archival Grain: Epistemic Anxieties and Colonial Common Sense* (Princeton, NJ: Princeton University Press, 2009), 32.

28. See Ann Laura Stoler, *Duress: Imperial Durabilities in Our Times* (Durham, NC: Duke University Press, 2016).

29. Simin Davoudi, "Planning as Practice of Knowing," *Planning Theory* 14, no. 3 (August 2015): 316–331, on 316.

30. Suzanne Tesselaar and Annet Scheringa, *Storytelling handboek: Organisatieverhalen voor managers, trainers en onderzoekers*, quoted in Iris De Boer, "Storytelling and Its Potential for Planning Practice" (bachelor's thesis, Wageningen University, 2012), 7.

31. Luuk Boelens, "Theorizing Practice and Practising Theory: Outlines for an Actor-Relational-Approach in Planning," *Planning Theory* 9, no. 1 (2010): 28–62.

32. Leonie Sandercock, ed., *Making the Invisible Visible: A Multicultural Planning History* (Berkeley: University of California Press, 1998), 2.

33. Yvonne Rydin, "Re-Examining the Role of Knowledge Within Planning Theory," *Planning Theory* 6, no. 1 (2007): 52–68.

34. Philip Allmendinger and Mark Tewdwr-Jones, eds., *Planning Futures: New Directions for Planning Theory* (London: Routledge, 2002); Philip Allmendinger and Mark Tewdwr-Jones, "The Communicative Turn in Urban Planning: Unravelling Paradigmatic, Imperialistic and Moralistic Dimensions," *Space and Polity* 6, no. 1 (April 2002): 5–24; Raphaël Fischler, "Fifty Theses on Urban Planning and Urban Planners," *Journal of Planning Education and Research* 32, no. 1 (September 2011): 107–14, https://doi.org/10.1177/0739456X114204; and Bishwapriya Sanyal, Lawrence J. Vale and Christina D. Rosan, eds., *Planning Ideas That Matter: Livability,*

Territoriality, Governance, and Reflective Practice (Cambridge, MA: MIT Press, 2012).

35. Daniel Pinson, "Urban Planning: An 'Undisciplined' Discipline?," in "Transdisciplinarity," eds. R. Lawrence and C. Despres, special issue, *Futures* 36, no. 4 (May 2004): 503–13

36. Gautam Bhan, "Notes on a Southern Urban Practice," *Environment and Urbanization* 31, no. 2 (2019): 639–54.

37. Shubhra Gururani, "Flexible Planning: The Making of India's 'Millennium City,' Gurgaon," in *Ecologies of Urbanism in India: Metropolitan Civility and Sustainability*, ed. Anne M. Rademacher and K. Sivaramakrishnan (Hong Kong: Hong Kong University Press, 2013), 118–43.

38. Donna Haraway, "A Manifesto for Cyborgs: Science, Technology, and Socialist Feminism in the 1980s," *Socialist Review* 80 (1985): 65–107; Donna Haraway, *Modest_Witness@Second_Millenium. FemaleMan©_Meets_OncoMouse™: Feminism and Technoscience* (New York: Routledge, 1997); John Law and Annemarie Mol, "Notes on Materiality and Sociality," *The Sociological Review* 43, no. 2 (1995): 274–94; Annemarie Mol, *The Body Multiple: Ontology in Medical Practices* (Durham, NC: Duke University Press, 2002); Helen Verran, *Science and an African Logic* (Chicago: University of Chicago Press, 2001).

39. Gary Downey and Teun Zuiderent-Jerak, "Making and Doing: Engagement and Reflexive Learning in STS," in *The Handbook of Science and Technology Studies*, ed. Ulrike Felt, Rayvon Fouché, Clark A. Miller, and Laurel Smith-Doerr, 4th ed. (Cambridge, MA: MIT Press, 2017).

40. Robert A. Beauregard, *Planning Matter: Acting with Things* (Chicago: University of Chicago Press, 2015), 57–75, 172–210.

41. Sean Sturm, "Terra (In)cognita: Mapping Academic Writing," *TEXT: Journal of Writing and Writing Courses* 16, no. 2 (October 2012), http://www.textjournal.com.au/oct12/sturm.htm.

42. Helen Verran, *Science and an African Logic*.

Census: New Hebrides/Vanuatu, 1967

Alexandra Widmer

Ni-Vanuatu[1] school teachers, I can imagine, might have been intrigued by the French and British[2] colonial authorities' request that they make the trip from their islands for a census enumerator training in the capital, Port Vila, in 1967. This being the first simultaneous census for the entire group of over 80 inhabited islands, prominent demographer of the Pacific Islands, Australian Dr. Norma McArthur had been planning the undertaking with a British colonial office civil servant, John Francis Yaxley, for a year. This was a lead-up time she found to be rather short, so she said that everyone involved would have to work hard to meet the deadline. In addition to developing relevant categories on the bilingual census cards, creating census regions that were possible to cover in a single day, and overseeing the logistics of getting census cards to all regions of the country, this preparation involved training dozens of Pacific Islanders to actually conduct the simultaneous enumeration of the population (of what wound up being 72,243 people). On the day of the census, each enumerator would visit every household in their assigned area. Every zone (of several areas) had a supervisor, a settler or missionary, who was tasked with ensuring the enumerators carried out their tasks.

Censuses are anything but neutral sets of facts. Among many political aspects, they produce racial and ethnic identities that are part and parcel of state forms of political administration.[3] Censuses have served particular ends in colonial circumstances,[4] not the least of which is being able to count a population, an action that "created the sense of a controllable indigenous reality."[5] It bears mentioning that earlier attempts at censuses were frequently rendered impotent by Islanders who did not participate. In the early twentieth century, French censuses were taken to demonstrate French dominance in the archipelago. Though they collected information down to the last vanilla plant the settlers cultivated, Indigenous people were not counted. In 1941, the British attempted to count Indigenous inhabitants to participate in an Empire-wide census, but this was only partially successful.[6] By 1967, the New Hebrides was the last remaining Pacific Island group without official census data to aid in planning for much needed public services and even a long-term view (at least in the British case) to decolonization.

McArthur and Yaxley prepared instructions for the enumerators that explained the rationale for the census. The Presbyterian and Anglican missions increasingly wanted to reduce their involvement in running schools and hospitals. Until this time, missions had been crucial to developing these institutions, particularly in the case of the British.[7] In the memo, the planners promoted the census in the name of planning colonial institutions in medicine and education. What is more, in the final segment, their instructions center the responsibility for the success of the census on the Pacific Island workers and admonishes them, "If you are lazy or not careful the census will be a failure. So please do your best."[8]

Such were the desired colonial subjects, capable of hard work and devoted to their task. What was less certain, at this point, was how this work ethic related to capitalist organization of social relations, and this was something that the census planners needed to capture. Ni-Vanuatu had a history of avoiding full-time, long-term wage labor in the twentieth century, preferring to grow their food in their own gardens, plant cash crops, develop their own co-ops, or work short-term contracts.[9] Having a reliable labor force had long been a concern of settlers, particularly British ones, as discussed later. The 1967 census was the first time that work was an important aspect of social reality that the census planners wanted to apprehend. The first census was far from the first attempt to count and categorize the inhabitants of the New Hebrides, as missionaries and some district agents had conducted preliminary work throughout the twentieth century, and even the late nineteenth century. However, these were basic head counts by village, taking note of inhabitants' approximate age and sex. Many are the early twentieth-century paper trails tracing relationships between officials and researchers that complain of Pacific Islanders' lack of interest in answering census questions.

The census planners in 1967 were the first to attempt to categorize all the inhabitants' relationship to a cash economy. Norma MacArthur was completely aware that the census imposed categorizations to simplify a more complicated social reality. She and her assistants revised the definitions of livelihoods more than once in conversations with Islanders and local authorities. Ultimately, two questions opened this topic: "What work do you do?"

(leading to the "subsistence sector") "Who do you work for and where?" (the "monetary sector"). The categorization of "work" in the "subsistence sector" here referred to food and shelter production from gardens, and not what 1970s second-wave feminists would later have called, "reproductive labor." The distinct categories of subsistence and monetary were to accommodate the fact that, while money was used throughout the New Hebrides, most Indigenous people, it was thought, were not dependent on cash income for survival but lived somewhere on a continuum between "pure subsistence" and "the monetary economy."[10]

There was a range of degrees of participation in the monetary economy even within the "subsistence" sector so McArthur had the enumerators further categorize livelihood in the subsistence sector as "villager, no cash crop" and "villager, gardener and copra maker." McArthur herself admitted these were arbitrary and subjective.[11] The copra (processed coconut) production activities (virtually the only export crop) could also be classified differently: Enumerators were told to distinguish between those who made copra for their own use or to fulfill communal obligations, and those who cut and prepared copra for sale.

Beyond "villager, no cash crop" and "village gardener and copra maker," McArthur did not develop further classifications for the subsistence economy because, as she wrote, "In view of the nature of subsistence and village agriculture there can be no division into grade or type of occupation as there is in those industries within the monetary economy."[12] The categories she assigned to the monetary economy were: "(i) professional, management and executive; (ii) supervisory and clerical; (iii) skilled and semi-skilled (combined because the low level of industrial development and in particular the absence of formal trade training make it impossible to distinguish the two groups with any degree of accuracy); and (iv) all other."[13]

McArthur categorized participation in the monetary economy in a hierarchy by the status of the occupation. This amounted to occupations associated with monetary economy being associated with categorizations of people's identities. So the broad majority of the population—Indigenous people—were associated with the uncomplicated subsistence economy without internal hierarchies, while the much smaller portion of the population were associated with economic activities that could be more or less complicated. It was likely well known by the census takers that the "subsistence economy" entailed hierarchies and skills, as researchers and missionaries had long documented the significance of pig husbandry for Indigenous ranked society and yam production for social reproduction,[14] but the census did not attempt to capture the specificity of these facts within the confines of its tables. As to why they did not ask these questions: the constraints on time and resources were considerable, and the planners likely suspected these economies and exchange relationships glossed as "subsistence economy" could not easily be monetized, taxed, or rendered into export goods.

And yet, with her broad definition of subsistence and monetary economic activities, McArthur made livelihood legible in a particular way. In particular, it meant that gender participation in economic activities was practically equal; 86.5% of men and 81.9% of women were economically active. This was remarkable! Before World War II, researchers frequently saw Pacific Islanders as a dying race, biologically vulnerable and culturally ill equipped for the modern world.[15] With McArthur's census, Pacific Islanders were strong, more numerous than expected, and able to cope, even prosper, with modernity. Indigenous people were not lazy or dying out but economically active. Indeed, Pacific Islanders carried out the enumeration.

"New Hebridean" as an identity category for a diverse group of Indigenous people, was produced in the census as an identity connected to livelihood. In this way, the first census in the New Hebrides would link Indigenous people to economic activities in a way that produced a self-evident category of Indigeneity as a countable and account relationship to the market economy. If Islanders in the New Hebrides had been asked what "Indigenous" meant for them at that time, I can imagine they might have used one of their own terms from one of over 130 Indigenous languages just as self-evidently, and it probably would not have been "New Hebridean" for most. In addition to a village or Island identity, another possible collective identity could have been *man ples*, someone of the place, that denotes genealogical attachments between people, land, and other non-human entities.

Had the census been conducted a mere five years earlier, the census planners would have had to contend more systematically with how to enumerate the Vietnamese plantation laborers, the descendants of contracted laborers, brought by French settlers in the 1920s. Back then, much to British settlers' frustration, British authorities would not permit the entry of indentured labor from their empire as they had done by bringing laborers from British India to nearby Fiji. By 1929, the Vietnamese (still referred to as "Tonkinese" in Vanuatu) numbered roughly 6,000, about 1/10 of the population.[16] The livelihood activities of the Vietnamese plantation workers would have been categorized in the lowest census category and least skilled form of labor—"all other." The small businesses that some operated in town would also need to be accounted for in the census. But, in 1963, the largest group of wage laborers in the New Hebrides boarded a ship for the homeland of their parents and grandparents, the Gulf of Tonkin in North Vietnam. This was the last of a few waves of repatriation. In the immediate post-war period, the potential of these laborers appears in the British file "Secret Returns to Secretary of State on (1) Political Feeling (Local). (2) Communism (Local). 1948–53."[17] The British Resident Commissioner requested that the British district agents keep tabs on Vietnamese laborers and Chinese settlers. If no concerns arose, a secret code word, "chang" should be telegraphed to the High Commission in Suva, from Port Vila, assuring those in the regional headquarters that all was quiet, which was usually the case. The British concerns were not completely unfounded, as some descendants of indentured laborers were organizing for their return to North Vietnam, and there were reports of the circulation of communist materials. According to some Vietnamese people in Port Vila who anthropologist Jean Mitchell interviewed in 2009, Ho Chi Minh signified "freedom, equality, and the nationalist project of decolonization that drew the Vietnamese back to their homeland."[18] In the 1967 census, the Vietnamese numbered only 395.

With a far-reaching definition of work attributed to the knowledge category of "subsistence," the census showed that people living in the New Hebrides were not lazy, isolated from capitalism, or incapable of living in the modern world. While this research was not directly aimed at winning the Cold War, it was connected to the aims of British and French colonial services and the Cold War–created South Pacific Commission[19] (which had commissioned and published McArthur's demographic work in the region), and displayed many of the same values of modernization in relation to capitalist development. Portraying the subsistence activities in as broad a way as possible might have allayed common Cold War concerns about revolutionary consciousness-raising among poor people. For "subsistence activities" would entail categorizing a group of people who lacked money not as poor but as the productive group (if not for colonial accounting) that they were. Islanders, who had often had access to land and a flexible attachment to wage labor, were rendered active, if not "careful," in the subsistence sector, rather than "lazy" as feared in Yaxley and McArthur's instruction memo. McArthur and Yaxley's numerical representation of Pacific Islanders formed a different narrative than the most common of the Pacific Islands as isolated, virtually uninhabited islands, a narrative which obscured the fact, as Ruth Oldenziel has argued, that U.S. Cold War technoscientific networks, made islands critical parts of their empire, a state of affairs that continues to this day.[20] "Subsistence" involved another set of categorizations used by policymakers and experts during the Cold War due to a new geopolitical threat whereby extreme poverty might lead to revolution. In an earlier period, planning for labor participation involved a transnational constellation of actors, from missionaries and colonial administrators to social reformers. In the postwar era, when the minds of planners were focused on decolonization and Cold War rivalries, labor in the South Pacific Islands became tied to the goals of independent state formation, ensuring capitalist or, in some cases, ongoing colonial hegemony and modern development interests.

It is common in the history of science to think of the Pacific Islands as being of interest to scientists as isolated scientific laboratories,[21] even while science contributed to a widening institutional web of influence and knowledge over the islands. These narrations of isolated islands, Sivasundaram suggests, sever the living material relationships of ocean, land, and human life.[22] Colonial planning that tried to account for subsistence (which is, after all, an assemblage of human integration into relationships among ocean, land, and non-human life) also produced reductionist accounts of human activities by fitting them into economic categories. The census economic categories, thus, could take expansive and complex relationships among humans, pigs, yams, and coconut groves (among others), and isolate them into categories of subsistence or monetary economically productive activities.

The story of a planning moment I have told here is not one that argues for abolishing the census. To borrow consumption metaphors from Brazilian postcolonial studies,[23] a stance might be not to eradicate colonial plans, but to disarm, ingest and incorporate them. As Bislama speakers in Vanuatu know, to "kakae" something is to eat it, materially or metaphorically, in a way that can incorporate it (rather than destroying it) and where consumption often incurs obligation and the regeneration of relationships.

Rather than purging, when it comes to the census, I suggest a metabolic approach to thinking about the transformation of colonial realities. Indeed, Islanders took multiple approaches to colonial engagements during this during this time period. Attesting to the importance of subsistence, Islanders pursued with renewed vigor land tenure issues unmentioned in the census planning: for instance, Jimmy Stevens' grassroots movement of land occupation in Northern Vanuatu. Meanwhile, more elite Ni-Vanuatu, like physicians Philip Ilo and John Kalsakau on the New Hebrides Advisory council run by the colonial authorities, argued for better land registration maintained by colonial courts to prevent settlers from claiming Ni-Vanuatu lands.[24]

By historicizing the knowledge categorizations in the census that solidified relationships between Indigeneity and subsistence in ways that privilege planning for waged labor over Indigenous values and ontologies, I join with Pacific and other Indigenous scholars and experts and hope we might unpack practices of categorizing subsistence and indigeneity in order to think more about relationships among humans, non-humans, and environments. This brings us closer to understanding the Pacific as a "sea of islands" rather than "islands in a far sea."[25]

I hope too, that by highlighting how colonial administrators connected "lazy" and "careful" to historically specific forms of knowledge like "subsistence," I can show that these character traits were part and parcel of the requirements of labor regimes that privileged economic productivity. The knowledge category "subsistence" had the potential to render erstwhile "lazy" people—a common colonial accusation of colonized subjects—as productive, even as it elided pigs and yams. In spelling out how categories and specific knowledges are made in/visible and un/known, we might move toward thinking, producing, consuming, and relating otherwise.

Notes

1. Ni-Vanuatu are the Indigenous citizens of Vanuatu, an archipelago of over 80 inhabited islands in the southwestern Pacific. While technically an anachronism here, as Vanuatu became a nation in 1980, it is a common practice to use this term in the historiography of the region.

2. The New Hebrides Condominium (1906–1980) was a joint colonial project of Britain and France.

3. For example, David I. Kertzer and Dominique Arel, *Census and Identity: The Politics of Race, Ethnicity, and Language in National Census* (Cambridge: Cambridge University Press, 2002).

4. Bernard Cohn, *Colonialism and Its Forms of Knowledge: The British in India* (Princeton: Princeton University Press, 1996).

5. Arjun Appadurai, "Number in the Colonial Imagination," in *Orientalism and the Postcolonial Predicament: Perspectives on South Asia*, eds. Carol A. Breckenridge and Peter van der Veer (Philadelphia: University of Pennsylvania Press, 1993), 314–36.

6. For more on the French and British censuses in the New Hebrides in the first half of the twentieth century, see Alexandra Widmer, "Making People Countable: Analyzing Paper Trails and the Imperial Census," in *Sources and Methods in Histories of Colonialism: Approaching the Imperial Archive*, eds. Kirsty Reid and Fiona Paisley (New York: Routledge, 2017), 96–112.

7. On mission in involvement in biomedicine in the New Hebrides, see Alexandra Widmer, "Genealogies of Biomedicine: Formations of Modernity and Social Change in Vanuatu" (PhD diss., York University, 2007) and education, see Keith Woodward, "Historical Note," in *Tufala Gavman: Reminiscences From the Anglo-French Condominium of the New Hebrides*, eds. Brian Bresnihan and Keith Woodward (Suva: Institute of Pacific Studies, 2002), 16–72.

8. New Hebrides British Service, "Condominium Census 1967. Instructions to Enumerators in Rural Areas and to Field Supervisors," 1967, C.D. 1 and C.D. 2,

MSS & Archives 2003/1, NHBS 7, series IX, file 2, box C3025765, Western Pacific Archives, University of Auckland.

9. Michael Allen, "The Establishment of Christianity and Cash-Cropping in a New Hebridean Community," *Journal of Pacific History* 3, no.1 (1968): 25–46; Jean Guiart, "Les mouvement coopératif aux Nouvelles-Hébrides," *Journal de la Société des Océanistes* 12 (1956): 326–34.

10. Norma McArthur and John F. Yaxley, *Condominium of the New Hebrides: A Report on the First Census of the Population 1967* (Sydney: New South Wales Government Printer, 1968), 57.

11. McArthur and Yaxley, *Condominium of the New Hebrides*, 57.

12. McArthur and Yaxley, *Condominium of the New Hebrides*, 57.

13. McArthur and Yaxley, *Condominium of the New Hebrides*, 57

14. John R. Baker, *Man and Animals in the New Hebrides* (London: Routledge & Sons, 1928); John R. Baker, "On Sex-Integrate Pigs," *British Journal of Experimental Biology* 2, no. 2 (1925): 247–63; Margaret Jolly, "The Anatomy of Pig Love: Substance, Spirit and Gender in South Pentecost, Vanuatu," *Canberra Anthropology* 7, no. 1/2 (1984): 78–108; Felix Speiser, *Ethnology of Vanuatu: An Early Twentieth Century Study* (Honolulu: University of Hawai'i Press, 1996).

15. See, for example, Patrick Buxton, "The Depopulation of the New Hebrides and Other Parts of Melanesia," *Transactions of the Royal Society of Tropical Medicine and Hygiene* 19, no. 8 (1926): 420–58; Alexander Carr Saunders, "Review of Essays on the Depopulation of Melanesia," *The Eugenics Review* 14, no. 4 (1923): 282–83; George. H. L. F. Pitt-Rivers, "The Effect on Native Races of Contact with European Civilization," *Man* 27, no. 2 (1927): 2–10; William H. R. Rivers, "The Psychological Actor," in *Essays on the Depopulation of Melanesia*, ed. William H. R. River (Cambridge: Cambridge University Press, 1922) 84–113; Stephen Roberts, *Population Problems of the Pacific*, (London: Routledge, 1927); Speiser, *Ethnology of Vanuatu*.

16. Miriam Meyerhoff, "A Vanishing Act: Tonkinese Migrant Labour in Vanuatu in the Early 20th Century," *Journal of Pacific History* 37, no. 1 (2002): 45–56, on 47.

17. Secret Returns to Secretary of State on (1) Political Feeling (Local). (2) Communism (Local). (Former number 52/95). 1948–1953, MSS. Archives. 2003/1. NHBS 17. Series III. File 8, Box: C3026184. NHBS 17: Records. MSS-Archives-2003/1-NHBS 17. Western Pacific Archives, University of Auckland.

18. Margaret Jean Mitchell, "'Comrades,' 'Trouble-Makers' and French 'Ressortissants': The Repatriation of the Tonkinese from New Hebrides to North Vietnam" (paper presented at Association for Asian Studies Annual Conference, Toronto, ON, March 16, 2017), 3.

19. Michael Howard, *Fiji: Race and Politics in an Island State* (Vancouver: University of British Columbia Press, 2011), 123.

20. Ruth Oldenziel, "Islands: The United States as Networked Empire," in *Entangled Geographies: Empire and Technopolitics in the Global Cold War*, ed. Gabrielle Hecht (Cambridge: MIT Press, 2011), 13–42.

21. See, for example, Henrika Kuklick, "The Colour Blue: From Research in the Torres Strait to an Ecology of Human Behavior," in *Darwin's Laboratory: Evolutionary Theory and Natural History in the Pacific*, eds. Roy M. Macleod and Philip F. Rehbock (Honolulu: University of Hawai'i Press, 1994), 339–70; Henrika Kuklick, "Islands in the Pacific: Darwinian Biogeography and British Anthropology," *American Ethnologist* 23, no. 2 (1996): 611–38.

22. Sujit Sivasundaram, "Science," in *Pacific Histories: Ocean, Land, People*, eds. David Armitage and Alison Bashford (London: Palgrave Macmillan, 2013), 237–62.

23. Carlos A. Jáuregui, "Oswaldo Costa, Antropofagia, and the Cannibal Critique of Colonial Modernity," *Culture & History Digital Journal* 4, no. 2 (2015): e017.

24. For more on this, see Alexandra Widmer, *Moral Figures: Making Reproduction Public in Vanuatu* (Toronto: University of Toronto Press, 2023).

25. Epeli Hau'ofa, "Our Sea of Islands," *The Contemporary Pacific* 6, no. 1 (1994): 148–61.

Charcoal: Dar es Salaam, Tanzania, 1973
Emily Brownell

Every day, thousands of worn, plastic gunny sacks of charcoal crisscross the murky boundary between Dar es Salaam and its hinterland. Covered at the mouth of the sack with a loose mesh of rope or palm fibers to keep overflowing pieces from escaping, men and women along roadsides or under the shade of trees sell the charcoal by the bagful, tinful, or even the handful. The presence of charcoal sellers in the interstices of the city might operate quite differently than the infrastructure of pipes and electrical grids that deliver other kinds of fuel, but they serve much the same purpose.[1] Charcoal is the main fuel source for the city and it always has been.

In the colonial period, peri-urban communities of agriculturalists made charcoal in sites scattered among the latticework of Dar's green periphery.[2] Operating both within and outside of colonial forest preserves, local communities were tacitly encouraged to make charcoal. Charcoal functioned as a tool of underdevelopment. The colonial state saw local forests as providing a "subsidy in nature."[3] As long as local trees provided cooking fuel as well as heat and light, the state could avoid expanding urban infrastructures for the delivery of fuel and electricity in African neighborhoods. In this way, Africans were obliged to always keep "one foot in the subsistence economy" by producing charcoal while they were otherwise obliged to increasingly stay out of forests, which "helped fuel accumulation for the state."[4]

Local charcoal producers preferred the trees of Miombo woodlands, which make up a vast part of the southern sub-tropic zone of the continent and abut against the edges of the city. Less dense than the region's forests, these woodlands were usually outside of the purview of state control. The Zaramo, who had customary ownership over much of Dar's rural regions have historically been the main merchants of the charcoal trade, though today the trade is more diverse. There is a hierarchy of trees within these woodlands when seen through the eyes of a charcoal producer. What in Kiswahili is known as *miombo*, scientists refer to as the *Brachystegia* genus of trees, and the region is home to a broad range of species within this genus. Some of these trees are denser wood, which is a good trait for charcoal. Others, covered in thorns, are left as a last resort. Beyond particular trees, charcoal production also followed along a spatial logic, hugging the contours of Dar's expanding arterial roads or the paths of peri-urban settlement where trees were cleared for cultivation or new villages.

After trees have been chosen for a burn, the cut wood is buried underground in a deep hole dug to serve as a kiln. After lighting a fire, the kiln is then covered with blocks of earth and grass, creating an environment where carbonization can occur with a limited supply of air. Over the course of a few days, the covered wood burns slowly under conditions of very high heat and low oxygen. In choosing the wood, monitoring the fire, and extracting the final product, charcoal makers—despite the inherently inefficient process—produce a remarkably efficient form of energy over the course of a few days.[5] In this act of transformation, the resulting carbonized product unearthed from the ground returns to the surface a shiny, black and impossibly lighter version of the trees cut and buried. What emerges is also an eminently mobile and easily portionable form of energy that is ready for use using few tools.

Charcoal production and its use transect the history of Dar es Salaam from the precolonial period to the present. This particular story, though, starts in the 1970s, a decade marked globally by a new focus on the "limits of growth" and the finitude of natural resources—particularly fossil fuels. Urban Africa's expanding use of charcoal came to sit at the center of a struggling national economy, international conservation efforts, and a hemispheric struggle to define what constituted a sustainable future.

When Indira Gandhi spoke at the first global environmental conference in Stockholm in 1972, she claimed that it was not industrial pollution but poverty that posed the largest threat to the environment. To address this, Gandhi urged the West to invest in providing the "Third World" with the technology necessary to develop. "On the one hand," noted Gandhi, "the rich look askance at our continuing poverty" while at the same time "they warn us against their own methods."[6] Without jobs, as well as the interventions of science and technology "we cannot prevent them [poor people] from combing the forest for food and livelihood; from poaching and from despoiling the vegetation.... The environment cannot be improved in conditions of poverty. Nor can poverty be eradicated without the use of science and technology." Gandhi sought the West's technical solutions as a way to mitigate what she described as the environmental predations of the poor.

A decade later when the president of Tanzania, Julius Nyerere, visited India to receive the Third World Award from Gandhi, he used his speech as an opportunity to express a very different sentiment, one that likely reflected the evolving debt crisis. Nyerere urged the assembled audience of 44 representatives of the "Third World," that it was the path dependency of Western technology and development that was in fact the root cause of environmental destruction in the Global South. Instead of looking to the West or using their technology, Nyerere asked that the audience turn inward, to "our own roots and our own resources." Over the course of the 1970s, Nyerere had become increasingly skeptical of building factories and infrastructures that were "technological lock-ins": these imported, "packaged" technologies could not be unpacked or easily fixed by locals due to strict rules as to where users must purchase their parts and sometimes even who could service them.[7] Western technology, once seen as the savior, was now the culprit of both underdevelopment and environmental crisis.

At the heart of this shifting conversation was the reverberating effects of the Arab Oil Embargo. In the years following the embargo, Tanzanian development had failed to be realized in the material terms first imagined. By 1982, Tanzania was spending 60% of all its export earnings on purchasing oil to the detriment of buying spare parts, new machinery, or fertilizer—many of these crucial imports were themselves tied up with the petroleum economy. Rising oil prices had also spiked the price of other "modern fuels" such as kerosene and diesel, which powered much of Dar's electrical grid. Despite these struggles, Nyerere and many other leaders of the "Third World" saw the oil crisis as a lesson in economic sovereignty as well as an opportunity to develop alternative fuels. OPEC's move to increase their bargaining power against the industrialized world both harmed Tanzania's economy and served as a powerful example of what must be done to create a new economic order.[8]

And yet the petroleum imports which ate up so much of Tanzania's foreign exchange only accounted for 7% of the nation's total energy use.[9] By the 1980s, 98% of the energy used by Tanzanians was derived from biomass, mostly wood and charcoal and the percentage was only increasing over time. The predominance of biomass as a source of energy across the "Third World" did not go unnoticed by Western environmentalists, themselves fueled by neo-Malthusian fears of the "Third World" and dwindling resources. As a result, scientists and activists proclaimed a second fuel crisis, known as the wood fuel crisis. Through creating new multinational institutions, the production of hundreds of scientific studies, and conducting a global campaign, environmentalists and scientists came to identify the collection of firewood and the production of charcoal as leading to imminent deforestation and desertification across the world.

Fear of wood scarcity was thus tied in complex ways to the rising price of oil and the specter of the end of hydrocarbons. The vision of what postcolonial development would look like in Africa did not go to plan; rising oil prices required rewriting a future that had been underwritten by cheap petroleum. Not only did the oil crisis prompt an increasing reliance on biomass in the Third World, but it also played a key role in creating the new notion of a global environment.[10]

East Africa's emergence into the crosshairs of the wood fuel campaign in the 1970s also resulted from a prolonged drought that struck both sides of the continent. The subsequent famine across the Sahel and down as far as Tanzania connected fuelwood use to concerns about widespread desertification, deforestation, and the despoiling of land for crops. While the drought subsided in a few years, campaigns against wood fuel continued. Fears of deforestation also underlined emerging policies of "sustainable development" when a second drought and famine hit the region just a decade later.

We see then that the anxieties around charcoal and other wood fuel use in East Africa coalesced due to a potent mix of ideas in ascendance: a new way of thinking about the environment as globally interconnected and imperiled, energy shortages, and a shifting approach to development that aimed to turn "small" into "beautiful." International organizations that shaped and circulated concerns over wood fuel connected these to a corollary concern that women were being left out of "development." Gathering and using wood fuel was frequently the work of women in these communities and thus it was their time that was increasingly tied up in gathering scarce supplies. As is so often the case, this crisis brought new objects of planning into being. Two conjoined objects came into view as the answer to the growing reliance on wood and charcoal. First, was the seedling. Tree planting efforts were initiated by organizations across the globe in the 1970s to address the concerning calculus of the "wood fuel gap theory." This theory sought to calculate wood fuel needs against available forest resources to figure out how short any community of fulfilling its needs. This persistent and growing "gap," scientists worried, would spark a snowballing deforestation crisis. Tanzania was no exception, as tree planting campaigns emerged across the country and millions were planted. In nearby Kenya, Wangari Maathai emerged as a leading figure in the environmental movement in part through her work encouraging women to plant trees, further consolidating the connection between women and environmental management in the developing world.[11]

The other object was the cookstove, also aimed at women. A dizzying array of prototypes have emerged in the decades since the fuel wood crisis in an effort to make a better, more efficient stove than the "three stone hearth." The promise of the improved cookstove was tripartite: it would use biofuel more efficiently, relieve the strain on women's time, and improve air quality. The search for this holy grail of development become an obsession for NGOs and engineers and also a core project of collaboration with which to enlist local communities.[12] You can follow the travails of these enterprises as they continued in the 1980s (and even today) in publications such as *Boiling Point*. In their 1987 issue, a project report describes a three-year pilot project in Morogoro, Tanzania to train women how to make and sell new stoves, in partnership with the Morogoro Women-Focused Afforestation Project.[13] Other issues explore the frustrating problems of finding the right mud and clay and how to fire the stoves in a kiln without have them crack.[14]

While these two objects—the seedling and the stove—became potential solutions to the fuel wood crisis, they also worked to frame local communities as the perpetrators of this crisis itself. Among other effects, this allowed for the privatization of forests for biodiversity and conservation, but also for the exportation of lumber. It also empowered international environmental organizations to interfere in Tanzanian development, turning to lending institutions to enforce an emerging environmentality and bypassing African leaders, who they had come to see as recalcitrant.[15]

But as charcoal became the culprit of a global crisis, it also became a source of potential salvation for Dar's residents. In the absence of oil and foreign exchange, urbanites were forced to craft their livelihoods with local resources at the center. Suffering dwindling wages as well as empty shelves, people sought relief in the production and consumption of goods that did not rely on the necessity of petroleum. For both urban consumers and peri-urban producers, charcoal was a way to survive rupture—whether it was infrastructural or financial. Just as a bag of charcoal took very little foreign inputs for locals to produce, it also required very little to use. Charcoal's most widespread use was cooking, where women used a basic stove called a "jiko." The jiko was made out of salvaged metal welded together over a charcoal stove in one of Dar's growing number of small industry garages. Selling food along major roads became a key way women supplemented dwindling family wages. And whereas "modern fuels" were only sold erratically or in certain locations, charcoal had none of the same constraints in distribution.[16] These were not new methods of survival, but they were newly essential to an urban economy in decline. In a city moving toward its periphery in order to cope with crisis, having ubiquitous distribution networks mattered.

While the Tanzanian state embraced tree-planting and a push for appropriate technology that could sever widespread rural dependencies on biomass, they didn't necessarily eschew charcoal. State afforestation efforts in Tanzania may have only been partially interested in the same things as Western environmentalists. As one study suggests critically, the Tanzanian state "may have realized that promoting afforestation is more profitable than halting deforestation."[17] Without any ready energy alternatives, planting trees was as much about fuel creation as about afforestation. In fact, the state sought to expand the charcoal economy overseas and sought out other ways that Tanzanian trees could contribute to foreign exchange, such as providing much-needed pulp supplies to India and Indonesia.[18] Trees were also envisioned as playing a more prominent role in domestic industries. Efforts were made across several sectors to transform factories to run fuelwood and pulp waste.[19] The Tanzanian finance minister, Cleopa Msuya, already in 1975 explicitly urged the Tanzanian state to cut the high costs of energy importation by focusing on the "production of hydro-electricity . . . and use of charcoal."[20]

The Tanzanian state was searching for a way to escape a global energy infrastructure that had shaped a future for the Global South as both environmentally and economically unsustainable. Urbanites, on the other hand, were engaged in their own forms of rebellion from an assumed order, as their access to forests became increasingly closed off due to conservation concerns and the liberalization of the Tanzanian economy that made land available to foreigners as well as the tourist industry. And yet while state and international actors did assert control over forests it was still virtually impossible to tax or monitor charcoal production. As Mitchell has noted about oil and coal, the material realities of extracting and transporting these different sources of energy generate diverse political possibilities.[21] Indeed, the very paths and forms of how charcoal was made and sold shaped its contestability and the nature of state control. Acting as an anarchic fuel source, the state had little ability to intervene in charcoal due to the dispersed nature of its production across the hinterland, stretching as much 50 to 100 of kilometers from the city. Across the past few decades, charcoal bans or attempts to tax it have been quickly rescinded. Perhaps this signals a triumph for local producers, but it also reveals the fact that there is no hope of a ready replacement for the majority of urban Tanzanians, which is its own form of infrastructural intransigence.

The bag of charcoal, the seedling, the stove: in the course of planning, certain things come to shore up the hopes and fears of the planners and the planned (who are planners of their own making). These objects can traverse across these communities, and as they move they can take up new charges and valences, shifting their meaning and creating new effects. Sometimes they become ways to enlist cooperation and other times they become hopeful paths out of the austerity of the plan.

Notes

1. AbdouMaliq Simone, "People as Infrastructure: Intersecting Fragments in Johannesburg," *Public Culture* 16, no. 3 (2004): 407–429, on 407

2. For more on charcoal in postcolonial Dar es Salaam, see Emily Brownell, *Gone to Ground: A History of Environment and Infrastructure in Dar es Salaam* (Pittsburgh: University of Pittsburgh Press, 2020).

3. Roderick P. Neumann, "Forest Rights, Privileges and Prohibitions: Contextualising State Forestry Policy in Colonial Tanganyika," *Environment and History* 3, no.1 (1997): 45–68, on 60.

4. Neumann, "Forest Rights, Privileges and Prohibitions," 63.

5. Using this process about 100 kg of wood produces only 8–23 kg of charcoal, see https://www.charcoalproject.org/is-a-woodfuel-and-charcoal-crisis-looming-for-tanzania/.

6. Indira Gandhi's Speech at the Stockholm Conference in 1972, see LASU-LAWS, "Indira Gandhi's Speech at the Stockholm Conference in 1972," LASU-LAWS Environmental Blog, http://lasulawsenvironmental.blogspot.de/2012/07/indira-gandhis-speech-at-stockholm.html, accessed May 15, 2017.

7. "Technology is not neutral," *Daily News*, May 11, 1977.

8. Julius K. Nyerere, "A New Order" (speech), Arusha, February 12, 1979, http://www.juliusnyerere.org/uploads/unity_for_a_new_order_1979.pdf.

9. Fanuel C. Shechambo, "Urban Demand for Charcoal in Tanzania: Some Evidence from Dar es Salaam and Mwanza," Research Report No. 67, Institute of Resource Assessment, University of Dar es Salaam, May 1986.

10. Timothy Mitchell, "The Crisis That Never Happened," in *Carbon Democracy* (New York: Verso, 2011), 188.

11. Maria Mies and Vandana Shiva, *Ecofeminism*, 2nd ed. (London: Zed Books Ltd., 2014).

12. Burkhard Bilger, "Hearth Surgery," *New Yorker*, December 13, 2009, https://www.newyorker.com/magazine/2009/12/21/hearth-surgery.

13. Anne Sefu, "Morogoro Fuelwood Stove Project Tanzania," *Boiling Point*, no. 13 (August 1987), http://www.nzdl.org/cgi-bin/library?e=d-00000-00---off-0hdl--00-0----0-10-0---0---0direct-10---4-------0-1l--11-en-50---20-about---00-0-1-00-0--4----0-0-11-10-outfZz-8-00&cl=CL2.20.8&d=HASH3ec3d846525090878a0abe.11&gt=1

14. "Reviews and Summaries: Morogoro Fuelwood Stove Project." *Boiling Point*, no. 10 (August 1986), http://www.nzdl.org/cgi-bin/library?e=q-00000-00---off-0hdl--00-0----0-10-0----0---0direct-10---4----ste--0-1l--11-en-50---20-about-materials+for+mud+stoves--00-0-1-00-0-0-11-1--0-0-&a=d&c=hdl&srp=0&srn=0&cl=search&d=HASH110a57bad34a66995a8aa1.15.2.

15. Arun Agrawi, *Environmentality: Technologies of Government and the Making of Subjects* (Durham: Duke University Press, 2005), and Stephen Macekura, *Of Limits and Growth* (New York: Cambridge University Press, 2015), 99.

16. Gerald Leach and Robin Mearns, *Beyond the Woodfuel Crisis: People, Land and Trees in Africa* (London: Earthscan, 2016), 190.

17. Richard H. Hosier, "The Economics of Deforestation in Eastern Africa," *Economic Geography* 64, no. 2 (1988):121–36, on 129.

18. Andrew Hurst, "State Forestry and Spatial Scale in the Development Discourses of Post-Colonial Tanzania: 1961–1971," *The Geographical Journal* 169, no. 4 (2003): 358–69, on 362–63.

19. World Bank, World Bank Report No. 4969-TA: Tanzania; Issues and Options in the Energy Sector, November 1984, 68.

20. wikileaks.org/plusd/cables/1975DARES02228_b.html, June 23 1975.

21. Mitchell, *Carbon Democracy*, 38.

COBOL: The Pentagon, United States of America, 1959

Benjamin Allen

In May 1959 a group of about 40 representatives from computer manufacturers, US government agencies, and computer-using private companies met at the Pentagon to begin planning the development of a universal language for business computing, one that could be implemented on every model of digital computer then in use. These May meetings led to the creation of a consortium named CODASYL, short for "Committee on Data Systems Languages," which would oversee the new language's development.

"Business computing," here, refers to the use of computers to perform relatively simple operations across very large datasets; this is in contrast to "scientific computing," which involves the use of digital computers to perform complex operations on smaller datasets. Typical tasks within business computing included payroll management, inventory control, and the calculation of actuarial tables; tasks that were central to executives' management of business processes and states' management of populations, but that held little intellectual appeal to the mathematicians and physicists working in the then-emerging field of academic computer science. The sharp disciplinary split between business computing and scientific computing was reflected in the attendance at the May 1959 meetings; Jean Sammet, who represented computer manufacturer Sylvania at the meetings, would later note at the first ACM History of Programming Languages conference that one, but only one representative from an academic institution was present.[1] Academic antipathy to COBOL persisted throughout the period of its use; consider, for example, that first-generation computer scientist Edsger Dijkstra openly referred to the language as a "disease."

There were two competing desires held by the participants in CODASYL. The most pressing need was heading off an impending proliferation of business programming languages. Grace Hopper's team at Remington Rand had developed FLOW-MATIC, the first commercially used business programming language, in 1955. In the four years after the release of FLOW-MATIC, other computer manufacturers and computer-using organizations had moved toward producing their own mutually incompatible business programming languages; IBM had started development on their COMTRAN language, Honeywell on FACT, and the US Air Force had completed their own derivative of FLOW-MATIC, called AIMACO. The participants in CODASYL resolved that the joint development of *one* business programming language that could run on all manufacturers' machines would be preferable to the independent development of mutually incompatible languages. Although each manufacturer would be foregoing the possible competitive advantage obtainable through writing a proprietary language optimized for their own machines, the industry as a whole would benefit from the ability to run more or less the same program on different models of computer, and likewise from the reduction in programmer training expenses that would result from using a shared language for business software. Achieving this standardization required the very rapid development of a new programming language implementable on all machines.

The CODASYL consortium's second need, largely incompatible with the first, was the development of a *good* standardized business programming language, one worth using in the long term, after the "stopgap" language had halted the threat of language proliferation. In the interest of developing both a quick language and a good language, the consortium set up three committees. The Short Range Committee—more often referred to in internal documents as the PDQ Committee, short for "Pretty Damn Quick"—would develop a language "good for the next year or two." Two other committees, the Intermediate Committee and the Long Range Committee, would be tasked with developing the more thoughtfully designed language to follow the stopgap language. The CODASYL Executive Group, with representatives from US Air Force, DuPont, Esso Standard Oil, and US Steel, governed over these three committees. Charles A. Phillips, a US Air Force Colonel and the director of data systems research within the Department of Defense, chaired the Executive Group. It is worthwhile to note that all the organizations participating in CODASYL were from the US. Although the plan was to create a *universal* programming language, the organization that would make this universal language shared a set of very particular interests— broadly speaking, this universal language would be produced by and for the industrial/governmental formation which Dwight Eisenhower would later describe as the "military-industrial complex." It is noteworthy that the overwhelmingly American character of business programming language development is not reflected in scientific programming language development; the first cross-machine scientific programming language, ALGOL, was developed in 1958 through international cooperation between American and European academics.

Neither the Intermediate nor the Long Range committees produced any meaningful work, aside from two noteworthy advisory memos from the Intermediate Committee. As Sammet discusses in her speech at the 1978 ACM History of Programming Languages conference, these memos recommended that the creation of a new language be treated as something outside the Short Range Committee's purview, that the Intermediate Committee should be put in charge of producing the first CODASYL language, and that all of the Short Range Committee's already-produced work should be scrapped. Charles Phillips of the Executive Committee opted not to respond to these memos until after the Short Range Committee had completed their specification.

COBOL would remain in widespread use for the next forty years. One estimate made at the turn of the twenty-first century held that of the roughly 300 billion lines of running code in the world at the time, 240 billion of them were written in COBOL.[2] Views differ as to whether or not this outcome had been foreseen in advance by the participants in the committees. Although Jean Sammet, who led the Short Term Committee, would later state that she had genuinely expected COBOL to remain in use for about eighteen months and no more, Betty Holberton, who participated in the Short Range Committee as a representative of the US Navy's David Taylor Model Basin testing facility, quickly saw that the expense of implementing COBOL would ensure that it would necessarily become more than just an interim language. As Holberton put it: "I asked Grace Hopper if she could send me a cost estimate for implementing the compiler. She estimated that it would cost Sperry Rand $945,000 and 4514 man-years of effort. In no way was this language going to be an interim solution. This language was it!"[3]

The work involved in implementing the language ensured it would remain in use for longer than just eighteen months. This relatively longer period of use would in turn produce something like a feedback effect—the longer COBOL saw use, the longer it would see use. As companies and governmental departments built up a body of COBOL code to manage their business processes and financial planning, and as those same organizations trained programmers to use COBOL, they became incentivized to continue to use COBOL for future projects. Although COBOL was never particularly beloved by anyone, the infrastructural inertia established and reinforced by these path-dependent feedback effects ensured that it would remain in use for decades.

Universal Plain English

The word "Common" in Common Business-Oriented Language bears two meanings. COBOL was meant to be "common" in that it was designed to be universally implementable on all digital machines, but it was also meant to be "common" in that it would allow programmers to write in "common language," meaning English.

A major concern among executive-level figures within organizations considering the computer automation of business processes was the relative difficulty of monitoring and controlling the work of programmers. In order to make this technical work legible to non-programmers—or, at least, to make it *seem* legible—COBOL was designed such that each valid COBOL statement could also be a grammatically correct English-language sentence. It was this feature of the language that won it the most scorn from academic computer scientists, since in their understanding the language's verbosity tended to mask, rather than illuminate, the underlying logical structure of the code. And, indeed, even early in the development of the language it became clear that making individual lines of COBOL code resemble English-language sentences was not sufficient as documentation for the program's behavior on the whole.

Consider that nearly every programming language provides some capacity for the programmers to leave comments within the program's source code that aren't interpreted by the compiler, but are instead meant to explain how the code works to other programmers who may later read or change the code. Initially the Short Range Committee had planned to *not* include a comment function in COBOL, since they believed that the English-like character of the code would be sufficient for it to stand as self-documenting. Midway through the fall of 1959 committee chair Jean Sammet prepared a report entitled "Points of Greatest Controversy," discussing difficult design decisions the committee had made in the process of language development to that point. The first item listed reads: "Allow the word 'comment' or some equivalent such as 'note' as a verb. This would be reproduced for the listing but produces no object coding."[4] Whereas the discussion of most of the design decisions in this document involve long lists of technical pros and cons, the discussion of this item simply reads: "Reason: people often want to do this."[5] This laconic statement reveals a tacit admission of the code's inability to document itself.

If the code's resemblance to English allowed readers to understand it, people wouldn't need or want to leave comments.

Nevertheless, maintaining the resemblance to English was a top priority for many of the language developers. In fact, when members of the Short Range Committee first floated the idea of allowing programmers to use mathematical symbols rather than English words to express arithmetic operations, the Sperry Rand team threatened to pull out of CODASYL altogether, seeing this option as a step backwards. Their sharply worded memo reads as follows:

> It is the clear aim of Sperry Rand to ultimately design compilers which accept good English as their input. It is not the intention of Sperry Rand to progress backwards by introducing mathematical symbols and banning adjectives, etc. Sperry Rand would like to support the development of a common data-processing language, but suggests that such a language will probably have to be developed by people familiar with business data-processing problems and not be mathematicians and programmers.[6]

Although this memo is unsigned, the blunt writing style suggests it was written by Grace Hopper, who did not herself serve on the Short Range Committee but who selected and oversaw Sperry Rand's delegation to it. Ultimately this dispute was resolved by allowing both English-language keywords and arithmetic symbols for arithmetic operations, but such that arithmetic symbols could only be used on lines starting with the keyword COMPUTE, which served as a warning to readers that potentially mathematically complex operations could follow.

Why was the English-like character of COBOL code so important, given that it did not actually allow programmers to write self-documenting code? Promotional materials used to advertise COBOL provide one lead. As I argue, COBOL's resemblance to English is best understood as a type of *rhetorical device*, meant to persuade managers that code was trustworthy, rather than to make code actually as readable as natural language.

In many ways, programmers were outside of, and seen as implicit threats to, corporate hierarchies; they were people who managed machines rather than human subordinates, but made major decisions; most of them were trained in mathematics rather than business; perhaps most threatening of all, many early programmers were women. An advertisement for RCA's implementation of COBOL placed in the July/August 1960 issue of business computing journal *Datamation* highlights managerial anxieties about the illegibility of programming, and by extension the illegibility of programmers. The ad states that "RCA's new COBOL Narrator utilizes universal plain-English language to express your business procedures [...] The English language material prepared for the data processor is easily read and understood. *As a result, direct management supervision of procedures and records now becomes practical and effective*" (emphasis added).[7] If vital business logic was to be entrusted to code written by technical workers, management would like some way to be able to directly supervise the work. Establishing the *impression* that managerial control was possible had nothing to do with the program logic, but instead with what it looked like. COBOL allowed non-programmers to understand code as something like a business memo—a familiar and approachable tool for the exercise of control over work.[8]

Sammet's comments about COBOL given in her talk at the 1978 History of Programming Languages conference's COBOL session supports this interpretation of the significance of the language's English-like character. As she explained: "Surprisingly, although we wanted the language to be easy to use, particularly for nonprofessional programmers, we did not really give much thought to ensuring that the language would be easy to learn; most of our concentration was on making it "easy to read" although we never provided any criteria or tests for readability."

In his *Seeing Like a State*, James Scott discusses urban planning in terms of the need for local knowledge in designing and carrying out central planning.[9] A city built on a planned grid, like Chicago or Manhattan, is relatively easily navigable without assistance from locals, since one can get a rough sense of the location of any place in that city from its address. Likewise, a grid city is easier for a centralized power to govern, since the universalizing grid removes the need for the planner to engage with local, particular knowledges. The logic of the grid facilitates tasks ranging from digging sewer lines to policing against street crime or insurrection, without the planner or their agents needing to rely on local guides and trackers, since most salient facts about the city can be seen from outside and from above. COBOL was advertised as offering non-programming executives something like a view from above of their employees' work; whereas reading code in earlier programming languages required knowledge of complex technical detail, COBOL would allow non-programming managers oversight and control over their employees' work.

English, French, German

Although COBOL isn't English, and although the resemblance to English doesn't *really* allow non-programmers to read and understand it, the resemblance does allow non-programmers to understand what category of thing a program is. Individual pieces of code aren't made more legible, but the concept of programming itself is. However, this rhetorical effect is specifically only provided to English speakers; the American organizations involved in the production of COBOL saw no interest in extending the invitation to use that the language's design offered to anyone but English speakers. Managerial disdain for programming languages that resembled natural languages *other* than English was deep, and predated CODASYL; in Hopper's first demonstration of the language that would become FLOW-MATIC—a language that, like COBOL, allowed programmers to write statements that mimicked English—she had prepared a version of the language compiler that could accept not just English-language keywords, but also French and German. In her keynote address at the first ACM History of Programming Languages conference in 1978, Hopper would describe the aftermath of this demo as follows:

> Have you figured out what happened to that? That hit the fan!! It was absolutely obvious that a respectable American computer, built in Philadelphia, Pennsylvania, could not possibly understand French or German! And it took us four months to say no, no, no, no! We wouldn't think of programming it in anything but English.[10]

To Hopper as a programmer, the difference between English-language, French-language, and German-language keywords was purely superficial; the underlying bit patterns manipulated through those keywords remained the same, so it was trivial to produce different versions of the programming language using keywords in different languages. To management at Remington Rand, an American computer taking input in German was an offense.

Undoing Babel

The hopes of COBOL developers for their language can be read in the design documents used by the Short Range Committee. Notably, the playful cover page given for the draft specification the committee produced in October 1959 showed them conceptualizing their attempt to produce a universal language for business planning in terms of a much earlier attempt to explain the power of a universal language for project management—the Babel legend as given in Genesis.

The selection from the King James translation of the Old Testament is carefully chosen to emphasize the power fantasy behind proposing a universal language for control. It ends with the following verse: "And the LORD said, Behold the people *is* one, and they have all one language; and this they begin to do: and now nothing will be restrained from them, which they have imagined to do."

The quote ends midway through the story, immediately before divine intervention confounds the language of the tower-builders such that they cannot understand each other's speech, and therefore cannot coordinate the completion of their city and its blasphemous tower. The presence of this selection on the draft report playfully suggests that COBOL would rewrite the end of the legend—that the new universal language for business use designed by the PDQ Committee could replace the lost universal language of Babel, thereby allowing humans to build metaphorical towers to the heavens. Planners armed with digital computers and COBOL would be able to create anything they could imagine, with nothing restrained from them.

The layers of irony present in this reimagining of Babel are complexly tangled. Although the quote ends at the point of Babel's greatest potential for success, the ultimate fate of Babel has a ghostly presence. The quote appears as an expression of hubris, by suggesting that this new programming language may allow us to fulfill the dreams of the builders of Babel. But it also stands as a warning—just as the dream of Babel's builders proved impossible, so too the dream of perfected centralized command and control offered by a universal business programming language might be impossible. Moreover, the terms in which the declaration of COBOL's universal power are declared are themselves quite particular; an appeal to an English translation of a story from Abrahamic religion. This is, though, perhaps appropriate. Rather than attempting to create a universal programming language that could adapt to local contexts, the strategy taken in the design of COBOL was simply to declare all local context irrelevant—English-like programming was universal, because every language but English was *verboten*. And yet, this resemblance to English is itself something of a decoy; rather than allowing managers to read and understand COBOL code, it simply allowed them to believe that they could read and understand COBOL code.

Perhaps the most significant irony, though, is that the members of the PDQ Committee were in a sense justified in including this cover page. The language that they would design over the course of the last half of 1959 would become not just the only widely used English-like business programming language, but the most widely used programming language full stop. As data processing with digital computers became more and more central to large-scale planning tasks over the course of the last half of the twentieth century, COBOL became the standard language for planning those plans. For the next 40 years, programmers working on business and government computer systems worldwide would use a language rushed through development, with a syntax designed not for readability, learnability, or ease of writing, but instead to reassure particular elite figures in American organizations that computers were trustworthy for use in business.

Notes

1. Jean Sammet, "Early History of COBOL," in *History of Programming Languages*, ed. Richard L. Wexelblat (New York: Academic Press, 1981), 199–201.

2. Gary De Ward Brown, "COBOL: The Failure That Wasn't," The Cobol Report, Object-Z Systems Inc., May 15, 2001, www.cobolreport.com/columnists/gary/05152000.htm

3. Kurt Beyer, *Grace Hopper and the Invention of the Information Age* (Cambridge, MA: MIT Press, 2009), 290.

4. Jean Sammet, "Points of Greatest Controversy," box 2, folder "COBOL—Development, Standardization," Frances E. Holberton Papers, Charles Babbage Institute Archives, University of Minnesota, Minneapolis.

5. Sammet, "Points of Greatest Controversy."

6. "Sperry Memo," box 2, folder "COBOL Correspondence 1950–1988," Holberton Papers, CBI Archives. RCA, "Now RCA Removes More of the Mystery from Data Processing," advertisement, *Datamation*, July/August 1960.

7. RCA, "RCA Removes More Mystery."

8. Sammet, "Early History of COBOL," 219.

9. James C. Scott, *Seeing Like a State: How Certain Schemes to Improve the Human Condition Have Failed* (New Haven: Yale University Press, 1998).

10. Grace Murray Hopper, "Keynote Address," in *History of Programming Languages*, ed. Richard L. Wexelblat (New York: Academic Press, 1981), 17.

Computing: United States of America, 1949
Benjamin Peters

Computing is an unsettling and unsettled planning process. Planning, at some level, projects deliberate difference into the future; computing, at almost every level, is an activity by which groups set out to make plans toward some end, sometimes called programs, and invariably (and this is key) end up being taken somewhere else in the process. (The final cause—namely, the teleology plus the catastrophe—of logical mechanisms and its social processes is never final.) This essay, a brief sketch of the keyword "computing" and its alternative history, harnesses the basic ambiguous ends of computing understood as a planning process and subsequently seeks to reinterpret that keyword to enrich how scholars and students think about the role of groups and sociomaterial history in the postcolonial study of computing and planning media, information technology, and science.

Computing has become an instrument of choice among privileged planners for the exercise, extension, and automation of the legibility of state and colonial power: in fact the act of computing often appears an arch abstraction blind to the very world of mortal bodies, contingent institutions, complex practices, and fragile ideas it seeks to process.[1] However, this essay seeks to reconsider the keyword "computing" in hopes of rediscovering and recuperating intellectual resources long dormant but still available in the twentieth-century history of computing fit for a world computing more inclusive, participatory, and even at times radically disruptive practices. Computing need not only be a resource of state abstraction and legibility; so too can it be a participatory practice by which the deliberate projection of difference into the future may be disrupted and recalculated. For a simple example, instead of imagining computing to be an automated process of abstraction and calculation, what if computing were about the women and men who try to solve problems with plans—and, in the process, built computers, developed programs, and cared for, maintained, and serviced the tools by which planning has become digital and automatable? What if, furthermore, the social nature of computing as problem solving were extended to and centered on all those whose lives are unsettled by the "solutions" of such computing plans?

To that end, this brief, speculative essay proceeds in two parts: first, a short reflection on the lexical possibilities of computing and, second, a historical sketch of how computing, as understood as social groups in action, cannot be separated from the more radical invitation to settle the disputations that computing creates in its search for problems to solve.

Let's begin with the lexical limits of the keyword itself. Namely, "computing," conventionally understood, usually refers to a technology field loosely organized around computer science; Denning and Martell insist that computing is in fact an eclectic "science field" organized around six common principles.[2] This essay separates the idea of computing from its false near-synonyms—the computer and computation. Of course the history of computing interacts with the well-trod histories of computation and computers, but it also, I propose, potentially stands apart by emphasizing neither computers nor computation,[3] technology nor science, neither hardware nor abstraction, but instead the keyword reclaims digital action as an open-ended group activity. In particular, building on and borrowing from a burgeoning scholarly literature,[4] computing history, understood as the history of groups trying to solve problems, finds itself largely incommensurate with the great men celebrations so frequent in both popular and scholarly literature. A history of computing need not call forth either a history of computers, nor need it preoccupy itself with a history of computation. Furthermore, an emphasis on small groups "computing" stands to challenge the Silicon Valley romance with its own mythos about garage startups and angel investors, while at the same time credentialing a world of small group practices long present in computing worlds—workshops, studios, committees, small conferences, breakout sessions, coding scrums, hackathons (see Irani's "Hackathon: India," in this volume), and other forms of small groups solving problems. In a phrase, computing, understood as a gerund distinct from computers and computation, accounts for, circumvents, and unsettles the individualist, gendered, and cognitivist biases common in most conventional histories. Understood on its own terms, the activity of *com* + *putare* means to settle together, and thus a closer analysis of computing demands a refreshed series of stories about how information technology worlds have set about settling problems as live groups.

Now for a second glance at the *comput*-root lexical neighborhood itself. The keyword "computing" presses into thinkable form several features for unsettling and stirring critical analysis that the material noun computers and abstract ideation of computation do not: computing is a gerund without a given subject that refers to an ongoing activity in the present tense, a verb in motion. But not just any verb in motion, but a gerundial form that leaves strategically vague the actors involved in settling and solving problems. From the prefix *com* one gleans that no problem can be settled or even framed until another has been involved. Computing, built on a preposition requiring another, is definitionally social and relational. The connotations of the root *putare*, or "to settle," are diverse, open-ended, and plural as well: "settling with" has the obvious sense of reckoning, judgement, and the balancing of accounts with another at the end of an economic transaction, although it can also have resonant connotations in English with the sifting through of analytic problems; as in, to settle a dispute. Perhaps most troubling is that the root of "to settle" implies the occupying or colonizing of a space (whether literal or conceptual) together. Like Rana's gripping history of the "two faces of American freedom," the term computing suggests a strategic, complicit ambiguity in which the other with whom one computes may be a fellow settler, another with whom one settles, another alongside whom settling happens, and, vitally, one who is settled upon.[5] (It is obvious, at a glance, why the term computing should stir anxieties!)

At the same time, this same ambiguity also opens another type of present-tense possibility: namely, that the term computing implies a group whose present-tense interactions constitute not only the resolution but the making and consideration of problems. There is nothing about the present-tense that requires the term "computing" to arrive at an end solution. The field of computing would do well to set aside its fixation on solutions except as an acknowledgement that that is what it cannot provide: to compute is to process problems, not to solve them. Which problems? Or more precisely, whose problems? The study of computing as the study of the processing or settling of problems, without solutions, breaks open a wider range of problems to consider—not just symbolic and automated processes, but also groups at work on sociomaterial problems, such as labor, bodies, mechanisms of settlement and extraction, inclusion and exclusion, language and cultural practices. The present-tense status of ambiguously defined groups in action, in turn, does not cordon off the small group into a well-defined silo or coherent variable, but instead invites consideration of how the live settling of group's problems—among other phrases that stand in for computing—has long involved conversations, resource sharing, and other means of problem solving far beyond the normal ken of computing powers in the corporate, military, and academic disciplines.

The gerund also renders invisible the legible liberal subject from the history of computing as planning that is its favorite and most dubious hero—usually styled as geniuses, innovators, or geeks, and almost always privileged, white, and male. Instead by turning over the work of computing to an ambiguously defined collective group, or some plural set of actors working together, the history and analysis of computing stands to name, make temporary, and constitute in their relations (similar to how Monika Kirloskar-Steinbach refers to the constitution of the citizen in the chapter "Constitution: India, 1950") the study of a broader set of group dynamics. Groups, in ways that Colin Coopman has examined in his history of racist real estate data policies,[6] invariably constitute membership identities by excluding others; they debate and negotiate the terms of their work; they function by the grace of often invisible material labor and infrastructure; they invite relational analysis that both narrows the empirical analysis of problem solving to relatively manageable and temporary meetings in time and space and grounds problem solving as a collective activity whose value is greater than any accounting of its end accomplishments or consequences. As anyone who has enjoyed the toil of group effort (such as this volume) knows, sometimes the greatest value of settling problems together lies entirely in the settling process, not in the problem solved. Computing not only permits, but the keyword lexically speeds, in this refreshed analysis, new possibilities for non-individualist critique and analysis of the history and analysis of how groups identify, select, compromise, resolve, inhabit, and defer baseline problems.

The proposition at the heart of this speculative lexical exercise—namely, how can the decentered, pluralized, and ambiguous collective subject at the heart of computing—applies just as well to the potential for planning studies. As Van Beurden, Irani, and others argue in their own ways in this volume, there are diverse, abundant, and vibrant traditions for considering how, anywhere people may gather, group problem settling practices are at work. Instead of understanding planning as the imposition of the state or higher authority upon the collective, let us seek ways, such as in computing studies, to reverse that—and find how people solving problems together anywhere invites a reconsideration of the very coordinates of the analysis: no longer is planning only about projection of power into the future, now it can be understood as a present-tense action for whom outcomes, while the primary conscious aim of the group, are in fact secondary to the consequences of that process for the others it settles and unsettles. No longer is planning only about projection of power into a space, it can be grasped as a collective activity that constitutes, in the present-tense moment, that power not in the solutions it brings, but in the collective summoning of diverse actors. No longer are "computing" and "planning" only words fit for colonizers (as, by contrast, "computers" and "plans" may remain); so too may these live group actions be descriptions of what constitutes community in the moment. (May we avoid the unspeakable portmanteau of "compunity," a community constituted in the act of computing.)

Now, a historical sketch of alternative computing small groups. In practice, the sketch of six small computing groups that follows falls far short of the radical potential of what postcolonial computing studies might do to counter computational hegemony; a proper history of the others who bear the burdens of computing's wrongheaded and mis-imagined planning logics would include countless communities in the wake of colonialism. Nevertheless, in bridging more conventional and unusual stories about computing, even this somewhat conventional sketch offers a broader, bolder historical outline than has previously been legible. Six groups follow: first, Josiah Royce's Summer Schools on the Scientific Method at Harvard 1913-1915; second, the "Lusitania" group and the Orthodox name-worshipping rituals behind modern set theory in 1920s Moscow;[7] third, the interwar Vienna Circle and Wittgenstein's poetic responses to it; fourth, the Macy Conferences on Cybernetics in postwar New York (including the culture sleuthing of the power couple Margaret Mead and Gregory Bateson);[8] fifth, the Ratio Club in London;[9] and sixth, the Dartmouth Summer Research Project on Artificial Intelligence,[10] among others. Each group owes more to social computing as a settler planning process than their better known contributions to the development of computers and computation.

Consider the strangeness and variety of previously hidden specific group practices and institutional networks behind the problems settled by these groups: in the foreground we see in the spotlight material objects common to most academic conferences—tables, doors, and transcripts as constants in most meeting halls. But in the background we also see embodied participants such as Margaret Mead, who broke a molar tooth in her enthusiasm for the group work at the Macy Conferences, religious mysticism as a source of mathematical creativity in atheistic Moscow, poetry as a performance of how language overflows logic in interwar Vienna, and countless other unexpected totems of computing. Lighting the same stage we see the familiar knowledge base backing the military-industrial-academic-complex emerge in the context of war and military research, such as the Russian revolution behind the fervent religiosity of the Lusitania group, the Great War and the collapse of the Habsburg empire behind the Vienna Circle,[11] Allied wartime research behind cybernetics and so many of its researchers (Bateson's disinformation campaigns, Alan Turing's Bletchley Park, Vannevar Bush's "endless frontier," etc.). These military motivations cannot be separated from the subsequent service their groups provided to the major research universities (Harvard, Moscow State, Dartmouth) and philanthropies (the Macy Foundation, the Alfred P. Sloan Foundation).

In short, these and other troublesome institutions compose the twentieth-century triple helix of communication, computation, and cognition, and their contributions often stray far afield from computer hardware and computation abstractly defined. Along the way, we encounter surprising non-analytic, non-computational ideas such as the central role of analogy and induction (or even "abduction," as Royce called it) in modern scientific method, the mathematical innovations in modern probability theory and set theory (Moscow), the vocabulary for describing communication as computation (Macy Conference), cognition as performance (Ratio Club), and cognition as computation (Dartmouth Artificial Intelligence). Central to this interweaving of the vocabulary of modern-day computing are the unresolved debates and the striking out of contradictions coursing throughout these twentieth-century groups that turn to computing—namely, that the very groups that pedestaled the popular myth of the isolated mind as the ideal processor did so precisely as institutionally privileged, deeply internally contested, non-reconciling groups.

These particular groups, of course, happen to be populated by supposed "great men" and their ideas in the past; however, nowhere in the group histories sketched here are these particular elements necessary. For example, cybernetics, a meta-science of system analogies (the organism is like the computer is like mediated society) has often been associated with Norbert Wiener while, in fact, much of his synthetic work drew its method from his introduction to the idea of analogies between systems (or "the method of concomitant series") as a philosophical method in his adviser's (and perhaps the world's only absolutist pragmatist) Josiah Royce's summer seminars on the scientific method at Harvard, where Wiener was a teenage doctoral student. More than a story of intellectual influence, however, the summer seminars stand out as a model for the small group behavior that Royce elevates elsewhere to the status of the "community of interpretation."[12] This small-group seminar, lodged firmly in Wiener's mind as a place where any philosophical stake could be sounded out, in turn inspired a series of generative small groups in, and beyond, the early history of cybernetics and modern computing. By comparison, the Lusitania graduate seminar in the early Soviet Union also features unusual group practices such as the name-worshipping Jesus prayer uttered by the seminar's founding mathematicians in the basement of an Orthodox cathedral in Moscow; for them, this prayer stood as a ritual and acoustic rehearsal of the idea that both God and infinities can be variously named but never measured, thus invoked into reality but not counted symbolically. In this, the Lusitania graduate seminar fascinatingly rejects the fixed symbolic idealism implicit in most understandings of abstract computation.[13] Ludwig Wittgenstein, at about the same time but 2000 kilometers to the west, responded to the Vienna Circle's embrace of that symbolic idealism and its focus on logical empiricism (or language that can process only either a priori true or empirically verifiable statements) by a significant social gesture—by turning his back to the group of philosophers and reading poetry out loud. It is perhaps fitting that the group most preoccupied with advancing computational principles was riven in practice by a bit of poetry. Ludwig Fleck, too, first articulated the social constitution and circulation of scientific truth claims in reaction to the Vienna Circle, ushering a long tradition ready to rethink computing as necessarily a relational activity.[14] The other case studies—such as the Macy Conferences, the Ratio Club, and the Dartmouth Conferences on Artificial Intelligence, remembered today for sparking cybernetic and artificial intelligence discourses, did so in opposition to intra-group pressures (e.g. the Ratio Club jokingly admitted only those whose interests in cybernetics preceded Wiener, meanwhile the AI conferences dedicated itself to articulating mental representation without cybernetic vocabulary). Notably the groups that ushered in the computing age defined their boundaries not by ideas or technology but in relationship (often of rejection) to other groups. Computing has never not been social. These details reveal the irreparable and contentious politics of small groups directing and inspiring the tools of computational planning then and now—who, not what, makes up a group is also what keeps it small.

Much in planning and technology today clings to a romance of the small: the garage startup, the window office, executive and faculty retreats, the decision room of generals and imperialists, and countless other cabals of planners. Such groups are remarkably smaller in terms of their demographics than they may appear at first glance: of the over one hundred scientists in these six examples, only a handful identify as not male. The elitism inherited by Western settler colonialism promotes and excludes members, shuts doors, locks rooms, arranges seats at tables, privileges certain speaking languages and identities, publishing and editing houses, and access to other resources. Yet, in a crowning bitter irony, it is often also these very small groups that have also championed a notion of the open mind as the preferred liberal subject that embraces liberty, autonomy, and the use of reason,[15] and in turn celebrates the kind of centrist, strategic, market, and mental openness. This mental openness in turn, because it is open, fits particularly well into the basic research discourses connecting the military-industrial-academic complex, refashioning the "open mindset" into a mobile, anonymous service to various military, commercial, and intellectual rationally strategic purposes. Thankfully scholars such as Mar Hicks, Lilly Irani, and Sreela Sarkar (all critical of the reach of techno-industrial planning on various ends of the British empire), are examining whose so-called open mind, or as Halcyon Lawrence develops, whose open voice, is privileged in computing's colonial romance with small groups of planners.[16] No small group can plan the world and yet Facebook (or "Boomerbook" as the rising generation sometimes puts it) has a mere 70,000 employees straining to plan and manage over three billion monthly users (roughly one employee per fifty thousand users).

In lieu of a conclusion, the history of computing is a history of self-celebratory small groups of failed planners; in particular, the concentration of ideologically strategic open mindsets enables a special kind of cognitive smallness that disables researchers from more fully considering the effects of their research, and in turns shepherds in the smallness of their idea-world to stand in for its completeness. Small groups, because they are small and capable of articulating core disagreements, promote the institutional self-selecting logic of elite status, the prestige, and the strategic selectivity, and the perception of self-coherence of the open mind. Thus, as Edwards has amply illustrated,[17] open minds thrive in distinctly closed world systems; perhaps computing then settles problems only so far as the system it operates within unsettles the world without.

The resulting problem besets both intellectual and institutional computing histories: when pressed into the service of strategic goals, open mindsets beget a sort of an analytic narrowness that characterizes much of the twentieth-century cognitive, communication, and computation revolution. Instead of allowing multiple competing regimes of evaluation to coexist, the open mind, once formulated in small groups in search of coherence, usher in strikingly narrow frameworks and communities that have kept the too often individualist-industrialist history of computing from previously considering the very relational interactions of the groups that make it up.

This essay seeks to lexically rethink the keyword "computing"—perhaps the keyword most synonymous with the automation and expansion of analytic power and state legibility in the twentieth century. Computing is definitionally social, contingently present tense, and ambiguous in the actors it attempts to settle. By sketching out a broader, bolder, more varied archipelago of small groups that launched the collective processing of computing problems together, this essay calls for a future in which the groups that plan the next chapter of this information age may be less small. In other words, this essay reimagines computing as a small group activity for self-consciously settling problems that—in its smallness (and often its rank blend of elitist sexism, racism, and classism)—often unsettles the very problem it thinks it has solved, resituating those problems onto others. Computing is another name for small groups that, often with a self-celebratory nod to the mythos of technical analysis, believe they have solved problems while actually creating new problems and shifting them onto others. So too does this fresh intellectual toehold into the twentieth-century history of computing look beyond the strategic, male open mind in the dashes of the military-industrial-academic complex to the others with whom computing must settle its (dis)putations. The future of computing should be its past—a call to settle the problems it has created worldwide for those whose lives were never included in its small groups. Perhaps by building more inclusive intelligence quorums, without the computational hierarchies of intelligence quotients and the materially classist advantages of so-called smart computer hardware, critical scholars of planning may unsettle the very neuro-hubris and mental machismo characterizing the history and present of computing planning discourse in the name of settling the often unacknowledged disputes computing has speed worldwide.

Notes

1. Audra Wolfe, *Competing with the Soviets: Science Technology, and the State in Cold War America* (Baltimore: John Hopkins University Press, 2013).

2. Peter Denning and Craig Martell, *Great Principles of Computing* (Cambridge, MA: MIT Press, 2015), 3-9.

3. Michael Maloneym, "The History of Computing in the History of Technology," *Annals of the History of Computing* 10, no. 2 (1988): 113–25.

4. Nathan Ensmenger, *The Computer Boys Take Over: Computers and the Politics of Discourse in Cold War America* (Cambridge, MA: MIT Press, 2010); Mar Hicks, *Programmed Inequalities: How Britain Discarded Women Technologists and Lost Its Edge* (Cambridge, MA: MIT Press, 2016); Thomas S. Mullaney et al., eds., *Your Computer Is on Fire* (Cambridge, MA: MIT Press, 2021).

5. Aziz Rana, *The Two Faces of American Freedom* (Cambridge, MA: Harvard University Press, 2011).

6. Colin Coopman, *How We Became Our Data: A Genealogy of the Informational Person* (Chicago: University of Chicago Press, 2019).

7. Loren Graham and Michel Kantor, *Naming Infinity: A True Story of Religious Mysticism and Mathematical Creativity* (Cambridge, MA: Harvard University Press, 2009).

8. Claus Pias, ed., *Cybernetics: The Macy Conferences, 1948–1953: The Complete Transactions* (Zürich: Diaphanes, 2016).

9. Andrew Pickering, *The Cybernetics Brain: Sketches of Another Future* (Chicago: University of Chicago Press, 2014).

10. Martin Gardener, *The Mind's New Science: A History of the Cognitive Revolution* (New York: Basic Books, 1985).

11. Allan Janik and Stephen Toulmin, *Wittgenstein's Vienna*, 2nd ed. (Chicago: Elephant Paperbacks, 1996).

12. Josiah Royce, *The Problem of Christianity* (1913; repr., Washington, DC: Catholic University of America Press, 2001).

13. Graham and Kantor, *Naming Infinity*.

14. Ludwik Fleck, *Genesis and Development of a Scientific Fact*, eds. Thaddeus J. Trenn and Robert K. Merton, trans. Fred Bradley and Thaddeus J. Trenn (Chicago: University of Chicago Press, 1979).

15. Jamie Cohen-Cole, *The Open Mind: Cold War Politics and the Sciences of Human Nature* (Chicago: University of Chicago Press, 2015).

16. Mullaney et al., *Your Computer Is on Fire*.

17. Paul Edwards, *The Closed World: Computers and the Politics of Discourse in Cold War America* (Cambridge, MA: MIT Press, 1997).

Constitution: India, 1950
Monika Kirloskar-Steinbach

The Indian constitution, which came into effect on January 26, 1950, marked the culmination of a long process of imagining a diverse Indian nation. Entrusted with the task of framing a constitution for undivided India, a body of elected officials that made up the Constituent Assembly formally took up its deliberations under colonial rule in December 1946. It continued its task through the partition of India and Pakistan until political independence was achieved. The constitution laid the foundation for a postcolonial, independent state, which was set to plan and organize the lives of a multicultural, multi-religious plurality of people.[1]

Undeniably, the structural and institutional continuities between the colonial state and the postcolonial Indian state are hard to overlook. However, this short essay consciously steps back from that trite narrative propped up on a metropole/periphery binary. It will foreground aspects that were specific to the making of the constitution in postcolonial independent India.[2] It will home in on how Bhimrao Ramji Ambedkar (1891–1956) constituted a postcolonial citizen of a civic nation in India and briefly place Ambedkar's ruminations within the broader framework of intellectual decolonization.

Ambedkar's biography is unique in many respects. Ambedkar was a double doctorate in economics from the London School of Economics (1923) and from Columbia University (1927). He became the most distinguished leader of the "Dalits"[3] in colonial India and went on to become the chairperson of the constitution's Drafting Commission.[4] His politically motivated conversion from Hinduism to Buddhism in 1953 led to a resurgence of such conversions in India.

Ambedkar's point of departure was his observation that there had never been a nation in colonial India. They "are only an amorphous mass of people."[5] Going by his own experiences as a member of the underprivileged Mahar caste, he assumed that there were no preexisting commonalities, which could serve as a binding force to unite people living on the territory of this state; pernicious and divisive casteism and social discrimination ruled. In addition, there had been a decided lack of a commitment to substantive principles in the populace. These observations led him to hold that for all social and psychological purposes, India was not a nation when political independence from the British became imminent. It was thus futile, and to some extent naïve, to believe that independent India should simply tread down the path set by other nations in crafting their respective states. And yet Ambedkar did believe that some sense of nationality was needed if postcolonial Indians wished to retain their political independence. But if conventional nation-making projects were inappropriate for the specific situation of postcolonial India, what could be a viable path? Ambedkar's simple answer reads: India had to become a *civic nation*.

A civic nation is for Ambedkar a group of people who pursue similar sociopolitical goals within a state-like organization. Their sense of collectivity as a nation arises in the course of time. Civic ties ensue through the pursuit of common projects. Through this pursuit, a sense of commitment to common goods develops.[6] Members of a civic nation, furthermore, are not conferred rights by their state because of given criteria like descent, ethnicity, language, etc. Rather, they are considered to be right-bearers because of their participation in the socio-political projects this collectivity seeks to materialize.

Ambedkar's outright rejection of conventional perspectives on the nation also had to do with his concern that notions of the nation in India used by other political thinkers were unable to surmount the divisions of caste, gender, religion and so on. A case in point was Bal Gangadhar Tilak's (1856–1920) conception of a nation. Tilak traced the formation of a nation to shared, common criteria like descent, religion, culture, territory, and so on. In his view, individuals become right-bearers due to their birth into such a nation. And as members, citizens have a duty to maintain the distinct individuality of their nation.[7] He proposed India as the "Hindu nation,"[8] arguing that the British must confer this right. In a further step, Tilak attributed to these citizens the natural right to freedom, predicated upon membership of the nation.

For his part, Ambedkar seems to have perceived the difficulties of putting such a nation at the center of independent India's social imaginary. Such a notion would have been unable to end the protracted domination of minorities, who although free from alien rule would continue to be oppressed by societal domination. But as long as the latter persisted, it would be impossible to sustain the political independence of all those living on the territory of independent India, he reasoned.[9] This explains why Ambedkar sought to develop and implement a meta-plan which would constitute citizens of a civic nation on India's territory, in the process also constituting a new society. This meta-plan projected a radically different future for India. It also opened up a new moral and conceptual space in the context of the Indian political imaginary.

Ambedkar's projection of another future for India is driven by his understanding of a civic nation. He sought to lay the ground for a new way of perceiving individual persons in the Indian context. For this purpose, he attempted to push for a state that unambiguously set all of its citizens on an equal footing. This state would endeavor to establish material conditions such that social equality could emerge. In his reasoning, individuals would be able to make use of their freedoms only when a minimal sense of moral equality obtained in the polity. The post-independent Indian state would, in other words, have to make an Indian right-bearer. To do so, it would need to develop the constitution as a "moral text" such that it would enable oppressed minorities "to acquire self-description as subjective agents."[10] Ambedkar's vision of such an agent seems to have been informed by the following basic assumptions: (a) The individual is an end in itself; (b) she has certain inalienable rights which must be guaranteed and secured by the constitution; (c) she does not need to trade in these rights for the obtainment of certain privileges; and d) the state will not delegate its power of government to private persons. In other words: The status of a human being itself confers a right on an individual to legitimately claim equal treatment and respect from her state.[11]

With this move, Ambedkar demonstrated why India should on the eve of its political independence set up a robust state, which commits itself to the procurement and establishment of its citizens' rights. This state should, in his view, be bestowed with powers, with which it can abet and guide citizens in their realization of equality and freedom, regardless of their contingent social positioning.

Furthermore, Ambedkar sought to ascertain the durability of a discourse on equal freedom in independent India. In his reasoning, this discourse would have to be made continually visible in the social imaginary. But how to achieve this? Again, he banked upon statist measures in accomplishing this goal, his idea being that in the process of making the constitution, the judiciary should explicitly set the tasks of the legislative branch. In this manner a "constitutional morality", which is to be shaped by the values of freedom and equality, would ensue. In the first debate on the constitution's draft in 1948, he stated: "It follows that only where people are saturated with Constitutional morality [. . .] one can take the risk of omitting from the Constitution details of administration and leaving it for the Legislature to prescribe them. The question is, can we presume such a diffusion of Constitutional morality? Constitutional morality is not a natural sentiment. It has to be cultivated. We must realize that our people have yet to learn from it. Democracy in India is only a top-dressing on an Indian soil, which is essentially undemocratic."[12]

We see how Ambedkar deploys statist measures to effectuate changes in the social imaginary of independent India. He writes into existence a civic nation, one which had not previously existed in this state. Having said that, Ambedkar was, however, quick to realize that such coercive top-down measures would, on their own, be unable to carry forward sustainable changes. Planning itself did not suffice. Statist measures aimed at developing a constitutional morality had to be backed up by society. The latter must support the moral and personal development of its members. Through its "moral and social conscience," a society must help the state to secure the rights of its citizens.[13] From Ambedkar's perspective, society does indeed have an incentive for doing so: its realization that general welfare stands in direct relation to the capability of all its members to pursue self-realization. However, the viability of this task depends on these ideas being prevalent in society. One way of rendering these ideas visible in the social imaginary would be that a constitution explicitly endorses these ideas and gives them a prominent place within its framework. Another possible way of foregrounding these ideas within the social imaginary would be to look for ways in which a sense of belonging to a "we" can arise through the participation in common practices. Ambedkar seems to have been equally keen on exploring this path.

Ambedkar appears to have been optimistic that practices through which constitutional morality is materialized would themselves have positive effects. Citizens would, in this view, in the long run learn to acknowledge the founding role of the constitution in, and for, independent India. They would respect each other's equality and freedom simply because the constitution of their state guaranteed these goods to them. This respect would be translated into practices of self-restraint, toleration, and respect for plurality. These practices would result in a sense of fraternity amongst them. In the process, a civic nation would take shape.

In Ambedkar's statement presented to the Indian Committee on Franchise in 1919,[14] we read:

> Men live in a community by virtue of the things they have in common. What they must have in common in order to form a community are aims, beliefs, aspirations, knowledge, a common understanding; or to use the language of the Sociologists, they must be like-minded. But how do they come to have these things in common or how do they become like-minded? Certainly, not by sharing with another as one would do in the case of a piece of cake. To cultivate an attitude similar to others or to be like-minded with others is to be in communication with them or to participate in their activity. Persons do not become like-minded by merely living in physical proximity, any more than they cease to be like-minded by being distant from each other. Participation in a group is the only way of being like-minded with the group. [. . .] Like-mindedness is essential for a harmonious life, social or political [. . .].[15]

Ambedkar seems to have fully realized that conventional understandings of a nation possess a highly exclusionary potential. They operate with a clear sense of boundary policing and remain inaccessible to those who purportedly do not, and cannot, belong to the nation due to supposedly objective criteria that are said to define the nation. Being highly skeptical about the ability of such exclusionary concepts to weld together a group of diverse people in India, he aimed for an inclusivist understanding of a nation and a society. In his view, independent India, which was setting out on its journey of making a new state, had no other viable choice if it intended to organize and structure the lives of a pluralistic, multi-ethnic and multi-religious body of people. As chairperson of the Drafting Commission, he pitched the idea of an Indian citizen, who would understand the need to finally end structural domination inflicted by colonialism, but also by the societal forces of caste, gender, and religion.

It would not be far-fetched to consider Ambedkar's civic nation as a meta-plan for making a new Indian citizen. It was, equally, an attempt at planning a new, diverse Indian society. This plan envisaged a society which would allow for, and perhaps actively encourage, the interaction of several groups. The latter would through time develop like-mindedness of interests and aspirations across group divisions. Arguably, such a notion of a society is tailor-made for the diversity of India's people.

Focusing on a project which arose in the wake of the political decolonization in India, this brief account analyzes how Ambedkar's project of a civic nation attempted to produce knowledge for a particular context. In discontinuing the trite metropole/periphery binary, this entry deliberately refrains from tracing intellectual influences on Ambedkar to authors and positions in Euroamerica, without suggesting that Ambedkar developed his ideas in a social or political vacuum.

Intellectual decolonization, whether in histories of planning or otherwise, can be effectuated only when we—perhaps inadvertently as an intellectual reflex—stop tracing every single concept in intellectual history back to a purported historical antecedent in Euroamerica. As Sudipta Kaviraj warns about academic retellings: "At every stage in the academic presentation [. . .], this subaltern presence and contestation [is] erased in the retelling, so that the historical *re*-presentation of this process is far more European than the process itself."[16]

Ambedkar's ruminations on a civic nation can contribute to the decolonizing debate in another way. One contentious point in the decolonizing debate is the idea of the knowing subject. For long stretches of time, this subject was modelled after the self-image of the dominant male, who in his knowledge projects placed himself in the center of intellectual space such that the presence of others in this intellectual endeavor simply paled into insignificance. He became the unmoving, disembodied maker of disinterested knowledge, who in his knowledge-making endeavors easily traversed the limits of space and time. Undeniably, critiques of the knowing subject can aid in destabilizing conventional portrayals of such a subject. But would they suffice? A thorough decolonization, it seems, would have to transform our current perception of this subject, too (and perhaps even of the privileging of a unitary knowledge). What would such a transformation look like? One possible way would be to move away from understanding knowledge as the contribution of isolated, individual subjects, whose rational faculties lead to the generation of knowledge. It would have to investigate the social dimension of knowledge-producing endeavors and increase the focus on the participation of all societal groups in knowledge production. Knowledge making is not the prerogative of a privileged few.

Ambedkar's meta-plan for constituting a civic Indian nation resonates with this view when it seeks to make (and materialize) knowledge for the Indian context without relying on extant European understandings of "true" nationhood. Furthermore, his understanding of a civic nation underscores within the Indian context how participation in common endeavors can itself fashion a community, and its self-perception.

Notes

1. The Indian constitution can be found at http://lawmin.nic.in/olwing/coi/coi-english/coi-4March2016.pdf (last accessed August 15, 2016).

2 As this binary will have it, knowledge is actively produced in the global metropole while the passive periphery simply allows this knowledge to be applied to it. Such a binary suggests that the periphery is, relatively speaking, still incapable of producing knowledge on its own and must therefore be epistemically dependent on the metropole. See Naoki Sakai, "Theory and Asian Humanity: On the Question of *Humanitas* and *Anthropos*," *Postcolonial Studies* 13, no. 4 (2010): 441–64. Relatedly, Upendra Baxi draws attention to how certain political practices, which arose during the Indian independent movement, continue to be undertheorized in political theory today. They neither sprang up from discursive practices germane to the European Enlightenment nor were they solely grounded in the extant traditions of political thought found in India. Their potential as "new episteme" are not, in his view, sufficiently explored; see Upendra Baxi, "Outline of a 'Theory of Practice' of Indian Constitutionalism," in *Politics and Ethics of the Indian Constitution*, ed. Rajeev Bhargava (New Delhi: Oxford University Press, 2008), 92–118, on 106.

3. This term literally means "downtrodden." Ambedkar is said to have translated this Marathi term as "broken men" for the first time in 1948 in *The Untouchables* (New Delhi: Amrit Book Co., 1948). Today, the term is used as a self-identification by members of the lower castes, who were—and continue to be subjected—to casteist domination.

4. For some of his own reminiscences, see his "Waiting for a Visa" (available at http://www.columbia.edu/itc/mealac/pritchett/00ambedkar/txt_ambedkar_waiting .html).

5. Bhimrao R. Ambedkar, "Annihilation of Caste: With a Reply to Mahatma Gandhi" [1936], in *Dr. Babasaheb Ambedkar: Writings and Speeches*, ed. Vasant Moon (1979; repr., New Delhi: Dr. Ambedkar Foundation, 2014), 1:23–35, on 50.

6. Ambedkar, "Annihilation of Caste."

7. Parimala V. Rao, *Foundations of Tilak's Nationalism: Discrimination, Education and Hindutva* (Hyderabad: Orient Blackswan, 2011).

8. More specifically, he seems to believe that this right is the natural right of the upper-caste Hindu elite of Maharashtrian provenance. See his reference to the "Kulkarni Vatan" (the nation of the Kulkarni caste) in a speech in 1918, see Bal Gangadhar Tilak, *Bal Gangadhar Tilak: His Speeches and Writings* (Madras: Ganesh, 1922), on 112; Rao, *Foundations*, 269, 288.

If Tilak's strict refusal to open up education for women and members of the lower castes is any indication, it is doubtful whether members of these groups would enjoy the same status in this nation as its privileged members. See the critical discussion in Rao, Foundations, 96–171. See also Bal Gangadhar Tilak, *Sri Bhagavatgita-Rahasya*, or, *Karma-Yoga Sastra*, vol. 1 (Pune and Bombay: Tilak Bros., 1965).

9. His draft of the "Constitution of the United States of India" submitted to the Constituent Assembly in 1946 explicitly states that one of the goals of the new state will be to "to make it possible for every subject to enjoy freedom from want and freedom from fear" (Ambedkar, "Annihilation," 389). See also Bhimrao Ambedkar, "States and Minorities" [1947], in Ambedkar, *Writings and Speeches*, 1:381–449.

10. Gopal Guru, "Liberal Democracy in India and the Dalit Critique," *Social Research* 78, no. 1 (2011): 99–122, on 101.

11. Ambedkar, B.R. "States and Minorities" in *Dr. Babasaheb Ambedkar: Writings and Speeches, Vol. 1*, compiled by V. Moon, 2nd Ed., (New Delhi: Dr. Ambedkar Foundation, 2014 [1946]), 381–449, 409; Bhimrao Ambedkar, "The Hindu Social Order: Its Essential Principles," in *Dr. Babasaheb Ambedkar. Writings and Speeches*, ed. Hari Narake (1987; repr., New Delhi: Dr. Ambedkar Foundation, 2014), 3:95–129.

12. Bhimrao Ambedkar, "Draft Constitution—Discussion: Motion re Draft Constitution" [1948], in *Dr. Babasaheb Ambedkar:* Writings and Speeches, ed. Vasant Moon (1994; repr., New Delhi: Dr Ambedkar Foundation, 2014), 13:49–70, on 60–61.

13. "The prevalent view is that once rights are enacted in a law then they are safeguarded. This again is an unwarranted assumption. As experience proves, rights are protected not by law but by the social and moral conscience of society. If social conscience is such that it is prepared to recognize the rights which law chooses to enact, rights will be safe and secure. But if the fundamental rights are opposed by the community, no Law, no Parliament, no Judiciary can guarantee them in the real sense of the word." Bhimrao Ambedkar, "Ranade, Gandhi and Jinnah" [1943], in Ambedkar, *Writings and Speeches*, 1:205–240, on 222.

14. Ambedkar submitted a written statement and was invited to present oral evidence to the Franchise Committee on January 27, 1919. The committee is commonly known by the name of its chairman, Lord Southborough. Its members were asked for their recommendations on political representation, both at the federal and provincial levels.

15. Bhimrao Ambedkar, "Evidence Before the Southborough Committee" [1919], in *Ambedkar, Writings and Speeches*, 1:243–78, on 248–49.

16. Sudipta Kaviraj, "Global Intellectual History: Meanings and Methods," in *Global Intellectual History*, eds. Samuel Moyn and Andrew Sartori (New York: Columbia University Press, 2015), 295–320, on 303.

Dam: South Korea, 1961
Aaron S. Moore

A concept from geography called "multi-scalarity," offers a promising gateway into the historical analysis of planning large-scale infrastructure projects. Through this prism, we can analyze how various constituencies created and negotiated multiple scales such as the global, national, regional, and local in the process of planning, shaping, and completing one particular project. By breaking down the process of planning and constructing projects by scale, one then begins to see infrastructure less in terms of durable, fixed objects and more as complex, dynamic assemblages of political, economic, technological, financial, ecological, hydrological, and cultural flows and networks. One also begins to tease out the specific power relations and mechanisms that go into planning large-scale infrastructure projects and by extension, the larger project of nation building. By unraveling these assemblages of flows and networks, one brings out the contingency, friction, and uncertainty involved in planning infrastructure and therefore, reveals other unrealized historical possibilities.

Take, for instance, the South Korean government's efforts from the 1960s to construct multipurpose dams as part of its "Comprehensive National Land Planning" policy—plans which sought to rapidly industrialize the nation, modernize agriculture, and ultimately launch South Korea into the ranks of the industrialized nations. As in most cases involving planning infrastructure projects in the developing world, an analysis of the flows and networks of capital, people, ideologies, and technology often reveals a complex history dating back to the colonial era when colonial planners and engineers pursued their own plans and projects for the metropole. Post-colonial governments often resurrected these earlier plans and conceptions, inherited earlier infrastructure projects, relied on colonial records and data, borrowed money from their former colonial rulers, or signed contracts with companies that had earlier operated in colonized areas. My argument is that unraveling some of the flows and networks operating at infrastructure planning sites in the developing world often uncovers important dynamics that are at times left out of histories of planning and development—colonialism and its post-colonial effects.

In the case of South Korea, these were the colonial and post-colonial power relations operating within flows and networks concentrated around specific dam sites that emerged out of entangled histories of Japanese colonial rule and the rise of the US Cold War order in East Asia. The broader project of decolonization within post-colonial nations such as South Korea remains unrealized without fully tackling colonial legacies of planning and development, which remained very much alive after independence in the post-World War II, Cold War era.

The Return of Japanese Ex-Colonial Engineers to Post-Colonial Korea

In August 1961, only several months after General Park Chung-Hee's coup overthrowing the short-lived Second Republic of Korea (1960–1961) (after widespread student and labor protests overthrowing the earlier Syngman Rhee dictatorship) Japan's Foreign Ministry suddenly received a telegram from the Korean Central Intelligence Agency requesting the visit of Satō Tokihiko and five Nippon Kōei engineers at the Korea Electric Power Company's invitation. Satō was vice-president of Nippon Kōei, Japan's largest development consultancy founded and staffed primarily by former colonial engineers, and had himself enjoyed a twenty-year career in colonial Korea building dams for the Japan Nitrogenous Fertilizer Corporation (Nitchitsu). During this time, he rose to become head of civil engineering for Sup'ung Dam on the Yalu River, the second-largest dam in the world at the time (after Grand Coulee Dam), which provided power for the wartime heavy industrialization of Manchukuo and Korea during the 1940s. Nippon Kōei's founder and president, Kubota Yutaka, also had a twenty-year career in Korea and was the main visionary behind building large-scale infrastructure, primarily dams, for industrial development throughout the Japanese empire. As chief engineer for Nitchitsu's Korean subsidiary, Kubota along with other engineers like Satō, who later joined Nippon Kōei, supervised the construction of colonial Korea's hydropower dam infrastructure that produced the near equivalent to mainland Japan's total energy production at that time. By 1942, Nitchitsu owned 36 percent of Korea's industrial fixed capital and was establishing factories and infrastructure in Manchukuo, Taiwan, Hainan, and China.

Under the fiercely anti-Japanese Rhee regime (1948–1960), few Japanese could visit South Korea in any official capacity. Grounded in a strong anti-Japanese sentiment among the population, Rhee feared renewed Japanese economic imperialism and therefore required strictly balanced trade with Japan. He also demanded a minimum of $2 billion in reparations as a condition for normalizing relations between the two countries.[1] The status of fishing rights in the seas between the two countries as well as Koreans living in Japan were also under fierce dispute. Thus, in this charged environment, the Japanese Foreign Ministry was taken by surprise and warned Satō not to discuss political and economic matters with his Korean counterparts since reparations and the colonial past continued to be such a sensitive topic.[2]

What followed was a whirlwind five-day tour of Korea's potential hydropower spots and recently begun dam projects. The Nippon Kōei team, several members of whom had worked in colonial Korea, observed first-hand the Japanese colonial infrastructure legacy that the South Korean government was trying to overcome. This included an acute lack of energy infrastructure to meet the country's needs, since the Japanese had built most of the dams and power stations in the more industrialized north while prioritizing agriculture for the south; the acute lack of engineering skills among Korean engineers also due to this colonial legacy, which relegated Koreans to lower-level technical positions; the half-finished Japanese colonial dams the Korean government was trying to finish and seeking expert advice on; and the overall effort toward "comprehensive national land planning" for the purpose of establishing economic independence and rapid industrialization (i.e., decolonizing the economy). Park Chung-Hee, himself a former officer in the Japanese Imperial Army and more open to Japanese-style central planning, based on his career in Manchukuo, had included many Japanese-educated and trained bureaucrats and engineers in his government, which presumably led to this surprise invitation of former Japanese colonial engineers. South Korea's post-colonial plans for rapid economic development in the face of a more industrialized, prosperous North Korea had ironically brought Japanese and Korean experts together, which in turn immediately conjured up ghosts of Korea's colonial past.

Nippon Kōei was soon hired to write an assessment of Korea's hydropower infrastructure and potential, which led to the "Republic of Korea Hydropower Investigation Report," submitted to the Korea Electric Power Company in September 1962. By then, Nippon Kōei had grown into a leading development consultant for large-scale dam projects throughout Asia (many of which were in the former Japanese empire), in places such as South Vietnam, Indonesia, Laos, and Burma, which by then were Cold War hot spots.[3] As many as fourteen of its top twenty managers had long careers in colonial Korea, Manchuria, and Taiwan, and these senior engineers pitched project ideas and planning conceptions rooted in their colonial experience to post-colonial leaders throughout Asia, who turned to the kinds of high-modernist projects Nippon Kōei was offering in order to pacify and unify their divided populations.[4] South Korea was another lucrative market for the company, and engineers continued to appeal to their past colonial experiences in Korea and advanced technical expertise in order to win dam construction projects from Park's new military regime.[5] Thus, with the rise of post-colonial development dictatorships such as Park Chung-Hee's South Korea, former Japanese colonial engineers began to thrive throughout Asia as they offered high modernist planning visions to leaders who themselves were quite amenable to highly visible infrastructure schemes such as dams that promised rapid industrialization and national prosperity to increasingly restless and frustrated populations.

Colonial Haunting: Building the Intellectual Foundation for High-Modernist Water Resources Planning

Ghosts of the colonial past—in particular, past colonial river studies and dam construction projects—immediately asserted themselves within the Japanese engineers' 1962 hydropower study. In assessing the feasibility of dam projects on the Soyanggang River for the Seoul metropolitan area, a dispute broke out about the accuracy of Japanese colonial statistics on Korea's rivers, which the Korean government relied upon as an important basis for river basin planning and dam construction. The extremely technical debate centered on streamflow statistics from the area compiled in the 1929 Korea River Investigation Report, one of the most definitive and thorough compilation of statistics on Korea's rivers. From this study, Japanese colonial hydrology experts developed the "Kajiyama formula" for measuring the streamflow of Korean rivers based on the unique characteristics of its weather patterns and river basin environment. Korean government engineers (with the blessing of American consultants and Bureau of Reclamation engineers who had earlier used this formula in their own Korean river studies during the 1950s) once again used this formula as a way to roughly calculate streamflow, an essential statistic for determining dam capacity, structure, and size, and therefore, electricity output and irrigation/flood control ability as well.

The cause of the calculation problem lay in the actual climactic conditions around the proposed dam site. Apparently, large floods during the summer rainy season in the Soyanggang area occurred in much shorter, concentrated time periods (a few days) than what was previously assumed as a sufficient time frame for calculating average streamflow during the colonial study—four to ten days. Hence, the Japanese colonial report calculated the average streamflow of Korea's rivers for a longer time frame (four to ten days) than what may have been needed to fully account for the area's peculiar hydrological and meteorological conditions. In other words, the colonial statistics did not fully capture the very erratic streamflow and weather conditions of the Soyanggang region in the summer, which was when streamflow was at its highest (as was energy demand and energy production). Thus, the colonial report's average streamflow statistics may have been lower than what was required—a higher average streamflow in turn would necessitate a larger dam and other structural changes to accommodate the intense, erratic, and concentrated rainfall in the region. Structural design in turn affected how much water could be released for energy production and irrigation (water use) or kept for flood control (water preservation).

The statistics therefore had to be revised to provide a more accurate average, or a margin of error had to be determined to account for the lower streamflow volumes recorded in the Japanese colonial report, which did not fully capture the unpredictable river behavior in the Soyanggang River region.[6] Moreover, newer data collected by the Korean government had to be incorporated to improve the earlier Japanese study. In the 1962 report to the Korean government, Japanese engineers acknowledged the issue; however, they asserted their familiarity with the colonial study and longtime experience of negotiating Korea's unpredictable river behavior to successfully build colonial projects in the past as a way to reassert their own scientific and technical expertise over their Korean counterparts. In this case, they made their own expert assessments about how Korean government engineers arrived at new statistics and formulas for streamflow measurement and volume calculation, and gave technical advice in their report accordingly.

Unraveling the Colonial: The Korea River and Hydropower Investigation Reports

This very technical post-war dispute highlighted the contingent, makeshift, uncertain nature behind planning high-modernist projects such as dams. Seemingly the inevitable product of the progressive, rational application of universal scientific principles and the systematic *subjugation* of nature, dams projected powerful ideologies of modernization that frequently mobilized and *subjugated* various peoples and environments to state designs. However, upon revisiting the earlier colonial Korea River Investigation Report of 1929 (as well as the 1930 Korea Hydropower Investigation Report, conducted simultaneously) and the scientific authority projected through hundreds of pages of graphs, charts, photographs, drawings, and maps of these reports, one begins to see dams less as durable, fixed objects that decisively transformed nature for human needs. As we shall see, both of these studies were the products of dynamic flows and networks of ideals, materials, people, natural forces, and capital that needed to be forced together into a tense, unstable whole. This dynamic assemblage of flows and networks then figured itself as scientifically authoritative data to be used for project planning in Korea even after liberation from Japanese rule by the post-colonial South Korean government. The 1962 dispute over accuracy merely punctured a hole in the edifice of high-modernist scientific authority—a hole that is worth examining to discover the inescapability of uncertainty and contingency in the shifting nature of dam planning as a whole.

Japan's 1929 River Investigation in colonial Korea was the first to measure unstable environmental flows of weather, geology, and hydrology, thereby rendering them visible for large-scale dam construction. It was partially the product of new conceptions of dam building being developed worldwide whereby the full capacity of rivers would be harnessed into large reservoirs to be then used rationally for electricity production, irrigation, flood control, industrial and urban development, and river transportation improvement. New data for this way of understanding river basins was needed to build the large-scale, concrete gravity dams based on this conception. Although the final report presented Korea's rivers as quantifiable and manageable in the form of officially compiled statistics (i.e., the result of the systematic, rational application of ready-made scientific laws), the whole process of rationalizing river flows in this manner was in fact quite unstable, imprecise, and involved the contingent assembling of various extra-scientific institutions and practices.[7] Together, the 1929 Korea River Investigative Report (conducted between 1921 and 1929) and the 1930 Korea Hydropower Investigative Report (1922–1930) provide insight into the types of practices used on the ground to understand river flows in order to create a new regime of certainty, which made hitherto unfamiliar environments transparent in the form of authoritative graphs, maps, and charts for hydropower and other high-modernist development. The new regime of certainty created by these two reports laid the basis for the construction of large-scale dams at the nucleus of "comprehensive national land plans" during the colonial and post-colonial eras as well. The reports lent these plans and projects an aura of authority, durability, and scientific progress, which in turn made the projects difficult to question and contest in both eras. Moreover, the socio-political aspects of planning became further disguised in the form of authoritative data and diagrams.

Korea's colonial Japanese engineers only realized Korea's hydropower potential while conducting these studies, when they began collecting data on the maximum and normal streamflows of the country's major rivers to determine if they were sufficient for building the high, concrete gravity dams necessary to meet the electricity demand for Korean industrialization and urbanization.[8] First, for the 1930 hydropower study, colonial engineers settled on several promising dam sites based on earlier, smaller-scale investigations as well as the Imperial Army Survey Division's topographical maps. They took into account a variety of considerations to measure erratic and unpredictable flows such as weather conditions, topography, geology, and estimated transmission and construction costs.[9] After forecasting electricity output at each location, the engineers did meteorological and hydrological tests as they discovered new sites and set up precipitation and streamflow measuring devices at the most promising ones.[10] Their budget, however, limited their ability to build the necessary number of stations or conduct the necessary tests.[11] For example, there was only one hydrometric station per 1600 square kilometers, which forced engineers to make generalizations based on sparse data about streamflow across vast, diverse river systems.[12] The final published reports hid these generalizations through what appeared to be organized statistical charts, giving the sense of transparent river flows and networks subject to manipulation and control.[13]

Obtaining the measurements was difficult because it required negotiations with and the bringing together of different, often conflicting colonial institutional networks to create new assemblages that made the investigations possible. For example, the Communications Division supervising the hydropower study had to conclude formal agreements with other government divisions such as the Public Works, Internal Affairs, and Productivity Bureaus to share their respective stations.[14] Koreans who were "somewhat illiterate" with a poor command of Japanese had to be trained to take daily precipitation and hydrometric measurements, as well as to submit monthly reports.[15] The 1929 River Investigation had very detailed "rules and regulations" on how exactly to measure streamflow and precipitation—the times of day for measurement, the location and positioning of instruments, what phenomena to look for and record, how to record different types of data, how to make various calculations based on that data, the proper forms to fill out and where to file them, and so on.[16] In short, a whole disciplinary regime was imposed on Korean technicians and lower-level Japanese colonial engineers. Local post offices and police stations were mobilized to oversee Korean observers. Poor weather was also a frequent problem, with floods washing away gauges or preventing teams from reaching remote stations. Korea's rivers were wide and sometimes doubled or tripled in width during the summer rainy season. State engineers therefore often had to conduct makeshift streamflow measurements or rely on local hearsay to determine maximum water levels.[17] Japanese engineers also turned to classical records from the Chosŏn Dynasty (1392–1897) of river measurement and river improvement projects for data — however, these were often framed as erratic, unsystematic, and haphazard, therefore justifying Japanese colonial scientific intervention.[18] In the end, despite the mobilization of Japanese colonial scientific expertise, earlier environmental flows and networks continuously disobeyed the logics of the new scientific knowledge that Japanese engineers were bringing into being. At subsequent dam projects, floods would continually plague engineers and make construction difficult.[19]

Finally, the need to determine a uniform streamflow coefficient that could calibrate varying streamflow gauges resulted in the makeshift construction of a gauge inspection facility in 1924. Using a sedimentation tank at a water treatment facility outside the capital Keijō (Seoul), they built a 48.5-meter long track for a special rail car equipped with a velocity gauge that reached into the tank. By moving the car at different speeds and taking readings, they arrived at a device coefficient that would then uniformly measure streamflow across the peninsula.[20] Thus, scientific devices designed to accurately measure the environment disguised the uncertainty, ambiguity, and fluidity within their very structure. Seemingly stable, scientific instruments too were merely the product of bringing together new flows and networks of knowledge.

In sum, the environment was not simply an inert, passive object outside of the above practices, ready to be transformed into definitive data and manipulated by Japanese technical expertise. It was a dynamic assemblage actively constituted by bringing into relation many unstable, extra-scientific processes such as disciplining Koreans to take measurements, making erratic rivers conform to new knowledge and devices, translating vast generalizations, classical records, and local hearsay into numerical data, and building makeshift facilities to anchor new forms of measurement. It was this haphazard, contingent bringing together of flows and networks consisting of various processes, knowledges, peoples, and institutions into the pristine form of transparent river and hydropower studies that slightly revealed itself in the 1962 dispute over the accuracy of Japanese colonial data.

Conclusion: Reassessing Developmentalist Planning in South Korea

However, in the 1962 dispute over the accuracy of hydrological data in the region of the proposed Soyanggang multipurpose dam, rather than questioning and unraveling the colonial scientific edifice as a whole to reveal the power relationships and networks involved and the fluid, changeable nature of planning in general (which would then create the potential for interventions into the planning process), South Korean and Japanese engineers merely confined the problem to one of improving statistical accuracy. In short, they accepted the basic presumptions, conclusions, and foundations of earlier colonial studies that epistemologically grounded the whole edifice of multipurpose dam construction and comprehensive national land planning.

The Park Chung-Hee regime revitalized the type of high-modernist planning of large-scale dam and river basin projects grounded in colonial river studies and carried out earlier under Japanese imperialism. For this task they required the services of some of the same Japanese colonial engineers and earlier flows and networks of colonial scientific knowledge. Korean engineers added their own layer of data, statistics, and graphs, thereby further lending credibility and durability to the overall intellectual edifice of dam construction and comprehensive national land planning.

This opened the gateway to the subsequent flood of Japanese capital and investment into South Korea upon normalization of relations in 1965, in exchange for the technology and expertise that they brought to enable economic self-sufficiency or what would become known as the "Korean economic miracle" of the 1970s. Japanese trading, construction, machinery, building materials, and consultant companies acquired a new market and even provided long-term loans and credits tied to purchases of their products.

The ghost of Korea's colonial planning past, however, could never be fully contained, as revealed by the 1962 dispute over data accuracy described at the beginning of this chapter. During the actual construction of Soyanggang Dam (1967–1973), for example, the contingent, haphazard nature of earlier colonial efforts would once again come into view as fierce summer floods frequently caused significant construction damage, Korean engineers disputed Japanese designs based on earlier colonial experience, and residents resisted the efforts of high-modernist dam construction in different ways. Moreover, beginning in 1981 Soyanggang's water gates had to be opened to alleviate unexpectedly high water levels threatening to overtop the dam and cause serious flooding. Typhoons in 1984 and 1990 created similar problems. Therefore, in 2004 four additional spillways had to be constructed to alleviate potential excess water.[21]

The vision of a controlled and transparent environment presented by Park Chung-Hee's government in his comprehensive national land plans, which in many ways revitalized Japanese colonial planning legacies, were continuously punctured in the post-colonial context of high-speed economic growth, and even afterwards. Although South Korea presented itself as completely self-sufficient in taking charge of its own economic destiny and overcoming the legacies of colonial dependence and underdevelopment, the reality was that the colonial Japanese economic, engineering, and planning legacy remained quite strong in the form of Park's "developmentalist state," as the nation underwent decolonization after 1945.

Since the 1980s, South Koreans have begun to assess the legacies of Park's high-speed, state-led economic growth policies. For example, civil society groups are questioning the benefits of comprehensive river basin planning through dam construction by demanding more investment in the regions that house the dams (rather than in urban and industrial areas), drawing attention to the particular cultures and economies that have been displaced by big dams, and opposing the construction of more large dams throughout the country.[22]

Unraveling the historical flows and networks that ground Park's high-modernist plans denaturalizes their authority and durability. More importantly, it brings to light colonial practices such as mobilizing populations for data collection and measurement, rendering local and earlier conceptions of water resource planning "irrational" and "unmodern," and assembling state institutions to perform and enforce scientific certainty and authority despite evidence to the contrary. Unraveling these colonial histories that undergird post-colonial planning projects can therefore contribute to the ongoing project of decolonization in South Korea.

Notes

1. Merideth Woo-Cumings, *Race to the Swift: State and Finance in Korean Industrialization* (New York: Columbia University Press, 1991), 85.

2. Satō Tokihiko, *Doboku jinsei gojūnen* (Tokyo: Chūō kōron jigyō shupan, 1969), 286–87.

3. As Japan's economy began to rapidly grow from the late 1950s and the US began to increasingly commit more economic and military resources to the worsening conflict in Vietnam, the US urged allies in East Asia such as Japan, Taiwan, and later South Korea to take up more of the burden of investing in development projects throughout Asia to help combat Soviet and Chinese influence there. Nippon Kōei was well positioned to take advantage of this American policy.

4. Shimizu Tomihisa, "Kubota Yutaka-Nippon Kōei to Chōsen-Betonamu," *Shisō* 14 (November 1973): 35. James Scott describes high modernism as the ideology

behind "monotonic schemes of centralized rationality" such as large dams that "straitjacketed" the human and natural words. He defines "high modernism" as a strong "self-confidence about scientific and technical progress, the expansion of production, the growing satisfaction of human needs, the mastery of nature . . ., and, above all, the rational design of social order commensurate with the scientific understanding of natural laws." Scott, *Seeing Like a State: How Certain Schemes to Improve the Human Condition Have Failed* (New Haven: Yale University Press, 1998), 4.

5. Nippon Kōei would become involved in as many as twenty-four dam construction and river basin planning projects in South Korea. Cho Kab-je, "Ch'ongdokpu gogwan dŭl ŭi kŭdwi," *Wŏlgan Chosŏn* (August 1984): 294.

6. Ilbon Gongyŏng, *Taehanmin'guk suryŏk chosa bogosŏ* ([Seoul?]: Ilbon Gongyŏng Chusik Hoesa, 1962), 2–61, 62, 2–68.

7. Richard White, *The Organic Machine* (New York: Hill and Wang, 1995), 76. For other work that focuses on how environments are "rationalized," see Timothy Mitchell, *Rule of Experts: Egypt, Techno-Politics, Modernity* (Berkeley: University of California Press, 2002), 19–53.

8. Kawai Kazuo, "Dai niji suiryoku chōsa to Chōsen sōtokufu kanryō no suiryoku ninshiki," in *Nihon no Chōsen Taiwan shihai to shokuminchi kanryō*, eds. Matsuda Toshihiko and Yamada Atsushi (Tokyo: Shibunkaku, 2009), 304. Earlier they had only collected data on minimum streamflow to ensure that the minimum amount of power could be provided for smaller-scale dams largely built for small factories, transportation cities, and urban lighting.

9. Chōsen Sōtokufu Teishinkyoku, *Chōsen suiryoku chōsasho dai ikkan (sōron)* (Keijō: Chōsen sōtokufu teishinkyoku, 1930), 22.

10. Chōsen Sōtokufu Teishinkyoku, *Chōsen suiryoku chōsasho*, 15.

11. Kawai, "Dai niji suiryoku chōsa," 314–315.

12. Chōsen Sōtokufu Teishinkyoku, *Chōsen suiryoku chōsasho*, 171.

13. Chōsen Sōtokufu Teishinkyoku, *Chōsen suiryoku chōsasho*, 15, 22, 171.

14. Chōsen Sōtokufu Teishinkyoku, *Chōsen suiryoku chōsasho*, 12.

15. Chōsen Sōtokufu Teishinkyoku, *Chōsen suiryoku chōsasho*, 75.

16. Chōsen Sōtokufu, *Chōsen Kasen chōsasho* (Keijō: Chōsen sōtokufu, 1929), Appendix, 57–79.

17. Chōsen Sōtokufu, *Chōsen Kasen chōsasho*, 161.

18. Chōsen Sōtokufu, *Chōsen Kasen chōsasho*, 2–14.

19. For more, see Aaron S. Moore, "'The Yalu River Era of Developing Asia': Japanese Expertise, Colonial Power and the Construction of Sup'ung Dam, 1937–1945," *Journal of Asian Studies* 72, no. 1 (2013):115–139.

20. Chōsen Sōtokufu Teishinkyoku, *Chōsen suiryoku chōsasho*, 194–196.

21. Many experts conclude that climate change has caused unusually high water levels in recent years. Thus, it would not be completely fair to fault colonial engineers for not predicting climate change. However, their focus on water use for industrialization, agricultural modernization, and urbanization over water preservation

for flood control and ecological preservation led to dam designs that might not have fully accounted for potentially high and dangerous streamflow volumes, as was the case at Soyanggang. See Kim Se Gun, "Soyanggangdaemgwa chiyŏkchumindŭrŭi ilssang - hŭrŭgirŭl mŏmch'un kang kŭrigo chŏngch'ŏ ŏpsi hŭrŭnŭn sam" *Sahoegwahagyŏn'gu* 54, no. 2 (2015): 26, 29.

22. For example, see Chŏng-uk Kim, *Na nŭn pandae handa: 4-tae kang t'ogŏn kongsa e taehan chinsil pogosŏ* (Seoul: Nŭrin Kŏrŭm, 2010).

Dodecahedral Silo: Spain, 1953
Lino Camprubí

Autarky is back in the political debate. Modern state planners across the political spectrum have aspired to self-sufficiency in certain key strategic areas, but the goal of an autarkic political economy is historically more specific. For all its precedents, the autarkic imperative reached its practical and theoretical highpoint in fascist regimes. That they depended on international flows and a context of colonial competition paradoxically shows economic nationalism to be an international movement. This chapter explores some varieties of autarky, including the "Christian concept of autarky" developed in Francoist Spain. It does so through a close attention to the material objects—such as a coal silo—that simultaneously embodied autarky and put it in motion.

One specific coal silo, strikingly shaped as a dodecahedron, acted as a functional representative of the Francoist plans for the Spanish political economy. Built in 1953, it was made of nationally produced cement, fed with nationally extracted coal, designed by nationally oriented engineers, and built by Spanish workers whose work and morality had acquired a new national meaning through the Christian concept of autarky and a new system of rationalization of labor. It fueled the new Costillares laboratory for the National Institute of Construction and Cement. The entire building was also a functional representative of the new Spanish political economy: prefabricated national materials assembled together according to a design that favored efficiency, economy, and creative functional aesthetics constituted an industrialized landscape that would host the latest research, experiments, and tests to industrialize the entire Spanish territory. This was no science fiction. As the title of the book from which this image comes suggests, it enacted an entire philosophy of construction.

Figure 1. The Costillares Coal silo functionally representing autarky in 1953. From Eduardo Torroja, Razón y ser de los tipos estructurales (Madrid: Instituto Técnico de la Construcción y del Cemento, 1957). Reprinted with permission of Archivo Histórico Instituto Eduardo Torroja de las Ciencias de la Construcción (IETCC).

The Costillares dodecahedral silo embodied the Francoist planning for autarky. However, in a sense, the silo was a failure. Its history gives as an entry point into the history of autarky (and its Spanish version: "the Christian concept of autarky") as well as into the limitations of autarky and its relationship to economic globalization.

Autarky, or self-sufficiency, is an extreme expression of state planning. Now, economic protectionism and even self-sufficiency are suddenly back in the public debate, after decades of scorn by economists, politicians, and scholars. The global financial crisis led to left and right anti-globalist responses across the world, reactions to the COVID-19 pandemic reduced world trade and closed borders, the climate challenge makes oil-supported commerce less feasible, and war reminds us of the dangers of dependency. The illusion of unlimited global flows has come to an end. And state-planned self-sufficiency has reemerged as a seemingly viable option. To gauge it, it might be useful to look back to a time, not that long ago, in which autarky was an acceptable economic theory and economic nationalism was an international movement.

Autarky is usually understood as the theory and practice of economic self-sufficiency developed in fascist countries during the early twentieth century. Nevertheless, the precedents are as many as they are telling. They are usually found in nineteenth-century Germany, with works like Johan Gottlieb Fichte's *The Closed Commercial State* (1800) and Friedrich List's *The National System of Political Economy* (1841). The latter suggested that strong tariffs were the only way to protect a latecomer to industrialization against more developed titans. Other nations had followed similar strategies. In an interesting example of what Ha-Joon Chang has called *Kicking Away the Ladder*, the United States soon turned from protectionism to forcing Tokuwaga's Japan to open its frontiers to Western trade.[1] That is from the demand side.

From the supply side, import substitution meant relying on national raw materials and production. European powers struggling for resources had pursued this goal through expansion and appropriation. Latecomers to the colonial game needed to gain knowledge about domestic resources and to devise new ways of exploiting them. Once again, science met political economy. According to historian Lisbet Koerner, Swedish self-sufficiency was one of the major drivers for none other than botanist Carl Linnaeus.[2] Since then, the goal of making the most of seemingly unproductive territories to reduce dependency on foreign goods has triggered important research programs. In the late nineteenth century, the major European powers played by the rules of laissez-faire and the scramble for Africa. But after World War I, those powers disposed of their overseas territories and turned to autarky.

In *Fascist Pigs*, Tiago Saraiva shows how Hitler, Mussolini, and Salazar were all part of larger apparatuses of people, animals, and things that came together in the efforts of colonial latecomers to secure for the nation resources and space.[3] In this context, science became a weapon against colonial oligopolies. Pigs, potatoes, and calculations produced at laboratories advanced the material realization of fascism at home and in outposts in Africa and the East. A transnational history of fascism shows that the international move to economic nationalism was a transnational ideology which depended on actual exchanges of knowledge, techniques, and at times even genetic materials. Autarky as a project depended on very specific international regimes of circulation.

Autarky in Spain was no different. In 1939 General Francisco Franco proclaimed the end of the Civil War (1936–1939) after the defeat of the Republican Army. Italian and German forces had contributed to his war effort and, while the regime remained neutral in World War II, its links to the Nazis and the fascists were obvious enough to cause a blockade by the allies in exports of fuels and other goods. After WWII, the allies politically isolated Francoist Spain and banned exports of other resources, such as fertilizers. In addition, the Spanish economy was badly hit, and the capacity to acquire goods in foreign markets was severely reduced. However, economic exchanges with different countries were sustained through the entire decade of the 1940s. Rather than a reality, autarky was an aspirational project held by different powerful elites of the regime. The explicit goal was to be less dependent on strategically important goods, from food to feed soldiers to oil to fuel military vehicles. It was also a "Christian" project.

Engineer, physicist, and Jesuit Priest José Agustín Pérez del Pulgar provided the earliest and most elaborate argument. He delivered a series of lectures in the late days of the Spanish Civil War (1936–1939) which were published posthumously as *The Christian Concept of Autarky* (1940).[4] There, he argued that monopolies of basic products had turned into "a political weapon more fearsome than an army." Most of the resources he mentioned were under the control of the British and French colonial empires. In a text rich with references to recent history of science and technology, Pérez del Pulgar provided various agricultural and industrial examples of how research on science and technology was capable of ending those abuses. The Nazi political economy, and in particular the production of synthetic rubber against imperial monopolies, was his most admired model.

Pérez del Pulgar concluded: "political and national autarky is dependent on economic autarky." The quote could come from the Führer or il Duce— but, let us not forget that Stalin, Churchill, and Truman also had a clear view of the political importance of economic independence. For Pérez del Pulgar, being politically independent was a prerequisite to develop a Catholic political economy, a system neither socialist nor purely capitalist that would allow for the Social Doctrine of the Church to materialize through the plans of an authoritarian state devoted to research and production.

This argumentation became the backbone of the official ideology of the Francoist regime: national Catholicism. As I have argued elsewhere, this doctrine united different political families within the regime and provided a common goal for scientific organizers and religious leaders.[5] The best example was the mutual nourishment of the newly founded Opus Dei and the Consejo Superior de Investigaciones Científicas (CSIC). The organizers of CSIC were among the first members of Opus Dei, and they provided this organization with funds, ideology, and contacts. They turned the CSIC into the regime's largest and better-funded organization for applied research. Applied research in the service of the national economy was, according to them, the way toward import substitution of raw and manufactured materials to reduce dependency on foreign economies.

The National Institute of Construction and Cement was one of the institutes of the CSIC. In the time in which the new laboratory Costillares was built, the Institute received more funds than any other center within the CSIC. Eduardo Torroja, its director, promised to deliver the tools for the country's transition to an industrial economy through the industrialization of construction and the transformation of the Spanish landscape. Torroja was already internationally renowned for his laminar structures, as well as for his presidency of the International Association for Pre-stressed Concrete, the material of choice for European postwar reconstruction. He was also the designer of the Costillares laboratory, including its coal silo.

In his *Philosophy of Structures*, he explained its design:

> The dodecahedral silo is used for coal storage at the Technical Institute of Construction and Cement in Costillares, Madrid. This kind of structure can be easily and cheaply built using prefabricated slabs, each identical for the whole assembly. In these polyhedral shapes, the play of shadows and light corresponds exactly to the contours determined by the designer. This gives the shapes a typical hardness and clearcut outline.[6]

Torroja's description contained his industrializing program in a nutshell. A designer like him could find new kinds of beauty through new arrangements of construction pieces prefabricated according to standards also set by experts like him. Prefabrication of concrete construction pieces would allow his Institute to exert its standardizing authority over private producers. It would also enable the control of workers through rationalization of labor. Fordism and Taylorism were now applicable to construction because the industry had been industrialized through new materials and new designs.

Like Pérez del Pulgar, Torroja thought there was something Christian about this industrial program. His argument echoed what others had already imagined in nineteenth-century England: deskilling labor is actually a step toward the spiritual realization of workers, because some day their tasks will be performed by machines, and workers will be free to pursue more spiritual endeavors. It may sound like a bad joke to us, but people like Charles Babbage or Eduardo Torroja were serious about it. It also justified their own position at the top of the chain of design, decision, and command.

Moreover, as in Pérez del Pulgar, the industrializing program would provide the basis for achieving economic independence for Spain. As the dodecahedral silo shows, Torroja favored coal as a fueling material for the nation. Although oil was more cheaply available, Spain had coal reserves that, if not of great quality, could help avoid oil imports and provide an energy resource shielded from outside influences. Cement and iron, the raw materials of pre-stressed concrete, were also available within Spain—and Torroja himself was for a time in charge of the regulating and planning national corporatist production and distribution of cement.

Research conducted at Torroja's laboratory also emphasized a self-sufficient political economy. The first field of research was cheap housing to host industrial workers. Torroja negotiated with other administrators and state planners the number of houses necessary to sustain economic growth, what cities should be prioritized, and which techniques adapted best to the Spanish economic and cultural context. The second field of research was large dams. Francoist engineers planned dams throughout the country to irrigate agricultural land and to produce the necessary hydropower to get machines going. Torroja's team designed some of these dams to maximize efficiency and resistance. They tested their designs through physical scale models. The *Costillares* silo was thus a functional representative of a project aimed at transforming the entire Spanish territory.

But political economies rarely function according to plan. The history of the coal silo also points to the failures and limitations of the project of economic autarky. Shortly after its inauguration, the price of imported oil made national coal uncompetitive. Several administrative bodies and powerful engineers like Torroja himself favored continuing to use coal and decreed it obligatory in state-owned buildings like Costillares. They hoped that together with hydropower, coal would help avoid the dangers of energy dependency. However, global markets ruin national plans, and cheap oil won the battle: sometime in the early 1960s, Costillares heaters turned to fuel oil and the silo remained as a decorative object.

Even in dictatorships, different groups of experts favor diverging technopolitical projects. In the case of Spain, each energy resource was associated with specific interest groups with their own plans for economic growth. Their interpretations of economic independence differed. Elites planning for the national economy invariably disagreed on the specific means of achieving growth. In Francoist Spain, engineers used the project of autarky to increase their positions of authority and their agency within the regime. And then they fought amongst themselves on how to interpret and implement their projects. When discussing autarky, the first question we should as scholars and citizens is *whose* autarky? Who is the sovereign when we talk about state economic sovereignty?

The discussion over which source of energy to favor was as economic as it was political.[7] Other state actors, including military and civil engineers, supported nuclear energy as an alternative to coal, hydropower, and oil. Unlike coal and hydropower, they argued, nuclear energy has no limits to growth. Unlike oil, it is "national." They counted on the reserves of Spanish uranium, which had been estimated to be much higher than they actually are. Regardless, the nuclear lobbyists used the 1973 oil crisis to include nuclear energy in the new redefinitions of energy plans for the country. Around those years, Spain became the major US client in nuclear material, including reactors and enriched uranium. This was made possible by cheap credit and other advantages provided by US firms under the direct supervision of the Secretary of State Henry Kissinger.[8] "The trick in the world now," said Kissinger in a different but related post-colonial context, "is to use economics to build a world political structure."[9] Imported uranium and oil replaced national coal and international markets took over national plans. But international markets are no less political, no less planned, and no less national than autarkic economies.

Self-sufficiency was far from the economic reality of a small country in need of foreign currency and imports. But taking it seriously as an actor's category reveals its role in shaping and fostering certain projects for retooling the political economy, territory, and commercial relationships with other countries. Through this example, economic nationalism emerges as a central category to understand the world economy. The usual opposition between nationalism and globalization obscures a co-evolution: on the one hand, economic globalization has been manufactured by colonial empires and powerful states seeking to shape and control the world economy and, on the other, the history of national economies is inseparable from global entanglements.[10] The current new era of nationalism does not reflect the end of globalization. It does, however, point to the emergence of new alternative globalizations and new economic world orders.

Notes

1. Ha-Joon Chang, *Kicking Away the Ladder: Development Strategy in Historical Perspective* (London: Anthem Press, 2002).

2. Lisbet Koerner, *Linnaeus: Nature and Nation* (Cambridge, MA: Harvard University Press, 1999).

3. Tiago Saraiva, *Fascist Pigs: Technoscientific Organisms and the History of Fascism* (Cambridge, MA: MIT Press, 2016).

4. José Agustín Pérez del Pulgar, *El concepto cristiano de la autarquía* (Madrid: Revista de los ingenieros del ICAI, 1941).

5. Lino Camprubí, *Engineers and the Making of the Francoist Regime* (Cambridge, MA: MIT Press, 2014).

6. Eduardo Torroja, *Philosophy of Structures*, trans. J. J. Polivka and Milos Polivka (Berkeley: University of California Press, 1967).

7. Lino Camprubí, "Whose Self-Sufficiency? Energy Dependency in Spain from 1939," *Energy Policy* 125, no. 2 (2019): 227–34.

8. Joseba de la Torre and María del Mar Rubio-Varas, "Nuclear Power for a Dictatorship: State and Business Involvement in the Spanish Atomic Program, 1950–85," *Journal of Contemporary History* 51, no. 2 (2015): 385–411.

9. For Kissinger's statement, see Lino Camprubí, "Resource Geopolitics: Cold War Technologies, Global Fertilizer, and the Fate of Western Sahara," *Technology and Culture* 57, no. 3 (2015): 676–703.

10. David Pretel and Lino Camprubí, *Technology and Globalisation: Networks of Experts in World History* (London: Palgrave MacMillan, 2018).

EMES Sonochron: Federal Republic of Germany, 1986

Martina Schlünder

They were green, black, red, and white and very small: 7 × 3 × 3 cm (3 × 1 × 1 inches). Nowadays they can only be purchased online from vintage platforms that advertise their small size using a picture in which the clock sits alongside a cigarette, which is quite a bit longer than the EMES Sonochron. In the 1980s, when portable time was still analogue, this alarm clock was one of the bestselling travel clocks in Western Germany. The main attraction was its small size and light weight: it was portable, fitting easily into any suitcase. Nothing fancy inside the clock – nothing of the upcoming, new quartz technology. Combining mechanical and electrical components that were carefully separated in two distinct parts, the clock consisted of an old-fashioned tiny mechanical clockwork with a small crown at its right side to adjust the time. The alarm was set by rotating the glass pane over the clock face. In the left side of the clock a piezo speaker powered by a battery supplied the alarm signal. By pulling apart the two halves of the clock, the word

A
L
A
R
M

appeared in the middle of the clock. And thus, it was set. By squeezing the two halves together, the word disappeared and the alarm was cancelled. Snoozing was not an option. Whereas clocks might measure time in more or less silent

Figure 2. Photo of the EMES Sonochron alarm clock by Pit Arens; reprinted with permission from Pit Arens.

ways, an alarm clock is different. It epitomizes (daily) planning, it actively intervenes and interrupts a human activity (usually sleeping). Urgency, emergency, a call to action are the subplots of the alarm clock, in this case a call to arms in alarming times.

In the middle of the 1980s, the EMES Sonochron turned into Germany's most politicized alarm clock. This timepiece embodied the planning moments of three different actors: a militant feminist group, Rote Zora, used it as a time fuse in bombings, the BKA (the Federal Criminal Police Office) as a material trap for apprehending the militants, and legal authorities (Public Prosecutor General) as evidence of terrorism.

Cologne, where part of each of these plans were realized, was a major city in the west of divided Germany, about 25 km (or 15 miles) north of the tranquil city of Bonn, federal capital and seat of government of the Federal Republic of Germany (FRG). In 1986, West Germany was a post-colonial country, a fact not familiar to all, although the German Empire had possessed colonies and settlements in Africa, Asia, and the Pacific between the 1880s and 1919.[1] However, the preoccupation with colonialism played a central role in the student movement of 1968 and the militant groups emerging from it. Their enthusiasm did not lead to an intensive examination of their own, German colonial history, but above all to an identification with the militancy and the guerrillas of the liberation movements of colonized and decolonizing countries. From there, one took inspiration for action in the struggle against one's own government, which was understood as still fascist, with Germany itself a colony of US imperialism.[2]

It was the common struggle against imperialist capitalism that united all revolutionaries (whether in Germany or Vietnam) and leveled all historical differences.

In 1986 it seemed that all revolutionary plans and counter-plans intersected in a single object, in the very same clock, since all of the involved parties delegated crucial actions to it. However, in this sociomaterial account of planning at hand, it is possible to investigate the clock as something other than a passive and stable object.[3] Sociomateriality is an anti-essentialist approach to matter and a way to overcome persisting dualisms in Western thinking like human/things, sign/matter, active/passive. The social and the material are not separated into different registers or categories, they belong together. Thus, objects do not exist outside the sociomaterial networks that enact them. Each of the three sociomaterial networks examined here (Rote Zora, federal police, and public prosecutors) enacted a different version of the alarm clock and delegated to it crucial planning moments and socio-political tasks. The alarm clock, as a multiple object emerged in different versions from these sociomaterial practices that entangled humans and nonhumans in chains of practices in which agency was distributed between its human/nonhuman participants. Therefore, the activity of planning in a sociomaterial approach is not limited to human subjects endowed with will and intentions, but involves material things as well, even if they do not possess these properties. In a sociomaterial network planning is distributed across all participants of the network.

Sociomateriality also works as an analytical tool against planning's modernist rationality and functionalism. A sociomaterial account undermines the hierarchical idea of planning as primarily abstract activity, and its secondary implementation, which can simply be carried out or delegated to instruments. Thus, it helps to break away from an understanding of planning as an ideal practice and the outcome of pure rational choices. It enables us to analyze planning as a much more material, performative activity heavily embroiled in affect, collective feelings, and emotions.

The alarm clock enacted different versions of time and time itself became a resource and a target of planning while it was also inevitably inbuilt into its muddled processes. Michel Serres's idea of time as a spatiotemporal process is an important resource to rethink the often too linear ways planning's temporalities are conceived of as being mostly simply directed to futures.[4] By following three different enactments of the alarm clock—as time fuse, material trap, and legal evidence—this essay will investigate the sociomaterial entanglements that let the EMES Sonochron become a participant in distributed planning.

Imagine you are a militant feminist, organized in a cell of the anti-imperialist group Rote Zora (Red Zora) engaged in the armed struggle.[5] Your political goal is to take the struggle of the liberation movements of the decolonized South to the metropolises of the North. For you, bombings are a legitimate tool against the much bigger structural violence of the state and capitalism—so long as bombings are directed against infrastructures and buildings, causing no harm to people. It is early fall 1986 and a plan emerges to bomb the Lufthansa headquarters in Cologne since the airline makes money from its implications in imperialist processes; Lufthansa profits from deporting asylum seekers and the booming market of sex tourism. You have to acquire the time fuse and, while last year you chose an EMES Sonochron that was easy to find, now you have difficulty spotting one in the shops. You wonder why as you work out how to buy one.[6]

Rote Zora emerged as a militant group in the aftermath of the 1968 student revolt, initially as the feminist wing of the Revolutionary Cells (RZ, or Revolutionäre Zellen). In contrast to the hierarchical structure and the clandestine culture of the better-known RAF (Red Army Faction), members of Rote Zora and the RZ lived ordinary lives: they had jobs and paid taxes. Internally, they were organized in hermetically sealed cells, preferring a decentralized, non-hierarchical organization that was modeled along the strategy of the communist resistance against the Nazis. Rote Zora broke away from the RZ in 1977 as an independent group with an explicitly feminist agenda: fighting for the full liberalization of abortion laws, against trafficking in women and sex tourism, against the increasingly restrictive legislation for asylum seekers and refugees. For Rote Zora, these were all international, anti-imperialist issues, and they always understood their actions as part of a global struggle that particularly involved the countries of the so-called Tricont (i.e., Africa, Asia, and South America). Their actions, even if they took place in Germany, referred for example to global population policy or to the global exploitation of female labor or trafficking in women. Rote Zora understood themselves as a link between anti-imperialist and feminist groups developing the concept of feminist counter-violence as a motivation for their attacks, which were not driven by aggression but justified as a form of defense against structural violence against women.[7] But open discussions between the feminist groups were rare and proved mostly controversial since most groups rejected violence in all forms. Academic feminists even charged Rote Zora with imitating the attitude and behavior of white males and their masculine cult of militancy.[8]

From the middle of the 1980s onward Rote Zora's activities took a more anti-scientific stance as they engaged in the struggle against new reproductive technologies and genetic engineering. This topic was one of the most controversial issues at this time in Germany, with a multitude of groups with different agendas engaged in it.[9] Rote Zora was driven by the need to stop—at the very beginning—the development of what one could call "total planability," forestalling a possible future nightmare of making all life forms, whether human, plant, or animal, controllable by manipulating their genome to render them exploitable through global capital.[10] Rote Zora planted several bombs in emergent biotechnology centers and blew up a laboratory of the Max Planck Institute for Breeding Research in Cologne to destroy their "political plants." These seeds, as Rote Zora claimed, were later patented and owned by the German chemical industry, maximizing their profits while at the same time destroying the livelihoods of local farmers in the Third World, exacerbating hunger and famine on a global level.[11] In their manifestos justifying the attacks, the militants argued against what they saw as a new wave of eugenics in global population politics and they linked their militant actions to a historical and moral obligation to resist the persistence of Nazi ideas and ideologies in medicine and science. Their ultimate aim was to stop research that exploited biology for profit and increasingly sought to make all forms of life amenable for total planning. Planning was tantamount to total control and became the target of Rote Zora's militant actions.

Between 1977 and 1995 Rote Zora planted about forty bombs at research facilities, airline offices, and textile companies.[12] For example, on August 15, 1987, eight bombs went off in one night all over West Germany in various branches of the Adler clothing chain. Adler was one of the first companies to outsource parts of its production to South Korea. There, women had gone on strike against the undignified working conditions, and Rote Zora wanted to bring this strike and the reasons for it to Germany as well. Unlike the RAF they avoided shootings, kidnappings, and car bombings. Instead, they delegated their agency to a tiny alarm clock used as a time fuse to detonate their explosives. It was its unique design and the materiality of the Sonochron that made it particularly attractive to Rote Zora: the separation of its mechanical clockwork from the electronic buzzer facilitated its subversion into a timer of an explosive—in just a few simple steps. Delegating the timing of the explosion to a converted alarm clock was risky. Red Zora scouted the place very carefully beforehand to determine at what time no one would be hurt, usually in the early morning hours. The sociomaterial network of the Red Zora did not always hold. Sometimes the timers failed and the explosive devices did not detonate. These then came into the possession of the police, who thus became aware of the importance of the alarm clock for the attacks and began to link the alarm clock with their own sociomaterial practices.

By turning the alarm clock into a time fuse, Rote Zora tried to synchronize events on different time scales, to link the temporalities of the global struggle against capitalism with the moment of a local explosion, and to connect a specific place, a lab, a shop, or Lufthansa headquarters with the temporalities and scales of their anti-imperialist agenda, with exploited peasants in India, striking women in Korea, or trafficked women from the Philippines. Initially untroubled by postcolonial reflections on Germany's colonial history and ignoring Germany's genocidal past that had its origins in its colonial empire, Rote Zora embraced the idea of a leftist global sisterhood. Facing up to historical, colonial, ethnic differences between the global sisters and examining their own racisms and culpability in this was a difficult and painful process that started rather late. In their last manifesto, shortly before parts of the group distanced themselves from militant action and disbanded, they considered "how much we ourselves are still caught in our utopias (. . .) by our 'own' Christian-colonial history, for instance (. . .) in our image of the liberated woman, in our belief to be already more 'developed' and freer than women elsewhere?"[13]

Imagine you are a policeman, working in Cologne. On a very cold day at the end of October 1986, you are called to help your colleagues to search an area around the headquarters of the Lufthansa airline company that has been partly destroyed in a bombing. Hundreds of your colleagues are involved: together for one and a half days you rummage through the rubble, mostly on your knees. Then you find what you have been looking for: a clock-face from a miniature alarm clock, less than three inches long. Four digits are engraved on it: 6457. You do not know it at the time but one year later this number would assume enormous political significance and become a central actor in a state security trial.[14]

Initially, the police ridiculed Rote Zora as "after-work" or "part time" terrorists. However, mocking them did not help to track down Rote Zora members successfully.[15] Moreover, their strategy of hiding in "plain sight" proved extraordinarily successful. BKA's traditional methods of investigation — including infiltrating revolutionary groups with undercover informants — failed dramatically. Aware of this, the BKA switched tactics and performed an impressive material turn: instead of infiltration of groups, they turned to matter, scrutinizing the toolkit used by Rote Zora in their explosives. The plan was to transform these tools into a material trap with which to apprehend its members. It turned out to be very difficult to track down the different elements of the bombs planted by Rote Zora since the devices were very simple in their design and used components such as batteries and wires that were mass produced and therefore not easy to trace—with one notable exception: the small alarm clock that was used as a time-delay device in the bombs. Although popular and widely sold in West Germany, the EMES Sonochron seemed, potentially, to offer a means to track down the militant fighters. Accordingly, between 1983 and 1985, the BKA planned and implemented the so-called "alarm clock program." After getting hold of the remaining stock of the EMES Sonochron clock, the police started imprinting four-digit numbers into each clock face, every number exactly three millimeters high: this rendered each clock identifiable and traceable. The Sonochrons were then returned to the company and sold only in a small number—about thirty pieces—at shops that had been fitted out with video cameras. Those working in the selected shops were contracted and trained in observational skills by the BKA to look out for customers purchasing the EMES clock who fit the profile of Rote Zora, that is to say, women between 18 and 45, who seemed alert and educated.[16]

But the well-oiled planning machine of the BKA stalled after the first bomb detonated at the Lufthansa airline headquarters in Cologne on October 28, 1986. It was only by chance that the clock face inscribed with a number specific to it survived the explosion. After identifying the shop where the clock was sold, the BKA was able to watch the sales transaction on video but the police had no clue how to identify and find the woman who purchased it. Only by watching a TV program about *EMMA*, Germany's top feminist magazine, did a member of the police recognize the journalist Ingrid Strobl as the woman who bought the Sonochron. Strobl was put under surveillance and her phone was wiretapped under the anti-terrorist laws from February 1987 to December 1987. Strobl was arrested in the course of a carefully planned manhunt—more aptly called a "womanhunt"—against Rote Zora.[17] More than 300 policemen searched private flats, a newspaper office, printing presses, photo labs, doctor's offices, and the offices of a group of feminists fighting against new reproductive technologies and genetic engineering. Investigations against 23 people were initiated, literature and newsletters were seized, two women were detained including Ingrid Strobl, and five more were hunted down and served arrest warrants.[18]

To this day it is unclear how much BKA's alarm clock obsession cost the German taxpayer. By inscribing state-authorized markers onto the face of the Sonochron clock, the BKA delegated the power of convicting terrorists to the materiality of an alarm clock. However, this sociomaterial network, in which the police had articulated themselves with marked alarm clocks, video cameras, and trained vendors, did not prove to be very stable. At first it was difficult to find the tiny clock face in the rubble so that it seemed impossible to link it to the shop where the clock had been sold. Another gap in the police's sociomaterial network was the link from the video recording of an alarm clock purchase to the identification of the purchaser. The network held only by luck and chance, by improvisations, by a series of contingencies. The spatiotemporal enactment of time embodied by the alarm clock also differed from the network of the Red Zora. Here, no time fuse connected places like Cologne with the South Korean city of Iri (today Iksan). Rather, the engraved markings on the clock face connected the manufacturer, police laboratories, stores, and the place of an explosion over an unpredictable period of time. Finally, BKA's obsession to link the clock with militant feminists transformed it into an iconic interface of feminism and terrorism.

Imagine you are a judge at the Criminal court for state security cases in Düsseldorf, close to Cologne. You have a reputation as a hardliner. Zero tolerance for terrorists and for those who sympathize with them. You fully support the amendment of §129a, the so-called "terrorist law" that treats the planning of terrorist attacks as if the deadly action had already happened. It blurs the line between a plan and its execution. Some of your colleagues are afraid that as in the Nazi era the state is on a slippery slope toward "political justice" (*Gesinnungsjustiz*) leading to the persecution of thoughts, opinions, and plans instead of (criminal) acts that are actually committed. You do not share these fears. You pack your EMES Sonochron in a dark green color into your briefcase as you always do when you travel. Tomorrow you will chair the trial against the feminist journalist Ingrid Strobl, who was caught on video buying the same type of alarm clock that was later found in the rubble of the bombed Lufthansa headquarters. You will take your clock to court and then you will make crystal clear that there are several ways of buying an alarm clock and that buying the EMES Sonochron nowadays in particular circumstances counts as an act of terrorism, especially if you are a feminist.[19]

§129a became the legal vehicle for the state's persecution of increasing numbers of its citizens. Introduced to the German penal code in 1976, §129a, aka Lex RAF (Red Army Faction), enabled prosecutors to detain individuals who might have been engaged in planning and preparing an attack but in fact, and crucially, had not committed it. It also made it possible to convict members of a terrorist group for crimes that they personally had not committed but which might have been carried out by a fellow member. Thus, the border between planning, plotting, and committing a crime was blurred. §129a and its application considerably intervened in civil rights, eroding the rights of suspects, undermining key principles of the law, such as the presumption of innocence. It also affected the rights to demonstrate, enabled particular forms of manhunt and of personal data collection, and it facilitated personal surveillance and wiretapping. Since the elements of crime under §129a were intentionally left underdetermined, more and more politically engaged people came under suspicion of sympathizing, supporting, or even campaigning for terrorist groups. This affected leftist bookstores but also bourgeois newspapers who might have printed one of the justification letters after a bombing.[20] During the persecution of Rote Zora the German Federal Attorney finally created the new term of *anschlagsrelevante* topics, i.e., topics that had triggered bombings and assaults. Now the prosecutors argued that activists engaged in these topics were already in a mental proximity to terrorists (*"geistige Nähe zum Terrorismus"*) which, in turn, justified arrests as if activists were already part of the armed struggle. Here again, the time between a future possible act and the present—where such acts were not committed but might have been planned—had been flattened.

Strobl's trial attracted a lot of attention. Critical leftist lawyers had campaigned and succeeded to install an independent committee observing the trial; a support group set up a newsletter reporting about the trial. Its name—"clockwork 129a"—indicated that a new sociomaterial network articulated and enacted the alarm clock, namely the supporters of the defendants and lawyers who critically opposed the extension of §129a. This network now competed against the sociomaterial network of the legal apparatus, which had articulated the alarm clock not as a material trap, but as legal evidence against Strobl.

Since neither the BKA nor the Public Prosecutor General could actually prove that Strobl was involved in preparing the bomb or placing it at the Lufthansa headquarters, the fact that she was videotaped while purchasing the clock was taken as a proof that she was actually a member of Rote Zora—which she vehemently denied. But the amendments of §129a allowed the state persecutors to treat her as a terrorist. In Strobl's trial the Sonochron clock appeared as a material witness: the state-registered numbers 6457, according to prosecutors, provided the direct link between the act of buying the alarm clock and the detonation of the bomb. While under civil laws, Strobl could only have been proven to have bought an alarm clock (which can hardly be considered a terrorist act), under §129a she had become the epitome of a feminist terrorist. In specific circumstances (e.g., being a leftist feminist engaged in topics that potentially could trigger a bombing), purchasing the EMES Sonochron would mean one was either a member or an active supporter of a terrorist organization.

The state made legal differences disappear. It denied the difference between a plan and its execution. The sociomaterial network of justice enacted the clock as a kind of time-sucker, since it extinguished the divergent temporalities between plan and execution: Buying an alarm clock was the same as setting off an explosive device. Thus, the state vehemently denied the planning moment itself, in which the risk of failure is also inherent. It delegated the legal evidence to the combination of purchasing an alarm clock while being a feminist. For the editors and authors of clockwork 129a, again it was not the purchase of an alarm clock, but rather the mishandling of the clock that revealed the unrelenting machinery of state and justice that was threatening and grinding the principles of independent justice—thus working like a time fuse of an explosive that would destroy the foundations of Germany's hard-won constitutional democracy.[21]

Militant groups like the Rote Zora emerged from the student revolt of 1968 and withered with the end of the Cold War. They were part of the postcolonial situation of West Germany, but they did not reflect it as such (which is also true for the police and the judiciary). Rather, they were characterized by the combination of ghostly post/colonial presences and absences. Rote Zora attempted to use the colonial liberation struggles of others as a means of identification and of mobilization of their own society. They understood their plans and actions not as postcolonial, but as a collective intervention, alongside colonial liberation movements and their later heirs in current struggles against US global imperialism. While being part of a plan to take the struggle of the decolonized liberation movements of the Global South to the centers of the North, the ticking of the clock that stood in the center of their planning unleashed a time of crises that was full of the unspoken ghosts and gaps of Germany's unrecognized genocidal past. Although the history of militant groups in West Germany was a specifically German way of dealing with their own colonial past, the question remains whether the utopian identification with colonial liberation movements and the non-awareness of one's own postcolonial situation did not characterize many more countries of the Global North.

The sociomaterial approach studies planning not as an abstract activity but as a sociomaterial practice. It therefore does not focus solely on the ideologies, the intentions, and purposes of Rote Zora and their counter-planning sociomaterial networks and then neglect the material objects and relations, it rather understands both the social and the material as constitutively entangled. By focusing on the alarm clock as an enactment of different sociomaterial networks, different versions of the clock emerged, as well as different forms of time and its spatio-temporalities, including its ghostly post/colonial gaps. The

EMES Sonochron throws into sharp relief that a plan never emerges alone, but is rather part of a constellation of plans half done, half dreamed, disrupted by counter-planning, or the need to react and improvise. Planning appears here not as the ultimate rational choice but rather as a distributed sociomaterial practice, a messy, material process, as contingent, as performative, never innocent, as a web of heterogenous actors, nonhumans and humans alike embroiled in a chain of distributed agency.

Notes

1. Surprisingly, even reviewers of this volume belonged to the circle who were not familiar with it, for the facts see Sebastian Conrad, *German Colonialism: A Short History*, trans. Sorcha O'Hagan (Cambridge: Cambridge University Press, 2012).

2. Petra Rethmann, "On Militancy, Sort of," *Cultural Critique* 62 (Winter 2006): 67–91; for a critical re-assessment of the "identification hypothesis," see Quinn Slobodian, *Foreign Front: Third World Politics in Sixties West Germany* (Durham: Duke University Press, 2012), 5–16.

3. Lucy Suchman, *Human-Machine Reconfiguration: Plans and Situated Actions* (Cambridge: Cambridge University Press, 2007); Wanda Orlikowski, "Sociomaterial Practices: Exploring Technology at Work," *Organization Studies* 28, no. 9 (2007): 1435–44; John Law and Annemarie Mol, "Notes on Materiality and Sociality," *The Sociological Review* 43 (1995): 274–94.

4. Serres compares time with a crumpled handkerchief: time is folded, full of gaps, discontinuities, unexpected turns and connections. It takes substantial sociomaterial work to make time flow continually. Michel Serres with Bruno Latour, *Conversations on Science, Culture, and Time*, trans. Roxanne Lapidus (Ann Arbor: University of Michigan Press, 1992), 60.

5. The name comes from the children's book *The Red Zora and Her Gang* by Kurt Held in which a young girl leads a gang of orphans who, in an anarchical and Robin Hood style, bring justice to a small Croatian town.

6. "Falsch bombadiert" [sic], *Der Spiegel*, February 12, 1989, 64–65.

7. Katharina Karcher, *Sisters in Arms: Militant Feminisms in the Federal Republic of Germany since 1968* (Oxford: Berghahn, 2017), 45–70.

8. Claudia von Werlhof, "Leserbrief an die taz (13.2.1981)," in *Die Neue Frauenbewegung in Deutschland: Abschied vom kleinen Unterschied; Eine Quellensammlung*, ed. Ilse Lenz (Wiesbaden: VS, 2008), 277–79.

9. Stevienna de Saille, *Knowledge as Resistance: The Feminist International Network of Resistance to Reproductive and Genetic Engineering* (London: Palgrave Macmillan, 2017).

10. *Früchte des Zorns: Texte und Materialien zur Geschichte der Revolutionären Zellen und der Roten Zora* (Amsterdam: Edition ID, 1993), 615–16.

11. *Früchte des Zorns*, 617–26.

12. Oliver Ressler, "Die Rote Zora," 2000, video, http://www.ressler.at/de/die_rote_zora/.

13. Rote Zora, *Mili's Tanz auf dem Eis: Von Pirouetten, Schleifen, Einbrüchen, doppelten Saltos und dem Versuch, Boden unter die Füße zu kriegen*, 1993, 22, http://www.freilassung.de/div/texte/rz/milis/milis1.htm; on the inherent antisemitism in the anti-imperialist agenda and the fine line between antizionism and antisemitism in the movement of the Revolutionary Cells, see Jan Gerber, "'Schalom und Napalm': Die Stadtguerilla als Avantgarde des Antizionismus," in *Rote Armee Fiktion* eds. Joachim Bruhn and Jan Gerber (Freiburg: ça ira, 2007), 39–84.

14. Oliver Tolmein, "Da haben alle mitgezogen," *konkret* April 12, 1989, 15; "Falsch bombadiert," 64–65.

15. "Gottverdammter Zufall," *Der Spiegel*, February 21, 1988, 95–96; *Früchte des Zorns*, 598–605; Eva-Maria Thoms, "Der Freiheit eine Falle," *Die Zeit*, February 3, 1989.

16. "Gottverdammter Zufall"; Tolmein, "Da haben alle mitgezogen," 15.

17. Saša Vukandinović, "Spätreflex: Eine Fallstudie zu den Revolutionären Zellen, der Roten Zora und zur verlängerten Feminismus-Obsession bundesdeutscher Terrorismusfahnder," in *Der Linksterrorismus der 1970er-Jahre und die Ordnung der Geschlechter*, ed. Irene Bandhauer-Schöffmann and Dirk van Laak (Trier: Wissenschaftlicher Verlag, 2013), 139–61.

18. Edith Lunnebach, "Der Weckerkauf und seine Folgen—'Beschäftigung mit anschlagsrelevanten Themen oder geistige Nähe zum Terrorismus,'" in *Staatssicherheit. Die Bekämpfung des politischen Feindes im Innern* ed. Helmut Janssen and Michael Schubert (Bielefeld: AJZ, 1990), 140–50.

19. Tolmein, "Da haben alle mitgezogen," 15.

20. Lunnebach, "Der Weckerkauf und seine Folgen," 140–50; Thoms, "Der Freiheit eine Falle."

21. In the first trial Ingrid Strobl was sentenced to a five years imprisonment for membership of a terrorist group, a sentence that was rejected by the Supreme Court. In a second trial, Strobl was convicted for assisting a bomb attack and sentenced to three years imprisonment. Since she had already served twenty-seven months in remand custody, for the remainder of the sentence she was placed on probation, see Lunnebach, "Der Weckerkauf und seine Folgen," 140–50; "Erstmal wegschließen," *Der Spiegel*, May 20, 1990), 68–73. Strobl's membership in a militant group was never clarified. Not long ago, she published a book, about the alarm clock purchase and her imprisonment. In it, she confessed that although she was not part of a militant group, she was aware when she bought the alarm clock what it was to be used for, something she had vehemently denied in court, Ingrid Strobl, *Vermessene Zeit: Der Wecker, der Knast und ich* (Hamburg: Edition Nautilus, 2020).

Famine: India, 1877

Anindita Nag

During a three-month tour of the Indian subcontinent, Julian Hawthorne, the special famine correspondent of *Cosmopolitan* magazine recounted his most "haunting experience" in an orphanage in Jabalpur in the Central Provinces in 1897.[1] He encountered the "saddest and grimmest spectacle known to modern times"—women and children begging for food or lying in despair beside an empty plate, or reduced to scavenging for scraps along with dogs. Hawthorne combined eyewitness accounts with vivid images of the starving to make the suffering as real, as immediate, and as disturbing as possible. His distant readers are meant to see what he saw, to hear what he heard, to feel what he felt. Hawthorne's account soon became the focus of public attention as emblematic of the famine experience, generating a newfound interest in famines as a human tragedy.

In describing his experience, Hawthorne drew on a language of sympathy instead of numerical data to create a persuasive image of hunger as a human condition rather than an abstract scientific problem. The point here was not just to establish that hunger was a human condition that afflicted real people but to connect his readers directly with the suffering of those people. By the time Hawthorne came to write about hunger, a new generation of crusading journalists has been increasingly drawing the attention of the British public to the crisis of famine in India. These men—Francis Merewether, Vaughn Nash, and William Digby—made their reputations by presenting news in more digestible forms through human-interest stories, serialized narratives, and the use of headlines, graphs, maps, and photographs.

These reports from the front line of the famine sparked a fierce reaction in Indian administrative circles, begging the question of what Britain's duties and aims were in India. Inevitably, the Indian administration closed ranks in self-defense followed by a vicious public attack by J.D. Rees, a former member of the Governor General's Council. "Why among the pictures of famine are only those representing the dark side reproduced in England?" complained J.D. Rees of the photographs published by special correspondents covering famines in India. "Why," he asked, "did we never see photographs of tens of thousands of people tolerably comfortable, and certainly not hungry, busily occupied in earning bread from the State, but only reproductions of poorhouses in which are gathered together the waifs, the strays, the halt, the lame, the blind, the aged, feeble, and infirm, the flotsam and jetsam of teeming Oriental Populations?"[2]

The attack on British rule was, however, wide-ranging and sustained and was carried out with patience and learning. In a speech delivered at Chandos Hall in London, Henry Hyndman, the founder of Britain's Social Democratic Federation (SDF) did not mince words about the atrocities that passed for famine relief in India, especially the use of famine funds to wage an unjust war in Afghanistan.[3] Drawing from the Famine Commission reports, census figures, and eyewitness accounts of journalists, Hyndman provided a litany of examples that demonstrated a concerted British policy of starvation and depopulation.

Indeed, as the display of famine became more systematic and more powerfully executed, it became clear that famines were hardly a neutral matter. Famine relief funds became causes with which the rich and famous of the day wished to have their names associated by participating in fundraising events across the Empire. For example, best-selling author Arthur Conan Doyle read from his work in London at an event in aid of the 1897 famine fund.[4] Similar fundraising events became common among the European community in India and were a means to endow the sometime frivolous social activities of Anglo India with a veneer of respectability.[5]

It is no accident then that throughout the nineteenth century, the calculated administration of poverty and hunger became critical to planning and devising new forms of statecraft. The classic illustration of this was the enactment of the New Poor Law of 1834 in Britain which subjected the poor to the punitive regime of the workhouse, emphasizing that only hunger could uplift the poor, by teaching them the virtue of labor. The poignant image of a skinny, neglected little boy asking for more gruel in Charles Dickens' *Oliver Twist* has now become a classic image of the hardships of the Victorian workhouse. The management of hunger took a devastating form in Britain's oldest colony, Ireland, and its largest, India. Viewed as "zones of famine," the poor and hungry of both colonies were deemed as lazy, improvident, and adherents of "superstitious" religion. In Ireland as in India, Victorian racism coupled with a dogmatic commitment to laissez-faire economics meant that the colonial power preferred not to intervene in the situation of scarcity.

In India however, the force of state planning with objective techniques for managing hunger was first felt. India lacked Britain's Poor Law or Ireland's permanent system of workhouses, which accommodated a million Irish victims from 1847 to 1849. New institutions and techniques were therefore developed to distribute relief and to discipline the poor in India. After all, India had long been the testing ground for British political economists and utilitarian social reformers who viewed famines either as acts of providence or checks on overpopulation by peasants unwilling to learn the disciplines of the market economy. This was especially the case after the imposition of direct rule in 1859, when the government embarked on an unprecedented expansion of the colonial state through a mastery of the territory, its people, and the products of their labor. The centrality of famine is undeniable in the colonial state's forward march toward improvement and progress in India.

It was at this time, too, that the British colonial government in India embarked on a series of attempts to set up a network of relief and labor camps, most notably during the famines of the 1860s and 1890s. Given the vast and unsettled nature of the colonial landscape, spatial control became an important component of famine operations in India. Famine camps embodied the promise of careful and conscious planning and emerged to distribute relief and discipline India's poor. Above all, the famine camps counted, classified, and ordered an itinerant population, making them visible to the colonial state bureaucracy. The image of a Madras famine camp portrays this notion of "order" typical of relief images during this period. The strong visual composition in this picture gives the viewer a sense of stability, the feeling that the crisis is well managed, even in overcrowded camps.

Although normally associated with totalitarian regimes, camps emerged and evolved as a recognized instrument of British colonial rule in which imperial agents experimented with new techniques for the control of mass populations. In fact, Britain had encamped entire ethnic groups: over 1.5 million Kikuyu in "rehabilitative centers" in Kenya, while concentrating half a million ethnic Chinese in enclosed "New Villages" in the Malay Peninsula. Camps have remained an integral feature of our global landscape—whether in transit center like Calais and Nauru, extra-judicial detention cells in Guantanamo Bay, or in refugee enclosures, such as the Zaatari camp in Jordan built for refugees fleeing the Syrian civil war. After all, as Aidan Forth points out, camps left a legacy not only of hardship and torment; Britain rendered civilian camps a legitimate institution of colonial warfare and welfare

Figure 3. "The Famine in India Distribution of relief to the sufferers at Bellary, Madras Presidency." *The Graphic*, October 20, 1877.

which was to be recycled by future colonial administrations throughout the twentieth century.

As the site in which the "state of exception" is given a permanent spatial arrangement, the famine camp became the paradigmatic space where imperial rule rested on powers exceptional to liberal democracy and constitutions of Europe. As such, the organization and occupation of space in famine camps were enshrined in two important presuppositions underlying British imperial practices: Malthusian theories of population growth and Jeremy Bentham's schema of pauper management. Adam Smith and Thomas Malthus were the first to establish the modern political economy of hunger. In the Malthusian paradigm, famine was seen as moral as much as a social and economic breakdown. Malthus understood hunger in terms of the moral failure of individuals to learn the disciplines of the free market.[6]

Despite Adam Smith's insistence that hunger was not a necessary spur to labor, that no moral stigma should haunt those who suffered as a consequence of markets that were everywhere in chains, the evangelical revival in nineteenth century political economy bolstered the less optimistic Malthusian view of hunger as a mark of moral failure.[7] If man's sinful nature ensured that population growth always outstripped the market's capacity to generate food, then hunger would helpfully eradicate the irredeemable and bring salvation to others. In this new market ethic, hunger was understood not as a problem of but as a solution to the ills of political economy.[8]

If Malthusian doctrine provided the initial language for famine relief, then the rationality under which the new relief administration would take shape in the camps bears the imprint of the program of pauper management projected by Jeremy Bentham in the eighteenth century. The colonial government turned to the Benthamite principle of "less eligibility and the means to put it into practice—the workhouse. Bentham believed that paupers should never be relieved of their poverty above the level of scanty means of supporting life. Following Bentham, the British administrators believed that only indigence fell within the province of public charity, while poverty remained "the natural, the primitive, the general and the unchangeable lot of man." Thus, the distinction between poverty and indigence defined more precisely the legitimate realm of governmental intervention into the administration of poverty.[9]

The Malthusian prescription for the poor is about allowing the laws of nature to operate which bind the poor man to the yoke of wage-labor. The severity of famines stimulated the idea of public works as a legitimate state response to drought and dearth, one that did not contravene laissez-faire orthodoxy. The administration therefore felt a strong urge to concoct a system by which "the proper recipients of public charity can be most effectively ascertained," and to ensure that resources were concentrated exclusively on that category.

Requiring the poor to work for relief would provide three checks on waste: First, the distance test whereby the famine victims would have to prove that they were in need by leaving their homes to go to the relief work; secondly, the task and wage tests whereby they would have to fulfill a task not less than 75% of that performed by laborers in ordinary times, and for a sum gauged to provide only enough to sustain life; thirdly, the residence test where beneficiaries are required to reside at the place of relief (i.e., a poor-house or worksite) and thereby forgo the presumed pleasure of ordinary social life; and, fourthly, they would be under Public Works Department (i.e., European) supervisors who would prevent slack discipline and be personally honest.[10]

The Madras Famine of 1877–1878 marked a moment of heightened state intervention in famine management. The colonial administrator Sir Richard Temple was dispatched from Delhi to supervise the famine administration in Madras. Meanwhile, the Viceroy of India, Lord Lytton, was preoccupied with the "Proclamation Durbar," a spectacular pageant held in Delhi to mark the crowning of Queen Victoria as Empress of India. Lytton believed dogmatically in the prevailing Malthusian ideas of political economy and refused to interfere with prices or allow restrictions of the market by stockpiling grains. Famine, according to Viceroy Lytton, was a natural corrective to "overpopulation."

When Temple was tasked with addressing relief measures for the Madras famine, his policies reflected the government of India's tighter priorities of economy, efficiency, and control. Temple enforced some draconian measures, including a cruel "distance test," whereby adults and older children were refused work within a ten-mile radius of their homes; and the notorious

Temple Wage, a supposedly scientifically derived food ration lower than the daily ration at Buchenwald, the notorious Nazi concentration camp. Temple cut rations down to one pound of rice per diem despite medical testimony that the peasants were now "little more than animated skeletons utterly unfit for work." The Temple ration of one pound of rice per diem was half what felons received in Indian prisons.

Actual camp rations were the source of considerable controversy and experimentation. A substantial contribution to the debate came from the Sanitary Commissioner of Madras, Surgeon-Major Robert Cornish. In articulating his criticism of Temple's policy, Cornish drew from his experience during the famine in the Bombay Deccan in 1864 when he chaired an inquiry commission appointed by the British Association for the Advancement of Science. Cornish argued that rice is deficient in one main nutritive element, namely nitrogen. The quantity of nitrogen in one pound of rice varied from 68 to 80 grains, while the quantity of nitrogen eliminated from the human body is estimated at about 200 to 240 grains per diem. To sustain life on a moderate amount of rice, it is therefore necessary to add pulses (beans), or some other kind of nitrogenous food.

Apart from its sheer deficiency in energy, Cornish pointed out that the exclusive rice ration without the daily addition of protein-rich food would lead to "slow and certain starvation." Temple justified the allowance of one pound of rice per diem with the argument that offering over-generous wages would "demoralise," and reduce people's inclination to industry. Moreover, he claimed that the majority of the famine dead were not the cultivating yeomanry, "the bone and sinew of the country," but parasitic mendicants who essentially had committed suicide.[11] Temple's greater aim was to look for ways of saving the state's money, remarking that it would be unjust to the public interest to exceed the minimum needs of the starving.

Social and cultural factors played a role in where British administrators fixed the limit of subsistence wage. Even further, the contours of satisfaction itself could be extrapolated from the general mores and habits of an entire population: Indians were deemed too numerous to receive systematic relief, and as Charles Grant argued giving further weight to the naturalization of Indian poverty, they lived in a country where climate and custom had combined to keep down the standards of wants among the Indian poor.[12]

Indians were famine prone and the Indian society possessed, as the Famine Commission of 1880 later put it, structures that were "admirably adapted for common effort against a common misfortune."[13] These included the corporate body of the Hindu joint family, but also longstanding relations of moral obligations and mutual assistance, as between landlord and tenant, master and servant, alms giver and alms receiver, "which are of the utmost importance in binding the social fabric together, and enabling it to resist any ordinary strain."[14]

The full blueprint of the administrative practice of measuring nutritional needs and food rationing in camps was codified in the Famine Commission Report of 1878–1880. Famine camps were a materialization of the rhetoric of political economy, where the colonial state pursued the twin objectives of "ordering" a dislocated indigenous population and disciplining its own bureaucracy. The conflicting demands of economy and relief meant that financial constraints often trumped basic medical and sanitary provisions in famine camps. The mandates of relief, discipline, sanitation, and labor provided a logic to the famine camps that shaped the attitudes of the officials involved. A cruel facet of this scenario is that official representations never focused on the colonial neglect that resulted in a crisis like famine. Famines in India thus reveal the "dark side of planning" with its links to state mechanisms of social control, constraint, and exploitation.

Notes

1. Julian Hawthorne, "India Starving," *Cosmopolitan*, August, 1897, 369–84.

2. J. D. Rees, *Famine Facts and Fallacies* (London: Harrison and Sons, 1901), 10. By 1902 the Colonial Office had established its Visual Instruction Committee to produce photographic evidence of Britain's improvement of its colonial territories that was then disseminated through lantern slide lectures in British schoolrooms, lecture halls, and libraries. See H. O. Arnold-Foster, ed., *The Queen's Empire: A Pictorial and Descriptive Record* (London: Cassell, 1897), x.

3. Established in the aftermath of the 1876 famine, the Famine Fund promised investments in technical infrastructure like rail and irrigation that would improve India's rural economy and allow grain to be moved from places of plenty to places of dearth. But a substantial portion of the fund was diverted to pay for the Second Anglo-Afghan War (1878–1880).

4. *The Times*, February 21, 1897, 10.

5. Mary A. Procida, *Married to the Empire: Gender, Politics and Imperialism in India, 1883–1947* (Manchester: Manchester University Press, 2002), 173.

6. T. R. Malthus, *An Essay on the Principle of Population, or a View of Its Past and Present Effects on Human Happiness: With an Inquiry into Our Prospects Respecting the Future Removal or Mitigation of the Evils which it Occasions*, 7th ed. (London: Reeves and Turner, 1872), 304. Malthus wrote little about India but he was a professor of political economy, teaching India's future administrators, at the East India Company's College at Hertford and then at Haileybury from 1805 until his death in 1834.

7. Although such views were contested by some, they became hegemonic during the Age of Atonement or the first half of the nineteenth century. See Boyd Hilton, *Age of Atonement: The Influence of Evangelicalism on Social and Economic Thought, 1785–1865* (Oxford: Oxford University Press, 1991).

8. Gareth Stedman Jones, *An End to Poverty? A Historical Debate* (London: Colombia University Press, 2004); Gertrude Himmelfarb, *The Idea of Poverty: England in the Early Industrial Age* (London: Faber, 1984); Mitchell Dean, *The Constitution of Poverty: Toward a Genealogy of Liberal Governance* (New York: Routledge, 1991).

9. Jeremy Bentham, *Works of Jeremy Bentham*, vol. 8 (1843), ed. John Bowring, quoted in Mitchell Dean, *The Constitution of Poverty: Toward a Genealogy of Liberal Governance* (London: Routledge, 1991), 175.

10. John Strachey, "Minute on the Famine in North-West Frontier Province," in *Proceedings of the Lieutenant Governor of Bengal during February 1866: Revenue Department*, February 1866, State Archives, Calcutta, West Bengal.

11. Strachey, "Minute on the Famine."

12. Memorandum on Poverty in India, Willoughby MSS Eur E 308/51, OIOC, British Library, London.

13. Indian Famine Commission, *Report of the Indian Famine Commission: Part. 1* (London: HMSO, 1880), 35.

14. Indian Famine Commission, *Report.*

Fertility Survey Workforce: Puerto Rico, 1949
Raúl Necochea López

Immanuel Wallerstein had a bold idea for Bernard Berelson, director of Columbia University's Bureau of Applied Social Research.[1] In a 1961 memo, Wallerstein decried the social scientific manpower shortage of "underdeveloped countries." Wallerstein's solution was for the BASR to run a program "which would, on a relatively rapid basis, turn out not scholars but workmanlike applied social researchers" in African and some Asian nations: able to design studies, understand statistical data collection, and supervise a field staff. More studies featuring local talent in some capacity, Wallerstein suggested, would boost the prestige of social science in poor nations, while expanding the evidentiary basis for studies by cosmopolitan researchers like himself. Available records do not show whether Berelson approved of Wallerstein's idea, though, with or without his approval, the latter built a prestigious career as the originator of World Systems Theory.

Despite that archival loose end, it is clear that Wallerstein's memo bubbles with assumptions about the worth and intellectual capabilities of people in "underdeveloped countries," and makes a virtue out of their instrumentalization, for the good of policymaking, of sociology, and of scholars such as Wallerstein. It is ironic that a theorist known for his critique of a global economic system in which some nations exploit others would defend the treatment of people in African and Asian countries as mere suppliers of labor and raw data for Third World governments and intellectuals in capitalist societies. But let's not make this about Wallerstein, for his position, while bold, was not original. Over a decade earlier, the BASR had already participated in a major project that produced a cadre of "workmanlike" surveyors in Puerto Rico: the Family Life Study (FLS).

The FLS (1948–1959) aimed to document popular ideas about reproduction and sexuality and to use such knowledge to persuade people to favor smaller families. Executing the FLS called for an epistemic synthesis of research on representative population sampling, commercial advertising, and the fledgling field of sexuality studies.[2] It also demanded generous public and philanthropic funding, and the vertical control of data by the project directors. This included the production of questionnaires, the deployment of surveyors to data collection sites throughout Puerto Rico, the centralized aggregation and analysis of field data, and the ultimate display of data as tables and bullet points that Puerto Rican decision makers could use. Surveyors were a key part of the FLS's intricate plan.

The FLS's main report argued that Puerto Ricans' attitudes towards the small family ideal could be made more positive, given sufficient understanding of local culture. The FLS made broad generalizations about how a lack of communication about sexuality between partners, the "modesty" of Puerto Rican married women, and their "respect" for their husbands' desires, stood in the way of contraceptive use by couples.[3] The FLS was the prototype for hundreds of fertility surveys carried out worldwide through the 1970s.[4] These "KAP studies" (known by this acronym because they aimed to uncover *k*nowledge, *a*ttitudes, and *p*ractices about reproduction and sexuality) took place in scores of countries across Latin America, South and Southeast Asia, Africa, and the Middle East that the US was courting as allies during the Cold War.[5]

In the late 1940s, US demographers and politicians worried about rapid population growth's potential to fuel poverty and social discontent in the Third World, thereby enabling communist inroads. KAP studies responded to this worry: those funding, crafting, and overseeing the surveys believed not only that population limitation ought to be a component of policies to modernize "underdeveloped" nations along Western lines and steer their populations away from communism, but also that fertility surveys provided the knowledge necessary to design plans to align popular preferences with the small family ideal.[6] Yet, fertility surveys did more than pull social science into the politically fraught sphere of population policymaking: they also created the need for a new workforce in the Third World, one that could mediate research patrons' thirst for quantifiable evidence and the wariness of people asked to bare details about their sexual and reproductive lives.

Since the early 1940s, prior to the onset of the Cold War, the US colony of Puerto Rico had been serving as a testing ground for the hypothesis that expert-crafted policies, such as those devised during the administration of President Franklin Roosevelt, could turn "backward" places of the globe into prosperous showcases for liberal democracy.[7] The University of Puerto Rico's

Centro de Investigaciones Sociales (CIS, Center for Social Research), established in 1944, played a pivotal role in producing the policy-oriented research government authorities desired. Between 1945 and 1955, the CIS zeroed in on four areas: "population growth," "industrialization," "distribution of economic benefits," and "social structure patterns," each comprising projects directed by prestigious scholars recruited from the US, including Lloyd Reynolds (Yale), John Kenneth Galbraith (Harvard), and Melvin Tumin (Princeton).[8] By the mid-1950s, projects dealing with population growth, including the FLS, consumed almost half (US$ 199,000) of the CIS's budget, nearly as much as projects dealing with aspects of industrialization (US$ 206,000), the most expensive category of research projects.[9]

Foreign scholars' lavish contracts at the CIS irritated Puerto Rican social scientists' sense of fairness. More importantly, though, the latter disapproved of the government determining the CIS's priorities. To them, the leading political party, the Partido Popular Democrático (PPD, the Popular Democratic Party), had allied itself with US capital to favor a colonialist agenda that extended to the social sciences, dictating the questions worth asking while neglecting others that could not be readily translated into rapid industrialization policies, such as research on education programs, women's employment, or migration.[10] Indeed, in 1948, Puerto Rico Governor Luis Muñoz Marín had launched an ambitious economic policy, Operation Bootstrap, that defined the public sector's primary role as that of supporting investment in whatever area US capitalists deemed promising.[11] Because rapid population growth could counter the benefits of economic reforms, however, Puerto Rican authorities viewed family planning services, and greater understanding of how people might use them, as necessary complements to economic measures.[12]

Puerto Rican social scientists' refusal to let the UPR become an enclave of US-led pro-industrialization research was partly influenced by the stance of the Partido Nacionalista (PN, Nationalist Party), which railed against Puerto Rico's status as an enclave of US capital. Since the 1930s, the PN deployed a fiery, Catholic, hispanophile rhetoric about Puerto Rican uniqueness, and led strikes and even attacks on the US mainland in pursuit of independence. Though censored, harassed, and persecuted, the nationalist minority still enjoyed a following among students and faculty at the UPR in the 1940s and 1950s.[13]

Historical records about the FLS tell us nothing about the response of the CIS to its critics at the university. Although CIS researchers may have been insulated from the broader UPR community, they depended on a cadre of native research assistants to bring their projects to fruition. Fertility research, which the FLS exemplified, was in its infancy in the US, and demanded months of training of subordinate personnel, which allowed them to come into their own as rising scholars.[14] On paper, the CIS agreed with the importance of substantial training, and portrayed surveyor work as an apprenticeship that "provides valuable training, which prepares these younger scholars to use their improved competence in graduate study, in research and teaching, or in government services."[15]

In reality, however, FLS surveyors were not groomed for research careers. Their training was completed in a matter of days, and their work was precariously bound to the completion of the FLS. Not surprisingly, most have left no archival traces. But what little exists shows that their interests went beyond family planning. Surveyor ES, for example, recorded a man's disapproval of US beauty contests that objectified women's bodies. MES was shocked by the 12-year-old daughter of an interviewee, who verbally assailed her mother for leaving her husband for a new man. And BEGL1 noted verbatim the simple link a woman made between her poverty and family size preference: "If I were rich I would have many [children]."[16] It is clear that the surveyors who were to zero in on the social barriers to contraceptive use were also moved by aspects of the lives of their countrymen and women, such as their marital strife, material deprivation, and disapproval of US customs, issues the official FLS did not address in its cultural diagnostics and policy advice. The latter focused narrowly on birth control popularization, instead of the broader gendered conflicts surveyors observed but did not delve into.

Considerations of cost and the governmental need for rapid delivery of recommendations determined the quick pace of data gathering, turning the day-to-day work into something akin to guerrilla sociology: scantily equipped but motivated surveyors deployed by the project's leader to visit locales for short bursts of questioning, followed by swift retreats to prevent communal backlashes against the indelicate interview topic. Former surveyor Angelina Saavedra recalled how she eluded confrontation with a man who was fuming after learning of the scabrous questions she had posed to his wife. Similarly, repeat canvassing of La Perla in San Juan was not possible because of the entire neighborhood's negative disposition toward the questionnaire. Word of mouth turned it into a topic of local censure after a single day of interviewing.[17]

The FLS began formally in 1949, with government funds and contributions from the Rockefeller Foundation and the Milbank Fund. The CIS recruited University of North Carolina sociologist Reuben Hill as its director, along with BASR sociologist Joseph Stycos and Harvard anthropologist David Landy as assistant directors.[18] While Landy did ethnographic research on child-rearing practices in a rural area, Stycos carried out fertility surveys in rural and urban areas.[19] Hill and Stycos's plan was to begin with a survey to learn about child-rearing styles, sex roles, courtship patterns, and attitudes and practices related to marriage and birth control. The project was to conclude with an experimental stage in which different forms of intervention would be tested out to determine how best to publicize the small family ideal, following the policy-shaping imperative of the CIS.[20]

Once appointed to the FLS, Stycos's first challenge was turning Puerto Ricans into reliable information-gathering surveyors. Between 1951 and 1957, he worked with 44 of them; the vast majority, 39, were women.[21] In addition to requiring surveyors to be relatable quick studies, the FLS screened for traits such as the ability to improvise, a most handy attribute of fieldworkers. (All I managed to speak with professed strong pro-PPD sentiments, rather than sympathy toward the PN.) Stycos allocated three weeks to prepare surveyors, through background readings and lectures, interviewing tips, and mock interviews. The three-week training fell vastly short of the minimum demanded by researchers engaged in similar studies in the US.[22] To Stycos, surveyors were ad hoc subordinates, rather than apprentices, whose task was simply "to receive the information from the respondent and to transmit it for analysis. For this function, the information has to be complete, accurate and not influenced by the interviewer himself."[23]

Surveyors, however, did not see themselves as mere conduits between the field and the FLS directors. Stycos and Hill, for example, insisted that surveyors stick to the script provided, emphasizing their affiliation with the prestigious UPR, to create an authoritative distance from which to compel complete and accurate answers to questions. However, in practice, surveyors must diminish social distances to make respondents comfortable.[24] This occurred most often when surveyors confronted the squalor and violence surrounding some interviewees' lives. Moved by pity and a paternalistic sense of responsibility for the poor, surveyors brought clothes for children or advocated on behalf of interviewees at health posts, for example. Reciprocally, surveyors accepted food, horse rides, and even invitations to Christmas pig roasts as tokens of gratitude for their visits.

In addition to enjoying their modest ability to improve the lives of interviewees, surveyors also described their rising self-confidence when addressing the delicate subjects of sexuality and contraception. Surveyors seemed to relish the challenge of getting others to bare details of their intimate lives, which they did by asking questions in roundabout ways, seeming empathic, cajoling answers from unwilling respondents, and praising their looks, homes, and children. As one surveyor put it, "every person can talk freely to anybody whom he believes to be serious and trustful in character, and especially who have proved to be friendly [to] him."[25] Whatever Hill and Stycos may have intended, then, FLS surveyors saw themselves and acted as more than data conduits.

As others have found, Cold War-bound social scientific projects sponsored by the US elsewhere in Latin America fostered the transmission of scientific knowledge, along with values of individualism and self-reliance, as markers of modernity.[26] The story of Puerto Rican FLS surveyors shows us that this process influenced not only the topics of research, but also, more immediately, the subordinate personnel that carried out portions of the research. Rather than being mere reporters of evidence, surveyors developed identities on the ground as modernizing agents in their own nations, to whom planned population growth was vital to national security and prosperity. However, the surveyors also made observations about people's lives, particularly gender-bound ones, that the official record eschewed. Moreover, surveyors' often flaunted the instructions Stycos had given them; yet, ironically, the FLS assistant director wound up praising their "unanticipated excellence" to the director of the CIS.[27]

The FLS was more than a pioneering social science project in a deceptively tame colony. Because of the extent to which it implicated native researchers, the FLS also directs our attention to the relations between US and Latin American personnel engaged in research. For US social scientists, the FLS reinforced the image of the non-US researcher as a professional sidekick, crafty but naïve, and lacking initiative. Others, including Wallerstein, seemed to embrace such an ethnocentric portrayal, with little thought to the deleterious effect this had on the cultivation of Latin American talent. Whether and how these self-serving biases changed over time, this tale does not tell.

The limits of the surveyor role were evident to Puerto Rican social scientists in the mid-1950s: it did little to strengthen individual analytical skills, connect surveyors to related research outside the CIS, or award them the academic credentials that could further their professional autonomy outside the confines of the CIS.[28] Thus, for surveyors, their relationship to the FLS was mainly one of exploitation, despite the opportunities they had to shape the day-to-day data collection and the pride and pleasure they derived from their jobs. Wallerstein's memo and Stycos's own writings were as tone-deaf to the sense of accomplishment surveyors derived from their work responsibilities, as they were to the political and cultural context in which surveyors became their employees. Those of us who work as educators, especially with aspiring researchers in less-developed nations, can learn from this and do better.

Notes

1. Immanuel Wallerstein to Bernard Berelson, 1961, "Special Training Program for Social Researchers from Underdeveloped Areas," box 106, folder 262, Bureau of Applied Social Research Papers, Columbia University Archives (hereafter "BASR"),

New York, NY. Established in 1944, Columbia University's BASR was one of the most prestigious social science research centers in the US, attracting a sterling roster of intellectuals that included founder Paul Lazarsfeld, C. Wright Mills, Kingsley Davis, Robert K. Merton, and Wallerstein himself. Corporate and government patrons (such as Lucky Strike, Ex-Lax, Kolynos, and the Department of State's anti-communist propaganda office) sponsored the BASR's communications research into the factors that modified political and consumer preferences.

2. Raúl Necochea López, "Fertility Surveyors and Population-Making Technologies in Latin America," *Perspectives on Science* 25, no. 5 (2017): 631–54.

3. Reuben Hill, Joseph Stycos, and Kurt Back, *The Family and Population Control: A Puerto Rican Experiment in Social Change* (Chapel Hill: University of North Carolina Press, 1959), 331.

4. Examples abound, including Sitamraju Balakrishna, *Family Planning, Knowledge, Attitude and Practice, A Sample Survey in Andhra Pradesh* (Hyderabad: National Institute of Community Development, 1971); John Caldwell, "Fertility Attitudes in Three Economically Contrasting Rural Regions of Ghana," *Economic Development and Cultural Change* 15, no. 2 (1967): 217–38; Ronald Freedman and John Y. Takeshita, *Family Planning in Taiwan* (Princeton: Princeton University Press, 1969); Joseph Stycos, *Human Fertility in Latin America* (Ithaca: Cornell University Press, 1968). See also John Cleland, "A Critique of KAP Studies and Some Suggestions for Their Improvement," *Studies in Family Planning* 4, no. 2 (1973): 42–47.

5. Frank Notestein, "The Population Council and the Demographic Crisis of the Less Developed World," *Demography* 5, no. 2 (1968): 553–60.

6. Kingsley Davis, "Latin America's Multiplying Peoples," *Foreign Affairs*, July 1, 1947, https://www.foreignaffairs.com/articles/central-america-caribbean/1947-07-01/latin-americas-multiplying-peoples; José Janer, Guillermo Arbona, and J. S. McKenzie-Pollock, "The Place of Demography in Health and Welfare Planning in Latin America," *Milbank Memorial Fund Quarterly* 42, no. 2 (1964): 328–45; Simon Szreter, "The Idea of Demographic Transition and the Study of Fertility Change: A Critical Intellectual History," *Population and Development Review* 19, no. 4 (1993): 659–701.

7. Michael Lapp, "The Rise and Fall of Puerto Rico as a Social Laboratory, 1945–1965," *Social Science History* 19, no. 2 (1995): 169–199. Under US control since 1898, following four centuries of Spanish domination, Puerto Rico drew attention to glaring colonial contradictions. For starters, it was no longer formally part of a European empire but was still ruled by the US Congress. Furthermore, Puerto Ricans were US citizens since 1917, but could not (and still cannot) vote in national elections. Finally, following the postwar wave of decolonization, Puerto Rico was granted the status of a "free associated state" of the US, a unique designation that defies the traditional periodization of decolonization into colonial and post-colonial stages, since the island occupies both stages simultaneously. For a timely critique of the odd yet revealing place of Puerto Rico in the history of colonialism, see the

essays in "Puerto Rico: A US Colony in a Postcolonial World?," issue, ed. Margaret Power and Andor Skotnes, *Radical History Review*, no. 128 (2017).

8. "Informe Anual del CIS, 1956–57," 4, caja C-48, fondo informes anuales, University of Puerto Rico Archive (hereafter "UPR"), Puerto Rico.

9. "Informe Anual del CIS, 1954–55," caja C-48, fondo informes anuales, UPR.

10. Luis Nieves Falcón, "Implicaciones Sociales de la Expansión Cuantitativa en el Sistema de Instrucción de Puerto Rico durante el Periodo de 1940 a 1960," November 5, 1963, caja 10, serie 17, sub-serie 7, cartapacio 20, Luis Muñoz Marín Archive (hereafter "LMM"); Eduardo Seda Bonilla, *Interacción Social y Personalidad en una Comunidad de Puerto Rico* (San Juan: Juan Ponce de Leon, 1964); Manuel Maldonado Denis, *Puerto Rico: Una Interpretación Histórico-Social* (Mexico: Siglo XXI, 1969); Eduardo Seda Bonilla, *Requiem por una Cultura* (Río Piedras: Edil, 1970).

11. Carlos Zapata, *De Independentista a Autonomista: La Transformación del Pensamiento Político de Luis Muñoz Marín, 1931–1949* (San Juan: Fundación Luis Muñoz Marín, 2003); César Ayala and Rafael Bernabe, *Puerto Rico in the American Century: A History since 1898* (Chapel Hill: University of North Carolina, 2007); Emilio Pantojas-García, *Development Strategies as Ideology: Puerto Rico's Export-Led Industrialization Experience* (Boulder: Lynne Rienner, 1990).

12. Iris López, *Matters of Choice: Puerto Rican Women's Struggle for Reproductive Freedom* (Piscataway: Rutgers University Press, 2008); Annette Ramírez de Arellano and Conrad Seipp, *Colonialism, Catholicism, and Contraception: A History of Birth Control in Puerto Rico* (Chapel Hill: University of North Carolina Press, 1983); Laura Briggs, *Reproducing Empire: Race, Sex, Science and US Imperialism in Puerto Rico* (Berkeley: University of California Press, 2002).

13. Marcos A. Ramírez, "La Universidad, La Colonia, y la Independencia," March 1, 1948, UPR, recopilación especial 1 "Reforma Universitaria 1942–1962," caja 1-2, cartapacio "Reforma Universitaria, ños 1942–1948"; Jaime Benítez, *Junto a la Torre: Jornadas de un Programa Universitario* (Río Piedras: UPR, 1962); Ivonne Acosta, *La Mordaza: Puerto Rico, 1948–1957* (Río Piedras: Edil, 1989).

14. Alfred Kinsey, Wardell Pomeroy, and Clyde Martin, *Sexual Behavior in the Human Male* (Philadelphia: W.B. Saunders, 1948).

15. "Informe Anual del CIS, 1950–51," caja C-48, fondo informes annuales, UPR.

16. The full names of these surveyors were not included in the archival source. BEGL1, "Report of interview X-3, February 11, 1952," 20; ES, "Report of interview X-3M, February 14, 1952"; MES, "Report of interviews with X-6, February 11–12, 1952," box 41, folder Puerto Rico Study, Reuben Hill Papers (hereafter "RHP"), University of Minnesota Archives, Minneapolis, MN.

17. Telephone interview with Angelina Saavedra, San Juan, May 16, 2011.

18. Millard Hansen to Reuben Hill, June 17, 1951, box 15, folder Progress Reports, Univ. of PR project, Family Study Center collection (hereafter "FSC"), University of Minnesota Archives, Minneapolis., MN. On Hill's specialty, the sociology of the family, see his biographical file at RHP.

19. See Millard Hansen to Reuben Hill, June 17, 1951; and Millard Hansen to David Landy, n.d, box 15, folder Progress Reports, Univ. of PR project, FSC, University of Minnesota Archives, Minneapolis, MN.

20. "Stages in a long-term project in family life," FSC, box 15, folder Progress Reports, Univ. of PR project, FSC, University of Minnesota Archives, Minneapolis, MN

21. See CIS annual reports from 1951 to 1957, caja C-48, UPR, fondo informes anuales, University of Puerto Rico Archive.

22. Joseph Stycos, "Further Observations on the Recruitment and Training of Interviewers in Other Cultures," *Public Opinion Quarterly* 19, no. 1 (1955): 68–78; Joseph Stycos, "Interviewer Training in Another Culture," *Public Opinion Quarterly* 16, no. 2 (1952): 236–46.

23. Kurt Back and J. M. Stycos, "The Survey under Unusual Conditions, 1959," 34, box 11, folder 18, J. Mayone Stycos Papers (hereafter "JMS"), Division of Rare and Manuscript Collections, Cornell University Library, Ithaca.

24. See telephone interviews with former CIS research assistants Teté Fabregás, San Juan, June 16, 2011; Socorro Martínez, Trujillo Alto, May 26, 2011; Angelina Saavedra, San Juan, May 16, 2011; and Nilda Anaya, Silver Spring, June 20, 2011.

25. BEGL1, "My experience as interviewer in the Family Life Project," 12, box 41, folder Puerto Rico Study, RHP, University of Minnesota Archives.

26. See Chapter 4 in Raúl Necochea López, *A History of Family Planning in Twentieth Century Peru* (Chapel Hill: University of North Carolina Press, 2014); Jason Pribilsky, "Developing Selves: Photography, Cold War Science and 'Backwards' People in the Peruvian Andes, 1951–1966," *Visual Studies* 30, no. 2 (2015): 131–50. On the historical and contemporary importance of intermediaries in the production of knowledge, see Simon Schaffer, Lissa Roberts, Kapil Raj, and James Delbourgo, *The Brokered World: Go-Betweens and Global Intelligence, 1770–1820* (Sagamore Beach: Science History Publications, 2009); Patricia Kingori and René Gerrets, "The Masking and Making of Fieldworkers and Data in Postcolonial Global Health Research Contexts," *Critical Public Health* 29, no. 4 (2019): 494–507.

27. Joseph Stycos to Millard Hansen, January 14, 1952, box 41, folder Puerto Rico Study, RHP, University of Minnesota Archives.

28. Anonymous letter to the Governor, "Sobre la UPR y el Rector, 1955," 84, caja 9, serie 17, sub-serie 7, cartapacio 15, Archivo Histórico, Fundación Luis Muñoz-Marín, San Juan, Puerto Rico.

Fertilizer: South Korea, 1952
John DiMoia

Along with sugar, cotton yarn, and wheat flour (the "three whites"), fertilizer was the focus of aid dollars to South Korea in the late 1950s. The aim was to assist the country in producing its own commodities, lending substance to a narrative of benevolent assistance leading to self-reliance. This assistance was motivated less by altruism and more by a desire to enroll the nation in a set of economic assumptions that would embed it within an American-sponsored network of economic activity in East Asia, or so-called Free World Asia. By tracking the production of commodities such as fertilizer, wheat, rice, and cement, it is possible to observe the workings of plans to change South Korea's consumer habits and living patterns, a set of micro-transformations corresponding to the larger structural changes taking place. Once an agrarian nation embedded within the Japanese Empire (1910–1945), colonial Korea experienced temporary division in 1945, with two Koreas then assuming their status as new nations in 1948. A violent war followed, the Korean War (1950–1953), leaving both North and South Korea devastated. In the postwar period, South Korea, which was reliant mostly on light industry and agriculture, became the recipient of multilateral sources of aid—a showcase. Only in the mid-1960s did the south start to recover more fully, joining as part of a dynamic, triangular trade relationship with Japan and the US, with the Vietnam War driving rapid economic growth.

In the late 1950s, the first sacks of chemical fertilizer to be produced at the Ch'ungju Fertilizer Factory (충주 비료 공장), a new chemical complex located in South Korea, began to appear, featuring a logo representing the site's stated mission. Each bag of chemical fertilizer featured prominently the emblem of a Korean farmer at his task driving a plow in the fields, as farmers had done for decades past, a common sight.¹ Animal labor, here taking the form of an ox, pulled the plow, accompanying the farmer in digging a series of furrows, creating a grid pattern by which to plant for the growing season. Linking the agrarian past to the immediate present, the logo symbolically captured a significant portion of South Korea's rural economy, the production of grain (rice, barley, corn) for the Korean peninsula, but with an added dimension.² For the first time since the Korean War (1950–1953), South Korean farmers had access to a domestic source of chemical fertilizer "made in Korea," a brand of urea fertilizer produced at a newly built South Korean chemical plant, signaling a technological leap in the handling of hydrocarbons, as well as anticipated gains in agricultural yield.³

Together these two aims (of producing chemical fertilizer [hydrocarbons] and increasing agricultural yields) reflect the extremely modest position of this post-colonial nation within the first decade of its post-Korean War recovery (1953–1961), as South Korea found itself lagging behind neighboring North Korea in terms of economy and the rebuilding process, locked in a bitter competition for legitimacy.⁴ Beginning in 1948, the south lost access to sources of fertilizer produced in the north, especially those deriving from the major chemical complexes based along the northeast coast at Hamhung.⁵ In this sense, Ch'ungju was deliberately framed as an ambitious bid for South Korean self-sufficiency, liberated from roughly a decade of costly fertilizer imports, a necessity given the inability to work out a satisfactory relationship of exchange.⁶ Moreover, this uncomfortable dependency was not a recent development, as the origins of chemical factories in the north derived from previous circumstances of Japanese colonialism (1910–1945), especially its prewar industrial ambitions, or "comprehensive development," dating to the late 1920s and early 1930s.⁷ With the Ch'ungju site, South Korea hoped to rewrite its own history, thereby gaining a measure of agrarian and technological autonomy, especially in terms of managing its regional relationships with neighbors Japan, North Korea, and China (Manchuria).

If the Ch'ungju story had resulted in a conventional "happy ending," there would not be much more to tell, as such a narrative easily satisfies the expectations of its developmental burden. In fact, there is a powerful, supporting narrative, "the miracle of the Han" (한강의 기적), a Cold War story of unlikely success, with South Korea emerging from the devastation of war to achieve economic "take-off" by the mid-1960s, at which point the country's growth rate increased dramatically, continuing well through the early to mid-1990s.⁸ However, this story has become subject to critique within the past two decades, especially in the aftermath of the Asian financial crisis (1997–1998). The major problem with this story is the descriptive label itself, which borrows from a previous West German economic transition of the 1950s, the "miracle of the Rhine" (*Wirtschaftwunder*) a comparable, constructed Cold War story of "success," one requiring elision of the uncomfortable, recent past.[9] Perhaps even more powerful is the extent to which this brand of nation-centered story erases the massive, intersecting series of aid structures providing South Korea with many of its basic commodities, a complex set of technical exchanges and aid packages keeping the nation afloat for much of the first two to three decades following independence.

In the immediate short-term, this diverse array of "gifts" came packaged with a corresponding set of economic and social plans, with the earliest of these emerging during the American occupation (1945–1948), holding modest aims.[10] Following the outbreak of the Korean War, the scale of these plans grew, taking on the added weight of new expectations. Even during the combat stages, UNKRA (United Nations Korea Construction Agency) began its work as early as 1951, with the Nathan Report (1952) soon appearing as the economic vision projected for a postwar Korea.[11] Robert Nathan previously worked on the US War Production Board in the early 1940s and on this basis claimed an understanding of how to mobilize limited resources for a better future. Nathan's projections, along with those of a host of other advisers, joined a chorus of those seeking to promote South Korea in its new role as "Freedom's Frontier," a showcase of Western values and practices bordering on the Communist world.[12] Prior history and the East Asian context held little importance, with economic "success," or at least the semblance thereof—translated to satisfying key metrics and indicators—required, given the extent of the symbolic and material investment. At the same time, these new expectations, imposed through plans from abroad, did not necessarily fit the needs of South Koreans, who expressed their dissatisfaction in a number of ways.

The Elision of Empire: Identifying Colonial Echoes

If one seeks to challenge or reject this brand of "miracle" story, or an account celebrating Korean economic success without providing sufficient explanatory context, what type of narrative work does a bag of fertilizer, or some similar commodity, accomplish? The idea that South Korea needed to produce its own chemical fertilizer derived from import-substitute schemes popular among developmental economists in the 1950s. A recently decolonized country needed subsistence items to recover, and should not spend its limited resources (foreign exchange reserves) on costly import commodities, meaning that a small pool of such items could be identified, and subsequently targeted for domestic production. In the South Korean case, the shortlist of these commodities included the "three whites" (sugar, cotton yarn, and wheat flour), along with chemical fertilizers, all items valued for their prominent role in industry and daily life.[13] In industrial terms, similar plans targeted a plate glass factory at Incheon, a cement plant at Mungyeong, and the fertilizer factory, our focus here.

However, this narrative lacks an account of interdependence, that is, a thorough accounting of South Korea's transformation from a late nineteenth century component part of an East Asian / Sino-centric economic system (trading mainly with China), to a Japanese colony, one central to, and integrated within a regional, industrial empire with expansionist aims. The drive to produce these commodities, in other words, sustains an idealized conception of a newly independent nation-state in the early Cold War, divorced from its own history, and one uniquely capable of functioning on its own and meeting its domestic needs.

Instead of accepting these terms, if we continue with the bag of fertilizer, it stands at the juncture of two very different sets of stories. In the postwar period (1953–1961), as well during the occupation preceding the war, this type of fertilizer served as a painful reminder of these webs of dependence, with the item purchased from Soviet representatives for almost three years (September 1945–May 1948) prior to the formal creation of a North and South Korea.[14] The postwar transition brought a different kind of dependence, this time, upon the US, with American imports subsidizing a significant proportion of the South Korean economy, an uneasy relationship that lasted until the early 1960s. The fertilizer proved not only costly, but also brought related structural changes with it, as it hinted at increased mechanization, a dramatic increase in energy demand, the strategic use of chemicals, and new types of agricultural priorities.[15] In effect, the creation of a domestic source of fertilizer production reduced one type of dependence (Japanese imperialism), but only by exchanging colonialism for another form of dependent relationship, this time embracing a capitalist, cash-crop economy, whereby the south would expose itself to regional and world market forces, while "protected" within the sphere of American interests.

If this was the situation in the mid-1950s, it was not entirely new, as colonial Korea had already undergone a similar process in the 1920s, with its agriculture subsumed and integrated within a regional empire of Japanese distribution. More specifically, the use of chemical fertilizers, deriving from industrial sites based along the Yalu River border with China, made this possible, with the transformation of the rural sector coinciding (1920–1940) roughly with wartime preparations.[16] As early as the late nineteenth century, as Japan affirmed its borders and colonized Taiwan (1895) and Korea (1910), it made food security a priority, especially the provision of rice for the home islands. The industrialization of the northern part of Korea provided a substantial source of hydroelectric power, which in turn could be channeled to fuel the electro-chemical industry, in this case based along the northeast coast (Hungnam).[17] The massive plants based here provided the nitrogenous fertilizer that covered southern fields—especially the southwest, in *Chŏlla-Namdo*— through the early 1940s, with much of this production going to feed the Japanese military and home islands.

In this context, the bags of fertilizer produced at Ch'ungju in the late 1950s were not radically different from those produced at Hungnam (NE) in the late 1920s, despite the claim to novelty. The critical difference lay in the type of economic network valorized by each of the respective fertilizer sources, along with the accompanying stories that could be told. In the first case, Japan crafted a narrative of "Asian" industrialization and colonial uplift within an imperial setting, and indeed, told this sort of powerful, if extremely problematic, story for colonial Korea and later for Manchukuo, the new "nation" created with the invasion of Manchuria in 1931. One of the celebratory volumes marking twenty-five years of colonial rule and showcasing Japan's work in Korea was titled *Thriving Chosen* (1935), and it eagerly depicted the infrastructure, resources, and sources of education made available to colonial subjects.[18] In contrast, the American claim roughly two decades later (1954) mobilized the rhetoric of freedom and independence. If North Korea served as the rhetorical foil, American aid to South Korea sought to craft a compelling narrative to contrast with its previous role as one embedded within the constraints of Japanese empire.

The emerging alternatives, whatever they might be, generally required even greater involvement with American capital investment and economic assumptions. The bags of fertilizer held the prospect of promoting agricultural yield, albeit within a system where Korean farmers would submit their results to the fluctuations of the world market. In the meantime, the transformation to come involved not only agricultural production, but also the Korean diet. As noted previously, the "three whites" constituted the focus of much of 1950s import substitution, with wheat flour serving as one of the major elements. Surplus American grain production was passed on to the world market through PL480, or the "Food for Peace" program, which both protected the US domestic market and gave enormous leverage to American interests.[19] In the Korean case, the import of American wheat served as a supplement, deferring the immediate need for rice imports; and in terms of consumption, fueled the growth of instant noodles (ramen), consumables made from this cheap source of wheat flour.[20] By the late 1950s, American aid was becoming a means of addressing economic barriers, integrating South Korea's economy within an American set of regional concerns, and dramatically changing daily consumption habits.

The last of these points links directly to changes in agrarian priorities, with accompanying shifts to economic planning in the early 1960s. From import-substitute policy, which managed to support a limited domestic industry, the United States shifted its emphasis in aid policy to project loans and grants in the early 1960s, requiring structural change on the part of the recipient nation. In terms of immediate impact, this meant that commodity items continued to arrive, with wheat flour and milk powder prominent for South Korea, but now accompanied by numerous attached conditions for enacting social change. In the dissertation, "Foreign Things No Longer Foreign," Dajeon Chung recognizes the enormously disruptive effects of imports upon Korean bodies, homes, farms, and the agrarian economy, prior to the end of PL480 distribution in 1972.[21]

If wheat was not always consumed as such, and indeed, it took considerable time for Western-style bread to make its way to the Korean market, the availability of this cheap surplus opened up new possibilities for trade. Moreover, the wheat could be sold or exchanged to obtain desired items, or comparable staples more suited to Korean tastes. This observation underscores the point that even with dramatic change, grains such as barley, corn, and rice remained favored crops, especially the last of these three. The role of wheat, in other words, was not that of a replacement, but one of a supplement, deferring the need for expensive rice imports. Through the mid- to late 1960s, South Korea continued to produce its rice supply at a deficit, despite ambitions for achieving self-sufficiency.

Rice for the Peninsula: "Unification," or Tongil Rice

In the early 1960s, sensitive to this very question, North Korea made an offer of food aid to its southern neighbor, a gesture of assistance, and equally, one derived from pride. The uneasy relationship between the two countries kept alive the dream of restoring the rice supply, and this time, the means proved to be agricultural science, and not simply the use of additional fertilizer, although that, too, played a role. In the late 1960s, South Korean agronomists affiliated with IRRI (International Rice Research Institute), a site in the Philippines, identified a new strain of rice that would better suit the Korean climate and its short growing season. Adopted first on a trial basis, and then soon receiving the government's enthusiastic support, the newly labelled "unification" rice became celebrated domestically as the product of modern science.[22] As with

Ch'ungju, less than two decades previously, the use of this product required farmers to make a series of transformations in their practice.

However, even before noting the enormously disruptive force of this new grain, IR8, or "Tongil" rice, as it was labeled, there was another critical factor: the unusual taste. Despite all the positives in terms of growth metrics, and its ability to increase yield, Tongil rice did not meet the basic requirement needed to make it a marketplace winner. Korean consumers complained that it tasted "chalky," and failed to purchase it in sufficient quantities. As with Ch'ungju, there was too much top-down investment to permit this development, and the new rice ultimately became a "success," at least in domestic terms. How did this happen? Government subsidies and price supports kept the cost low in the market, which translated into benefits for both consumers and farmers. Of course, a relatively closed market economy and the mechanisms of an authoritarian state also helped to spread the message that this government-supported product was the best available option. Through these artificial means, Tongil became a major staple crop in the 1970s, enabling an increase in the food supply.

In Tae-ho Kim's version of the story, Tongil neatly coincides with the rise and subsequent decline of Park Chung-hee's power, especially in the 1970s, when he assumed complete authority.[23] Even as the south finally achieved self-sufficiency in rice production, the dubious achievement had to be supported through government subsidies. Moreover, the emphasis on this particular variety led to a dangerous reliance upon monoculture, leaving crops vulnerable in the event of weather events or outbreak of disease. As with other forms of aid, Tongil's "success" therefore represents a highly qualified development, one contingent upon numerous supporting infrastructures, an enthusiastic state, and a compliant population under an authoritarian regime.

Transition to Exports: Vietnam War Mobilization

When bags of fertilizer were first distributed in the late 1950s, South Korea was operating under a controlled import-substitute model of economic planning. This style changed in the early to mid-1960s, with a shift to export-oriented industry (EOI), the manufacture of consumer goods and other items that could be exchanged with partner nations in new markets. A driving factor contributing to this development was American intervention in the Vietnam War (1965–1973), circumstances creating an immediate need for commodities and transport, especially construction materials, and serving as an incentive to multiple nations in the region, especially Japan, South Korea, Taiwan, and

Thailand. If the ROK continued to produce fertilizer in larger amounts, and to import cheap wheat, it now provided manpower and consumer goods to South Vietnam, a close trade partner since the normalization of relations in 1956.[24] If wheat brought the nation into a reconfigured economic relationship with the US, cement would further transform its relations within Asia, and with its own domestic rural communities. In terms of the timing, the first exports of Korean cement began in 1964, just as the lead-up to Vietnam was accelerating.

As with fertilizer and wheat, American aid and its associated forms of planning heavily promoted import substitution, and this approach only changed with the dramatic growth of Japan's economy. Where once South Korea found itself within a dependent colonial supply chain, a heavily paternalistic system, it now found itself in a newly imagined economic relationship with a "new" Japan, following 1965. This partnership retained its colonial echoes and tensions, but now South Korea could negotiate space for itself, and more importantly, compete with Japan for Asian markets, especially in East and Southeast Asia. This last point surprised Japan, especially in the early 1970s, when Korean cars, ships, and cheap steel first began to make inroads into Japan's market share. Prior to this, the first Korean export product to carry the label "Made in Korea" abroad—wigs (made of natural hair), textiles, and other light manufacturing goods—took advantage of hand-intensive, mainly female labor, just as colonial Japan had done in the 1930s.[25]

Emerging at about the same time, cement became a desired export item largely because of these reconfigured economic relations, and in particular, because of the growth of markets within Asia. Korean cement had gone primarily to meet domestic needs during the first decade of rebuilding, especially for scarce construction and building supplies, but American participation in Vietnam had a dramatic effect on a number of regional economies. Moreover, as part of the normalization agreement with Japan, South Korea lobbied strongly for access to American procurements and supplier contracts. This effort meant the first international jobs for large numbers of Korean laborers, and logistics and supply contracts for Korean firms, again, providing international experience for the first time. In particular, the Korean cement market saw more than half of its output redirected to South Vietnam, where it would go toward the construction of new airbases, temporary housing, and built structures.[26]

Interestingly, a portion of this cement also went to the Korean domestic sector, in this case, targeting rural communities as part of the *Saemaul Undong* ("New Village Movement") campaigns of the early 1970s. In this scheme, thatch-roofed agricultural communities received building supplies (cement, roofing materials) and farming pedagogy (seeds, equipment), thereby altering farming practice and daily living habits. If South Korea underwent dramatic change during the three decades covering the 1940s through the 1970s, calling that change a shift to a "capitalist economy" is inadequate for capturing the underlying complexity. What really happened was an enormously disruptive series of transformations that brought Korea from an internal, almost exclusively East Asian market, to a colonial, heavily industrial one, and finally, to one poised between the "new" Japanese postwar state and American "Free World" ambitions for the East Asian region, still very much part of the dynamic governing Northeast Asia today. In turn, the commodities discussed here have served to penetrate markets, break down cultural, linguistic, and economic barriers, and ultimately to do much of the work of reconfiguring land use, the types of labor and lifestyles available, and to redefine South Korea's self-concept in relationship to a larger world. In embracing this style of economic success, South Korea also shifted to a hyper-capitalist mode of production, one still contentious among Korean scholars as the society currently experiences increasing economic stratification.

Notes

1. 忠肥 / 忠州肥料工場運營株式會社 (Seoul: 忠州肥料工場運營株式會社, 1963). The logo appears on the cover of this volume, published to celebrate the project's completion.

2. Tae-ho Kim, "통일벼"와 1970년대 쌀 증산체제의 형성" (New rice "Tongil" and the technology system of rice production in South Korea in the 1970s) (PhD diss., Seoul National University, 2009). See also Juyoung Lee, "Chemical Fertilizer and the Making of 'Scientific' Farmers in 1960s South Korea" (paper presented at Association for Asian Studies 2022, Graduate Student Paper Prizes, Northeast Asia Council, Hawai'i Convention Center, Honolulu, HI, March 26, 2022). Melany Sun-Min Park, "The Truss and the Cave: Architecture, Industrial Expertise, and Scientific Knowledge in Postwar Korea, 1953–1974" (PhD diss., Harvard University, 2020) covers the architecture and design of the plant.

3. *Review of Contracts* Dated May 13, 1955, and March 27, 1959, with McGraw-Hydrocarbon, a Joint Venture for the Construction and Operation of a Fertilizer Plant in Korea, Department of State; Report to the Congress of the United States by the Comptroller General of the United States (GAO: Washington, DC, 1960). Details of the plant's planning and construction stages are located at the UN Archives (UNARMS).

4. There was a similar type of ideological competition going on between the two Germanys, the two Vietnams, and between Taiwan and the PRC. In particular, the Korean-German dynamic, especially the South Korean-West German relationship, has become the focus of a growing body of literature.

5. Cheehyung Harrison Kim, *Heroes and Toilers: Work as Life in Postwar North Korea, 1953–1961* (New York: Columbia University Press, 2018). See also Kim's dissertation, "The Furnace is Breathing Work as Life in Postwar North Korea," (PhD diss., Columbia University, 2010); Rüdiger Frank, *Die DDR und Nordkorea: Der Wiederaufbau der Stadt Hamhùng in Nordkorea von 1954–1962* (Shaker Verlag: Aachen, 1996). See also Tae-ho Kim, "김태호, 리승기의 북한에서의 '비날론' 연구와 공업화: 식민지시기와의 연속과 단절을 중심으로." (Seoul: Seoul National University MA Thesis, 2001).

6. Between September 1945 and the spring of 1948, the two zones of occupation (north, south) exchanged electricity and fertilizer, among critical commodities, on a cash basis. This policy ended with United Nations-sponsored elections. "Electric Power Supply in the Republic of Korea (Korea, 1949–1970)," box 17, 4–7, Walker L. Cisler Papers, Bentley Historical Library, University of Michigan, Ann Arbor, MI. Electrical reconstruction work would begin in 1954, led by Pacific Bechtel. See Richard Finnie, *Korea's New Energy: the Construction of Three Steam-Electric Plants* (San Francisco, CA: Bechtel, 1956).

7. Shitagau Noguchi and the firm Nihon Chisso probably represent the most well-documented part of this story of industrial expansion in the north and into Manchuria.

8. "Take-off" as used here represents the language of Walt Rostow. For the reception of Rostow and modernization theory in the South Korean academy, see Tae-Gyun Park, "W. W. Rostow and Economic Discourse in South Korea in the 1960s," *Journal of International and Area Studies* 8, no. 2 (December 2001): 55–66.

9. Young-Sun Hong, *Cold War Germany, the Third World, and the Global Humanitarian Regime* (Cambridge: Cambridge University Press, 2015). Park Chung Hee's visit of December 1964 to West Germany is often cited as inspiration for his fascination with the country.

10. USAMGIK (United States Military Government in Korea) oversaw the southern zone from September 1945 until independence in August 1948.

11. Robert R. Nathan Associates, UNKRA (United Nations Korea Reconstruction Agency), *Preliminary Report on the Economic Reconstruction of Korea* (Washington, D.C.: Robert R. Nathan Associates, 1952).

12. Theodore Hughes, *Literature and Film in Cold War South Korea: Freedom's Frontier* (New York: Columbia University Press, 2012).

13. The "three whites" formed the basis for the first post-independence business activity that would later become some of the first chaebol.

14. United Nations elections in spring 1948 marked the beginning of a formal separation, as the north chose not to participate.

15. These changes are consistent with those expected in countries associated with the "Green Revolution."

16. Gi-wook Shin and Michel Robinson, *Colonial Modernity in Korea* (Cambridge, MA: Harvard East Asia, 2001).

17. Aaron S. Moore, *Constructing East Asia: Technology, Ideology, and Empire in Japan's Wartime Era*, 1931–1945 (Stanford, CA: Stanford University Press, 2013).

18. Government-General of Chosen, ed., *Thriving Chosen: A Survey of Twenty-Five Years' Administration* (Keijo: GGK, 1935).

19. Nick Cullather, *The Hungry World: America's Cold War Battle Against Poverty in Asia* (Cambridge: Harvard University Press, 2013).

20. Katarzyna Cwiertka, *Cuisine, Colonialism, and Cold War: Food in Twentieth Century Korea* (London: Reaktion Books, 2013).

21. Dajeon Chung, "Foreign Things No Longer Foreign: How South Koreans Ate US Food" (PhD diss., Columbia University, 2015).

22. Tae-ho Kim, "Miracle Rice for Korea: Tong-il and South Korea's Green Revolution," in *Engineering Asia: Technology, Colonial Development, and the Cold War Order*, ed. Hiromi Mizuno, Aaron S. Moore, and John DiMoia (London: Bloomsbury, 2018), 189–208

23. Park suspended the constitution following the 1971 election.

24. Ministry of Public Information (Republic of Korea) *Korea and Vietnam* (Seoul: Ministry of Public Information [Republic of Korea], 1967).

25. Janice Kim, *To Live to Work: Factory Women in Colonial Korea, 1910–1945* (Stanford, CA: Stanford University Press, 2009).

26. Joseph Halevi, "The Accumulation Process in Japan and East Asia as Compared with the Role of Germany in European Post-War Growth," in *Post-Keynseian Essays from Down Under: Volume II*, ed. Joseph Halevi, G. C. Harcourt, Peter Kriesler, and J. W. Nevile (London: Palgrave McMillan, 2016), 355–67.

Grid: New York, United States of America, 1972

Robert J. Kett

In 1972, visitors to the Museum of Modern Art's exhibition *Italy: The New Domestic Landscape* were ushered through a series of architectural environments, among them an installation conceived by radical Florentine architectural collective Superstudio.[1] Looking onto a platform of white gridded tiles, the visitor to Superstudio's intervention was surrounded by walls of mirrors, emplaced within an infinite gridded expanse. Set into one of the room's corners was a hub with a series of prop-like plugs splayed around it. Reproduced at regular intervals in the illusory, latticed landscape, the node signaled a connection to a less tangible, informational grid. A short film on a nearby monitor offered a window onto the futuristic vision for which the chamber served as a kind of anticipatory stand-in. *Supersurface: An alternative model for life on the Earth* imagines a techno-utopian future devoid of architecture, a world where all people are nomads and the grid provides regular, regulated services. Here, "a new mankind, free from induced needs, can survive with the help of the grid and the plugs. A new society based no longer on work, nor on power, nor on violence, but on unalienated human relationships."[2]

Superstudio's fascination with the grid, evident in *Supersurface* and a series of related projects, offers a vector for a broader consideration of the grid as both a latent infrastructure and a ubiquitous pattern. Within the broad repertoire of organizational programs available to designers and planners, none has proven as ubiquitous or seductive as the grid. The grid is at the heart of Western epistemology and aesthetics, informing perspective, the dominance of orthogonal forms, and models of space as a continuous, extensive surface. Through its apparent simplicity, the grid promises clarity, encompassment, and universality. For its wide extension through our material and conceptual

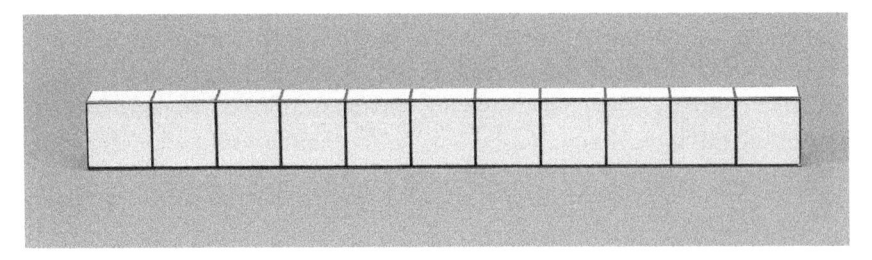

Figure 4. Superstudio, Device for a measuring rod, 1968. Veneer, 3 × 33 × 3 cm. Musée National d'Art Moderne/Centre Georges Pompidou, Paris, France. © Gian Piero Frassinelli.

heritage, the grid amounts as much to a cosmology as a geometry, "a mediatization of space from which hardly anything can escape."[3]

The grid is a meta-plan so mundane as to go everywhere unnoticed, a paradox consistent with its power to bridge and blur classic dichotomies of Western thinking: the material and virtual, the practical and the conceptual, the particular and the universal. This dual nature of the grid explains its durability as a tool of first resort for those seeking design and planning solutions as well as its unique power as a colonial technology. Superstudio's engagements with the grid unmask persistent historical dynamics in the grid's use in projects of design and territorial incorporation, but also signal the grid's transformation under an emergent networked society. In embracing a technology deeply implicated within processes of colonization, enclosure, and control, Superstudio explored the possibilities for liberatory reframings of tools of domination.

After Architecture

Supersurface appeared at MoMA six years after Superstudio's founding in 1966, and forms part of a larger effort to develop new "patterns of behaviour" for contemporary humanity beyond the scope of traditional architecture.[4] Superstudio's radical architectural practice was conspicuously devoid of architecture; the group eschewed construction in favor of developing speculative proposals as well as product designs for ambiguous objects whose underdetermined nature promised to catalyze everyday life. Through these efforts, Superstudio forwarded a vehement critique of design as usual. As founder Adolfo Natalini argued,

> If design is merely an inducement to consume, then we must reject design; if architecture is merely the codifying of the bourgeois models of ownership and society, then we must reject architecture; if

architecture and town planning is merely the formalization of present unjust social divisions, then we must reject town planning and its cities . . . until all design activities are aimed towards meeting primary needs. Until then, design must disappear. We can live without architecture. . . . Architecture is one of the superstructures of power.[5]

Much like establishing an architectural studio in order to "live without architecture," Superstudio's repeated use of the grid speaks to the group's systematic effort to work through and negate the forms, habits, and technologies at the heart of architectural practice. The grid offered a particularly evocative avenue for such a critique given its long status as a standard environment for architectural design.

In the wake of early experiments in gridded vision and inscription in Renaissance Europe, the grid's use as a design technology had been formalized in seventeenth- and eighteenth-century France. Descartes's 1637 publications *Discourse on Method* and *Geometry* gave a scientific account of physical reality built upon the grid and its axes and coordinates. This use of the grid as a model for a new Western rationality prompted its adoption as a method of enlightened scientific design. Following his appointment to the École Polytechnique in 1795, Jean-Nicholas-Louis Durand authored a new method of architectural practice based not on traditional techniques of verisimilar rendering but on a diagrammatic "code of points, lines, and planes to be organized on the newly introduced graph paper."[6] As Durand argued, the geometric plans generated through this gridded structure "fix ideas" and "communicate them afterwards," constituting rational, mobile, and durable artifacts of architectural intent.[7] The invention of such technical standards helped to establish the grid as an abstract, pervasive, and generative substrate, a fundamental armature for design as a mode of prospective action facilitated by representation and planning.

The modernism of the twentieth century launched a critique of the academic styles that Durand's techniques supported. However, this new architecture continued to inhabit the space of the grid and the graph paper page. The persistence of the grid and its implication within totalizing and colonial schemes has proven one of the primary targets of critics of modern architecture. Tom Wolfe's sardonic popular history of modern architecture lays the blame for inhumane skylines made up of "mean cubes and grids" at the feet of Gropius, Le Corbusier, and American acolytes of the International Style.[8] As Vidler notes, authors from Victor Hugo to Lefebvre and Focillon have pointed to abstract geometry as the primary cause of a fundamental "architectural alienation," a homogeneity and sterility caused by the "too easy translation of the new graphic techniques used by the modern architect into built form."[9]

It was this sterility, and its association with modern forms of capitalism and consumerism, to which Superstudio sought to respond. As Peter Lang notes, in outlining the failures of such modern architecture, Superstudio "did not so much seek to challenge the conventions of representation, or what Martin Jay classifies as 'Cartesian perspectivalism' . . . as they sought to operate to subvert the principles of architecture within this convention."[10] In so doing, Superstudio embraces the grid's role as a planning technology at the heart of architectural practice, employing it to extreme effect as a means of laying bare the structuring assumptions of an impoverished modern architectural imaginary and exploring the grid's fundamental ambiguity.

Every Point the Same

Supersurface was not the first work in which Superstudio recruited the grid as a substrate for a new kind of living and as a discursive tool in a parodic critique of the architectural status quo. *The Continuous Monument* (1969), the group's most notorious proposal, deployed hulking, gridded megastructures to cover entire swathes of cities and landscapes, foreseeing a "world rendered uniform by technology, culture, and all the other inevitable forms of imperialism."[11]

In *Supersurface*, the group continued to embrace the grid's pretense to universal spread, projecting a future with "always larger zones of the earth being developed, homogenous, and inhabitable." *Supersurface* fuses the Cartesian spatial field with the emergent grid of the technosphere. Adrift in a world without architecture, the grid becomes a flexible habitat for its motley residents, providing customized climates, access to informational networks, and a radical freedom of choice in the shaping of individual and collective lifestyles. For Superstudio, the grid "is to be understood not only in the physical sense but as a visual/verbal metaphor for an ordered and rational distribution of resources."

Across their projects, Superstudio embraced uniform infrastructures and platforms as a means of allowing subsequent freedoms of experimentation, customization, and choice. As they argued, to develop new "patterns of behaviour," "you supply the rules of a game to be played with all kinds of objects, or containers that can be filled with all kinds of things." Totalizing gestures such as the gridded worlds of *Supersurface* and *The Continuous Monument* are framed as enabling constraints, providing "a stage for a continuous performance or, in other words, a place for happenings, a place for the be-in."[12]

In describing *Supersurface* in MoMA's catalogue for *Italy: The New Domestic Landscape*, the collective celebrates the prospect of a world where "every point will be the same as any other," a world beyond the need for urban plans or architectures of social distinction, a world for wanderers.[13]

Yet despite the free, nearly hedonistic imagery of some of *Supersurface*'s nomadic residents, the grid's extension beneath these scenes of experimental living is at times more menacing than liberatory. Superstudio's animated depiction of the grid's ongoing extension across a variety of landscapes, meant to signal the growth of an infrastructure for a new form of living, simultaneously recalls the mandated grid plans of colonial settlements or the more abstract gridded mapping tools which have long underwritten projects of territorial incorporation and control. Charles Jencks noted something similar in his comments on *The Continuous Monument*, arguing, "the idea is a mixture of 'Fascist' 'total urbanization' (as they call it) and absolute egalitarianism. Everyone has exactly the same room, or the same white square gridiron, which is used for all functions."[14]

A more sympathetic reading of the grid as a universal spatial standard can be found in the work of Rem Koolhaas. Inspired in part by the work of Superstudio, his famous book *Delirious New York: A Retroactive Manifesto for Manhattan* (1978) performs an archaeology of the city's grid plan as a playground for speculative development and as the engine of a new kind of urbanism. In Koolhaas's reading, the grid plan is the primary engine of New York's exuberant, ecumenical urbanism: "The Grid's two-dimensional discipline creates undreamt-of freedom for three-dimensional anarchy. The Grid defines a new balance between control and de-control in which the city can be at the same time ordered and fluid, a metropolis of rigid chaos."[15] While in Superstudio's vision such fluidity and chaos are achieved through an end of architecture, for Koolhaas, this freedom is afforded by the skyscraper. With the skyscraper "cracks became evident in [the grid,] the edifice of neutralizing power." As skyscrapers rise, the grid is unable to fully regulate program, style, or form. And since no architectural statement can exceed the space of the gridded block, the encompassing logic of the grid proves a limit to other attempts at totalizing development.

However, Koolhaas' project is premised upon a celebration of the energy of American capitalist development, a phenomenon whose spread triggered the very anxieties that prompted Superstudio's recruitment of the grid as a means of abolishing architecture. And as critics like Sennett have noted, it is impossible to separate the proliferation of the American grid from the projects of settler colonialism and territorial incorporation.[16] Superstudio's repeated use of the grid seems to occur both because of and in spite of these associations. In engaging so deeply with the grid, the group articulates a satirical mode capable of highlighting the grid's historical and ongoing use in coercive plans but also a more experimental, perhaps optimistic attitude, which seeks potentials for less determined livelihoods in the midst of totalizing systems.

Little Squares

While *Supersurface* dramatizes the grid's power as a colonizing force and universalizing cosmology, it also leaves open the possibility that it may support experiments in living which extend beyond the remit of any plan. This possibility is acknowledged most obviously in illustrations of the individual, experimental practices which the grid is shown to support, but is also reinforced through Superstudio's materialization of the usually abstract grid. In the project's film and photomontages, the grid does not disappear as in the plans behind a finished work of architecture but mingles with more traditional materialities as it becomes a stage for provisional practices of nomadic living. In manifesting the grid amidst more quotidian objects and bodies, the project works to collapse the grid's ability to both signal the general, standardized, anonymous, and sterile and to give location to the particular, embodied, and provisional. In *Supersurface*, it is impossible to forget the grid; universalizing infrastructures are made apparent and subjected to the unpredictable effects of material practice.[17]

Superstudio is not alone in recognizing and exploiting the ambiguous, double nature of the grid. As Rosalind Krauss argues, in spite of its apparent tyranny, the grid manifests a fundamental "indecision about its connection to matter on the one hand or spirit on the other."[18] It is a technology for communicating abstract aesthetic ideas but only has effects through its instantiation in design artifacts, buildings, artworks, and other materializations. The grid can both implicate us within phenomenal space and alert us to space's virtualization, simultaneously rendering "projection as territory and territory as projection."[19]

Superstudio's materializations of the grid did not begin with *Supersurface*. The group's early projects included a series of experiments in architecture and furniture design born out of the grid, including a series of architectural sculptures (*Histograms*) as well as a line of minimalist furniture covered in gridded laminate (*Quaderna*). Echoing their tendency to work *through* conceptual and technical legacies, the group describes their early work in "little squares" as "a journey into the realms of reason, a journey with many lateral exits left open, and which at that time we did not wish to use, but which we have since used to get out of the cubical houses of disciplines and reasonable science." They also note the powerful effect their material experiments had upon the grid's abstract status: "the experiments amply showed that somewhere the links were slackening and spatial accidents had altered the archetypal grids so that nothing could return to its Platonic position. . . . we began to think of other grids with subtle non-Euclidean corners."[20] The realizations gleaned from these early manipulations and deformations of the grid help us understand how Superstudio saw decolonial possibilities in a universalizing, colonial technology.

Life between "A and B"

Supersurface, while framed in a speculative and parodic mode, has proven a remarkably prescient diagnosis of human life in a networked society. In many respects, we are now nomads on Superstudio's connected grid, accustomed to navigating networks and expectant of individual choices and customized modes of living. While built upon gridded cartographies, contemporary digital wayfinding technologies afford point-to-point navigation more akin to the associative movements of Superstudio's nomads than the intersectional plotting of previous forms of gridded mapping, echoing the collective's dream of a life of free movements between points "A and B."

However, as we are learning in the age of the network and the cloud, to live on the grid beyond traditional architecture introduces new considerations concerning design, systems, and control. Today, Superstudio's discourse of liberated life on the grid is echoed in the marketing imaginaries of Silicon Valley conglomerates as much as in radical architectural imaginaries.

More broadly, the grid's very success has resulted less in a unified world picture and more in a globe of layered networks of such complexity that it cannot be effectively imaged or comprehended. While contemporary digital platforms might on the surface resemble the all-encompassing spatial/informational grids envisioned by Superstudio, they are crucially lacking in the uniformity the Italian visionaries foresaw in their totalizing constructions. Today's grids are layered, interlocking, fragmented, (in)compatible, proprietary, and cloistered, and often take on decidedly messier material form.[21]

To some extent, these dynamics mitigate the menace of a single, centralized grid underpinning human life. However, they also underscore the questions of design and power at work in the creation and management of shared infrastructure. If we are in fact the nomads of Superstudio's networked speculation, who is designing our grid(s)? And if we cannot escape the grid, is Superstudio's dream of a universal grid as a platform for equitable and experimental common life achievable?

As Lang has noted, there are "degrees of deception" at work across Superstudio's practice. Speculative proposals are forwarded with a sense of satire and irony, often meant more to prod and provoke than to serve even as ideal design proposals.[22] *Supersurface* and the studio's other gridded experiments generated literal contexts for an exploration of alternative worlds—not utopian worlds where totalizing schemes are perfected, banished, or overcome, but sites full of strategies for making and displacing plans. As Superstudio argued, "The important thing is to keep on asserting ourselves, to go on making our mark on things. The important thing is to 'be there.'" In an age where learning to ask questions about the design of shared infrastructure is more important than ever, Superstudio's projects may still offer a useful provocation.

Notes

1. Over the course of its existence, the collective's membership included Adolfo Natalini, Cristiano Toraldo di Francia, Gian Piero Frassinelli, Alessandro Magris, Roberto Magris, and Alessandro Poli.

2. *Supersurface: An Alternative Model for Life on Earth*, 1972.

3. Bernhard Siegert, *Cultural Techniques: Grids, Filters, Doors and Other Articulations of the Real* (New York: Fordham University Press, 2015), 120.

4. Superstudio, "Evasion Design and Invention Design," in *Superstudio: Life without Objects*, ed. Peter Lang and William Menking (Milan: Skira, 2003), 116–17, on 17.

5. Adolfo Natalini, "Inventory, Catalogue, Systems of Flux . . . a Statement," in *Superstudio: Life without Objects*, eds. Peter Lang and William Menking (Milan: Skira, 2003), 164–67, on 167

6. Anthony Vidler, "Diagrams of Diagrams: Architectural Abstraction and Modern Representation," *Representations*, no. 72 (Autumn 2000): 1–20, on 9.

7. Durand quoted in Vidler, "Diagrams," 9.

8. Tom Wolfe, *From Bauhaus to Our House* (New York: Picador, 1981), 3.

9. Vidler, "Diagrams," 8.

10. Peter Lang, "Suicidal Desires," in *Superstudio: Life without Objects*, eds. Peter Lang and William Menking, (Milan: Skira, 2003), 31–51, on 44.

11. Superstudio, "The Continuous Monument: An Architectural Model for Total Urbanization," *Superstudio: Life without Objects*, eds. Peter Lang and William Menking (Milan: Skira, 2003), 122–47, on 122.

12. Superstudio, "Evasion Design," 117.

13. Superstudio, "Description of the Microevent/Micorenvironment," in *Italy: The New Domestic Landscape*, ed. Emilio Ambasz (New York: Museum of Modern Art, 1972), 242–51, on 247.

14. Charles Jencks, *Modern Movements in Architecture* (New York: Penguin Books, 1985), 56.

15. Rem Koolhaas, *Delirious New York: A Retroactive Manifesto for Manhattan* (New York: The Monacelli Press, 1994), 20.

16. Richard Sennett, "The Neutral City," chap. 2 in *The Conscience of the Eye: The Design and Social Life of Cities* (New York: Alfred A. Knopf, 1990).

17. For more on materiality and entropy in the work of Superstudio, see Lucia Allais, "Disaster as Experiment: Superstudio's Radical Preservation," *Log*, no. 22, (Spring/Summer 2011): 125–29.

18. Rosalind Krauss, "Grids," *October* 9 (Summer 1979): 51–64, on 54.

19. Benjamin Bratton, *The Stack: On Software and Sovereignty* (Cambridge: MIT Press, 2015), 84–85.

20. Superstudio, "Histograms," in *Superstudio: Life without Objects*, eds. Peter Lang and William Menking (Milan: Skira, 2003), 114–15, on 114.

21. See, for example, Bratton, *The Stack*; Nicole Starosielski, *The Undersea Network* (Durham, Duke University Press, 2015).

22. Lang, "Suicidal Desires," 46.

Hackathon: India, 2012
Lilly Irani

Hackathons are a practice of planning, rendered experimental, emergent, and probabilistic, if planning is a practice of anticipating and producing representations of possible futures. Private firms, public institutions, and civil society organizations have taken up hackathons as a way of engaging publics in prototyping possible futures. I trace transformations in Indian planning from independence through the Modi era. I show how hackathons emerge as an organizational form symptomatic of the movement of development out of the exclusive domain of the state, into the hands of NGOs, private companies, and state actors working in a liberalized regime. I then show how the practices and subjectivities of participating in planning as hacking foreclose more sustained forms of democratic engagement and investment.

Hackathon organizers invite programmers, designers, and others with relevant skills to spend one to three days addressing an issue by programming and prototypes. Organizers offer a space, power, wireless internet, and often food. Participants bring their computers, their production skills, and their undivided attention. Participants form work groups, explore ways to address the focal theme, and push towards a "demo"—a piece of software that supports storytelling around future technologies and use. At the end of a hackathon, those who managed to build demos might show them off and speculate about their futures with judges or conference goers. These demos were subject less to the question of "will this work" than "could this work." And they stood as proof of the capacities of the entrepreneurial team as much as, or even more than, the probability of the proposed solution.[1] After the hackathon, the team might promise to continue the work, or might just shake hands and say goodbye.

These hackathons were organizational devices to invite participants into exploratory, experimental planning. Facebook and the Bill and Melinda Gates Foundation, for example, hosted a hackathon to call on volunteers to build education applications atop Facebook's computing infrastructures. Facebook's plan was to grow its revenue streams and render itself indispensable to human social relations; BMGF's plan was to develop silver bullet interventions through "creative capitalism." How those plans would be accomplished—and how they could be accomplished with public legitimacy—was utterly unclear. The hackathon allowed Gates and Facebook to invite participants to dream education in the form of apps; by emphasizing voluntarism and personal agency, the events downplayed the terms of inclusion and the long-term planning horizons those terms served. Other organizations that solicited voluntaristic hacking included the World Bank and Infosys, an Indian IT service company. Their hackathon challenged volunteers to code demos to address "global sanitation" problems. More recently, the Government of India has offered up education datasets at an OpenEducation AI hackathon sponsored by IBM, Amazon Web Services, Google, and Indian education startups. Media scholar Sandeep Mertia has studied a range of hackathons in Delhi that open up "smart cities" planning and produce an emergent, new kind of expert—the "civic data scientist."[2] Hackathons proliferate as a space that allow firms to explore hires, investment, and ideas that might not otherwise readily emerge within the culture of a firm. Hackathons enable governments and corporations to expand their networks of partners, their imagination of plans, their public legitimacy, and their influence of public imagination.[3] How did hackathons become an instrument of Indian development planning? Hackathons seem to be a global phenomenon, but their function in Delhi could not be read as identical to that of Silicon Valley, though it draws its legitimacy from there.

Postcolonial Developments: From Five-Year Plans to Managed Speculation

Delhi at the time of my fieldwork seemed a development boomtown. Planning had given way to speculation; public sector companies and projects had largely given way to public-private partnerships expected to yield profit for investors.

Since before independence, Delhi has been a center of development planning and calculation to modernize Nehru's "needy nation."[4] The Planning Commission, or Yojana Bhavan, produced Five-Year Plans to model the nation as an economic system, plotting inputs and outputs, targets and controls. The Commission employed statisticians, planners, and economists as technocratic experts who set bounds on and goals for India's private and public sector, away from the "squabbles and conflicts of politics" to express the rational will and consciousness of a unitary nation.[5] Political economist Kalyan Sanyal characterized this state as an agent of "planned accumulation": "the task of the state was the promotion of capital formation by engaging in development planning, that is, by designing, implementing, and monitoring plans for the expansion of the modern industry."[6]

After liberalization, five-year plans gave way to market governance to facilitate the movement of capital investment and the growth of public-private partnerships.[7] The central government privatized national industries and exposed domestic capitalists to competition from foreign corporations. Some elites chafed at the changes; the state and high-tech capital's advocates responded that India's entrepreneurs were ready to compete globally without the protections offered by import-substitution policies.[8] Less cautious, high-tech capital pushed for the changes.[9] Five-year plans began calling on India's "entrepreneurs" to step into the void as providers of products, services, and infrastructures. Private sector consultants rotated through World Bank and public sector posts. Planning no longer resided securely in the state.[10] Over two decades, the Planning Commission moved from a language of planning to one of "enabling"—a commission that, following World Bank guidelines, minimized state intervention, allowed market institutions to distribute resources, and acted as a check on excesses.[11]

The enabling state opened a range of opportunities for private actors to experiment with development. Chief Ministers of Indian states shuttled to global cities to sell investors on economic zones and development projects. By 2004, Goldman Sachs directed global investors to the potential of emerging markets in Brazil, Russia, India, and China. Michigan business school professor C. K. Prahalad sparked a wave of private sector interest in developing products and services for the poor with his book *The Fortune at the Bottom of the Pyramid*.[12] The 2013 Companies Act mandated that large companies donate 2% of their net profits to corporate social responsibility (CSR) projects, including nonprofit health, education, housing, and welfare initiatives as well as technology incubators.[13] These CSR laws were one way the state and India's wealthiest industrialists compromised to restore the legitimacy of companies in a decade of increasing inequality.[14] These shifts generated a wave of investment in development ventures by Indian nationals and expatriates alike. Indian development had become an economic opportunity, not just a responsibility, obligation, or national task.

As the government shifted from controlling industry to enabling it, it required a new legitimating discourse. Entrepreneurship allowed elites to narrate every Indian not as a mouth to feed, but a potential leader of enterprising development. At the 2011 World Economic Forum, more commonly known as Davos, the CEO of one of India's largest banks discussed "India's Inclusive Growth Imperative" with representatives from Amnesty International, the IMF, the press, and the Government of India. Beyond the glittering dinners and expert panels, news of political unrest haunted the conference — from protesting farmers in West Bengal to the revolutions of the Arab Spring. "How do we share the growth?" the banker Kocchar asked, referring to the India's growing income inequality. She offered a proliferation of public-private partnerships as a solution. "In India, it is the entrepreneurial spirit that has contributed to a lot of growth," Kocchar announced. "How can PPP [Public-Private Partnership] work together in every field? Health? Education? Expanding the sources of employment? *It* [development] *is a responsibility for everybody and an opportunity for everybody.*"[15]

A responsibility for everybody. An opportunity for everybody. Planning and development were no longer the duty of a state to the people who put it in power. They were a diffuse opportunity for entrepreneurial citizens to enrich themselves while caring for the social body.[16] In 2015, Narenda Modi closed India's venerable Planning Commission. In its place, he announced NITI Aayog would act like a "think tank" to support an "entrepreneurial ecosystem." The move formalized a shift long underway in liberalizing India. We were all postcolonial planners now.

"Hacking for Change" from the Design Studio

I spent fourteen months with one group of middle-class Indians who answered this call, leaving jobs in Indian and multinational corporations to form an innovation consultancy I will call DevDesign. Their projects typically entailed ethnographic research, design, and planning for products and services for "the bottom of the pyramid" — water filters, community toilets, or digital learning platforms, for example. They did those planning projects for multinational philanthropies, NGOs, and corporations, sometimes even working with Indian municipal or central government officials. They put significant work into turning other Indians on to more socially aware forms of enterprise. They supported co-working spaces. They staged pop-up photography shows on the streets. They staged an annual conference that gathered workshops in arts, nonprofit management, social marketing, and design. These were the varied forms of knowledge needed to plan for a neoliberalizing community and nation, from culture, to technology, to business model.

DevDesign staged the hackathon as part of a festival celebrating design, development, and entrepreneurship. The hackathon was one of a set of workshops where fellows—accepted applicants—worked on short-term projects with NGOs, theater groups, and urban planners. The first morning of the hackathon, I ambled into the studio at 9 a.m. and the studio cook placed biscuits and small cups of black tea on the glass conference table. I knew Vipin, the convener, and Prem, an acquaintance I had invited to join the hackathon. Two other young men—engineers at the same consultancy in Bangalore—arrived together. Three others included a designer from Pune, a software engineer from Gurgaon, and a designer from London; all but the last where Indian. Vipin started the conversation by asking why we had volunteered to spend five days at the studio hacking for "open governance"—the theme of the hackathon—with people we mostly did not know.

Dev, a young web developer from Bangalore, explained that he wanted to see if he could improve the functioning of the Indian state. He explained that with "all the complaining" he heard about how the government "doesn't work," he saw this as "a chance to see if we can make a difference." Nikhil, an ex-startup founder turned software consultant, wanted to introduce government officials to the virtues of technology design more generally: "This could be just jamming the door. Getting the technology in. Ease their lives a bit with technology. So then in the future, even if it is not [this demo], they'll be more receptive [to other technology initiatives]." For Prem, a Marxist legal anthropologist, the world of software production promised a more immediate and direct way to make social change than was available to him in his writing. Prem also drew on his fieldwork on land rights activism among landless people in Uttar Pradesh as a source of insight into the design problem. Krish, another software consultant, spent his free time reading Donna Haraway, Rosi Braidotti, and painting. He had bicycled across western India and recounted those experiences as we debated potential purposes for the software.

None of us imagined the software seeds we would build here as a source of profit. Vipin hoped that he could get a philanthropic grant so he, Krish, and Nikhil might take a salary for building out the system after the hackathon. I had been part of a volunteer-run software project for years. If the software could generate income to sustain its operation, that would be great but remunerative futures were multiple and none of them paid very well. They fit, however, alongside a dense and vast complex of NGOs and social enterprises that worked along with the Indian state to organize and care for India's dispossessed.[17] I note this because analysts who explain the drive for innovation as the hegemony of neoliberalism miss the ways that participants in these projects may not subscribe to an economized or market-exchange mediated vision of the social. Few of us at the hackathon thought about our work in economized terms. The CSR sector, its networks of NGOs, and corporations and philanthropists who funded and sustained them, relied on our varied politics to ameliorate dispossession as long as we did not fight capitalist dynamics systemically or too vigorously.[18] The power of the hackathon, then, was that it invited us to translate our political hope into technical and speculative planning labor—labor that produced experimental plans and vehicles for investment by the for-profit or nonprofit sector.

We had 5 days to create the demo. The discipline of the demo forced us to orient to planning as a process of pragmatic compromise and working with already extant resources, relationships, and infrastructures.

We began by familiarizing ourselves with the domain. Vipin had recruited a friend at Parliamentary Research Service who guided us towards Parliamentary standing committees as a site where we could inform legal deliberation through the software we would design. Most of us had experience making software, but few of us had knowledge of legal process. We read through and critiqued a recent Road Safety Bill draft to put ourselves in the shoes of possible law-reading users. We learned about parliamentary procedures. Vipin pushed a stack of books on "Open Government" and e-Government, exclusively based on American case studies, to me and told me to skim for anything "that interested" me.

These activities were interwoven with expressions of time anxiety. Someone, most often one of the software engineers, would ask us to sketch a production schedule. How long could we talk about the law? Could we scope the time of debate to assure ourselves that we could produce "the demo"? As we negotiated milestone deadlines, Vipin pushed post it notes around the board representing the timeline leading up to the festival. This collective visualization of time forced us to work backwards from the demo, bounding the time to build components, preceded by negotiating what we could do that we wanted to do, preceded by where we were now—understanding anything about the problem to begin with.

Fairly quickly, major differences emerged in how Prem and Vipin understood politics to work. Vipin expressed technocratic fantasies of a website that could link dispersed Indian experts with state planners and politicians—a kind of Quora for the development state, as he described it. Vipin saw the law as a kind of code that set incentives through punishment; fix the law, fix the nation. Prem, on the other hand, had studied the implementation of the Forest Rights Act and told stories of how the law moved through activists, district officials, and landless Adivasi on the ground. The law as text was little match for the contingencies and power plays in which it was invoked. Prem, and many of us with him, did not share Vipin's faith in elite experts substituting for the politics of the poor.

These tensions led Prem and Vipin into a heated debate and many of us sided with Prem. Working with and through Prem's ethnographic cases, our interactions that followed were peppered with the subjunctive: "what you could do" and "what if we." Taking advantage of Vipin's absence for a few hours, we developed a concept called Jan Sabha, inspired by the Jan Lokpal, that would allow organizers to document face-to-face deliberations of poorer constituencies around central government issues. The hackathon, it seemed, could accommodate a more progressive politics. But, Prem warned us, it would require "some REAL footwork" to get "on the street" and work with existing organizations thinking in terms of political participation. As the sun sank deeper in the sky, we realized we had little time to reach out to NGOs or activist networks. We had little time to understand their information practices or to build trust with them. We could not even promise maintenance of any demo that came out of a potential collaboration.

That week, we weren't on the street. We were in the studio. The time, tools, and skills in the room were geared towards prototype work, not "footwork." Even the kinds of prototype work we could undertake was limited by the political economies of internet production in a country where few had direct access to the internet. Krish, a software engineer, explained to us that in the long term, the project could get into rural areas through interactive voice response phone systems, rural kiosks, or SMS-based systems. "In Andhra, there's a women's radio station," he told us, "The scope of what we want to envision is THAT. What we implement in five days is probably a website." The skills in the room were of the web; web tools were those most at hand for urgent hacking. He continued, "So we're going to go to a conversation where we'll chop off everything. Cut. Cut. Cut. Cut. But if there's a master document that accompanying this chopped up little thing..." he trailed off. Learning to demo was learning to work with the infrastructures that were already there. In the urgency of the hackathon, we could not invest in creating new infrastructures, maintaining old ones, or contesting the forms of the infrastructures that supported and disciplined our futures. Those plans and pitches which were out of bounds were accused of being "unviable."

The hackathon, then, was more than a way of bringing civil society into processes of planning. The hackathon also manufactured urgency to squeeze planning into a set of pitches—tests to see which futures could attract public, private, or personal investments. This urgency displaced deliberation in favor of inspiration; it privileged tacit knowledge and prior expertise over stretching one's matters of concern or care.[19] In expert discourses of business management and high-tech production, this urgent, experimental attitude was called "a bias to action." This is not just my description, but an actor's category originating in McKinsey management consultants Peters' and Watermans' work on how to manage corporations in the face of the failures of rational, predictive, linear models. The world, they argued, was one of complexity and rapid change. They advised that managers ought to quickly research, implement, experiment, and learn rather than run into "analysis paralysis." The "bias to action" they advised made it into job postings not only for the Delhi design studio, but even for Google. It was a central temporal and emotional ethos of innovation.

More than a regime of being-in-time, champions of "the bias to action" elevated it above other figures of democracy and planning. In joking moments, entrepreneurial Indians mocked other kinds of Indians: overly intellectual Malayali men who could find "six sides to a cube," Bengali men in *adda* satisfied to talk deeply, or academics attuned to political dilemmas over action. Collaborative design meant getting input from many kinds of people, but not letting the project run aground over the political. Planning, for these middle-class Indians, ought not be about long deliberations, extensive consultations, and resourceful investments in execution. In entrepreneurial form, as in the hackathon, planning had become a set of experiments, generated from social and tacit knowledge and ideologies, and subjected to processes of editing and selection. Hackathons were instruments of planning rendered entrepreneurial.

The form of innovation we practiced in the hackathon was more Schumpeterian than Edisonian. Historian Thomas Hughes famously called Edison a heterogeneous engineer—someone who not only invented a lightbulb but had to generate an entire infrastructure of electrical production, transmission standards, and power lines around the technology.[20] Hackathons, instead, presumed the world was full of infrastructure and the challenge was to articulate technologies that could ride those rails to create new kinds of value. These infrastructures were those forged with the growth of the internet, first under militarized internet research, then corporate telecommunications research, and then the open-source communities and internet industries that flourished across the US and Europe.[21] These infrastructures, in other words, accreted through the investments and for the needs of US and European technological use. This Delhi hackathon, like many others, encouraged participants to find use values for these infrastructures in other parts of the world. This was economist Josef Schumpeter's vision of the entrepreneur—the one who found new arrangements of existing resources, relationships, and techniques to organize novel forms of production. They did not necessarily labor, manage, or maintain. They made novel combinations out of existing skills, labor, and techniques.

Hackathons and Postcolonial Speculation

In the *Economist* in 2013 a Rolex ad promises that "anyone can change everything." The "anyones" are, crucially, people of color rather than familiar white saviors. Social enterprise promises a world without poles, where Global South elites can be presented as a grassroots, south-south achievement. Colonial anthropologists worked for companies or states to produce knowledge about difference in service of governmentality and extraction.[22] Knowledge constructing tradition, caste, and tribe, for example, helped render "terra incognita" navigable, exploitable, and governable. Today, entrepreneurs appeal to a wider set of patrons—financiers, philanthropies, government agencies, and companies. They too draw on applied anthropology and resource maps. But they also search themselves, their communities, and their resources to construct opportunity. Julia Elyachar argues that the idea of the "Bottom of the Pyramid" as a source of innovation, or "next practices," emerges in part out of critiques of development as universal, top-down expertise.[23] Entrepreneurs—in this case, children of India's governing classes—appear as another postcolonial resolution, pitching development from the middle-class up. Elided in the promise of this connected world of changemakers are the relations of exploitation—colonization, labor exploitation, resource extraction, and its newer extractive formations—that make certain plans "viable" and certain populations the named objects of care. These practices of planning as hacking and experiment direct attention away from the social relations, material resources, agendas that make certain forms of transformation more imaginable and more viable in the manufactured urgency of entrepreneurial time. Though "anyone," the ad tells us, can change "anything," the flexible skills of pitching, prototyping, and "what if" talk are new markers and practices that divide the modern and backwards, governing and governed, in entrepreneurial, experimental, investable form.

Notes

1. Smith characterizes demos in information technology cultures as a way of telling polyvalent stories about use, locating demos in a tradition of scientific demonstration. In cultures of entrepreneurialism, investors and workers commonly assume that projects will change or "pivot" as market conditions change. Thus, demos are as

much about signaling the capacity of the demo builders as they are about selling the technical solution itself. Wally Smith, "Theatre of Use: A Frame Analysis of Information Technology Demonstrations," *Social Studies of Science* 39, no. 3 (2009): 449–80, https://doi.org/10.1177/0306312708101978.

2. Sandeep Mertia, "FCJ-217 Socio-Technical Imaginaries of a Data-Driven City: Ethnographic Vignettes from Delhi," *The Fibreculture Journal*, no. 29 (2017), https://doi.org/10.15307/fcj.29.212.2017.

3. Hackathons can also have other purposes. In open-source cultures, they function as a site of joyful coding and co-present code maintenance (Coleman). They can also become sites where publics critique technical infrastructures by hacking a system, or a public pooling of archival labor. Andrea Muehlenbach, "Building an Archive of Vulnerability: #GuerrillaArchiving at #UofT," *EDGI*, 2017, accessed January 2, 2018. http://flolab.org/wp19/building-an-archive-of-vulnerability-guerrillaarchiving-at-uoft/.

4. Srirupa Roy, *Beyond Belief: India and the Politics of Postcolonial Nationalism (Politics, History, and Culture)* (Durham: Duke University Press, 2007).

5. Partha Chatterjee, *The Nation and Its Fragments: Colonial and Postcolonial Histories* (Princeton: Princeton University Press, 1993).

6. Kalyan K Sanyal, *Rethinking Capitalist Development: Primitive Accumulation, Governmentality and Post-Colonial Capitalism* (London: Routledge, 2007), 170.

7. Stuart Corbridge and John Harriss, *Reinventing India: Liberalization, Hindu Nationalism, and Popular Democracy* (Cambridge: Blackwell Publishers, 2000), 120.

8. Latha Varadarajan, *The Domestic Abroad: Diasporas in International Relations* (New York: Oxford University Press, 2010).

9. Stanley A. Kochanek, "The Transformation of Interest Politics in India," *Pacific Affairs* 68, no. 4 (1995): 529–50.

10. Historian David Ludden argued that as states have lost their disciplining power to markets, they have lost their leadership role in development as well. Territorial boundaries, he argues, no longer define the "participants, populations, and priorities" in the development process. David Ludden, "Development Regimes in South Asia: History and the Governance Conundrum," *Economic and Political Weekly* 40, no. 37 (2005): 4048.

11. Planning Commission (Government of India) *Eleventh Five Year Plan*, vol. 1 (New Delhi: Oxford University Press, 2007), 223–24.

12. C. K. Prahalad, *The Fortune at the Bottom of the Pyramid: Eradicating Poverty through Profits* (Upper Saddle River, NJ: Wharton School, 2005).

13. Ekta Bahl, "An Overview of CSR Rules under Companies Act, 2013," *BusinessStandard India*, March 10, 2014, http://www.business-standard.com/article/companies/an-overview-of-csr-rules-under-companies-act-2013-114031000385_1.html.

14. Lilly Irani, *Chasing Innovation: Making Entrepreneurial Citizens in Modern India* (Princeton: Princeton University Press, 2019); for inequality, see Atul Kohli, *Poverty Amid Plenty in the New India* (Cambridge: Cambridge University Press, 2012).

15. "India@Davos: Batting for Inclusive Growth," *NDTV*, January 29, 2011, video, https://www.ndtv.com/video/news/the-big-fight/india-davos-batting-for-inclusive-growth-189493.

16. Irani, *Chasing Innovation*, 27.

17. Sanyal, *Rethinking Capitalist Development*.

18. In *Rethinking Capitalist Development*, Sanyal argues that an assemblage of NGOs and international agencies care for the populations rendered surplus by capitalist dispossessions—the vast informal sector—as a way of maintaining the legitimacy of postcolonial capitalism among the masses rendered surplus by it.

19. Bruno Latour, "Why Has Critique Run out of Steam? From Matters of Fact to Matters of Concern," *Critical Inquiry* 30, no. 2 (2004): 225–48; Maria Puig de la Bellacasa, "Matters of Care in Technoscience: Assembling Neglected Things," *Social Studies of Science* 41, no. 1 (2011): 85–106.

20. Thomas P. Hughes, "The Evolution of Large Technological Systems," in *The Social Construction of Technological Systems: New Directions in the Sociology and History of Technology*, eds. Wiebe E. Bijker, Thomas Parke Hughes, and Trevor Pinch (Cambridge, MA: MIT Press, 1987), 51–82.

21. Janet Abbate, *Inventing the Internet* (Cambridge, MA: MIT Press, 2000).

22. Kavita Philip, *Civilizing Natures: Race, Resources, and Modernity in Colonial South India* (New Brunswick, NJ: Rutgers University Press, 2004).

23. Julia Elyachar, "Next Practices: Knowledge, Infrastructure, and Public Goods at the Bottom of the Pyramid," *Public Culture* 24, no. 1 (66) (2012): 109–29.

Kishikishi: Belgian Congo, 1956
Sarah Van Beurden

The object that provides a point of entry into the history of planning in this essay is a kishikishi, a figurative sculpture from the Pende culture of the southwestern Kasai region in today's DR Congo. Locally, the kishikishi were mounted on top of chiefs' huts as a symbol of the permanent protection of the chief by the spirits.[1] Several of these rooftop finials have also found their way into the collections of museums in the Global North, where they are viewed as ethnographic artifacts, but also as art objects of considerable financial value. In this chapter, I examine what a particular kishikishi made by the Pende sculptor Kaseya Tambwe Makumbi and located in the collection of the Institute for National Museums in Kinshasa can teach us about the intended and unintended consequences of colonial planning. The object's history intersects with the history of a network of colonial artisanal workshops. Planned as spaces for the preservation of so-called traditional life and artistic production, these workshops instead became spaces for the popularization of stylistic innovation. As part of a centralized plan for the control and exploitation of a colony-wide cultural economy, they became sites of tension between the preservation politics that gave rise to them, and the innovation and commercialization inherent in the idea of a planned cultural economy. As such their history helps question the cultural hegemony of the colonial plan.

In 1974, Nestor Seeuws, a Belgian employee of the Institute for National Museums located in Kinshasa, the capital of Zaire (formerly Belgian Congo), visited the workshop of Pende sculptor Kaseya Tambwe Makumbi in Kasai and bought a kishikishi sculpture for the museum. Kaseya was a well-known sculptor, whose reputation reached beyond the confines of the Pende community, and whose work was already present in several museum collections in

Figure 5. Sculptor Kaseya Tambwe Makumbi with a kishikishi he made for the National Institute of Museums in Zaïre (today's DR Congo), 1974. (Photo by Nestor Seeuws, Courtesy of IMNC).

the Global North. The Royal Museum for Central Africa in Belgium acquired a kishikishi made by Kaseya Tambwe in 1949, for example. The museum's then-director, Frans Olbrechts, considered it "one of the most important acquisitions" the museum had made that year and described it as a statue that deserved a place of honor in the collections.[2] The reputation of Pende arts made them much-desired objects for museums in the Global North, along with several other central African artistic cultures. A representative collection of African art would not be considered complete unless it contained some examples of their sculpture. Olbrechts, Seeuws, and their colleagues considered Kaseya Tambwe's work as representative of an "authentic" African art—as the product of a timeless artistic tradition that had not been "tainted" by the influence of colonial modernity. Ironically, as we shall see in this entry, Kaseya's kishikishi's were nothing of the kind.

The Plan

The workshop Seeuws visited had been part of colonial network of workshops established during the 1950s as part of a plan for the development and exploitation of a Congolese cultural economy. Kaseya was one of many Congolese artisans and artists working in these spaces to further their own reputations and expand the market for their work. These workshops were part of a planned *politique esthétique* created by the members of a Belgian governmental committee given the task of preserving "authentic" Congolese art and artifacts: La Commission pour la Protections des Arts et Métiers Indigènes, or the Commission for the Protection of Indigenous Arts and Crafts (COPAMI). Founded in the 1930s, COPAMI's aim was to reinforce and advance the political and economic interests of the colonial state by controlling Congo's cultural and artistic production. The committee was made up of Belgian artists, high ranking former colonials, museum curators, and university professors who relied on the anthropological and art-historical knowledge of the era to shape their plans for a new cultural economy, one that could reconcile the image of the colonial state as the benevolent custodian of Congo's artistic traditions with its more fundamental identity as an extractive regime.[3] The *politique esthétique* thus embodied the Commission's politics. In the document, committee members outlined plans for controlling the quality of art education in the colony (which put the Commission into direct conflict with the missionary sector in the colony, since the latter controlled all education); the establishment of museums in the colony for the conservation of objects in loci so that they could serve as inspiration for contemporary artisans; and the establishment of state-controlled craft workshops and sales counters. It is the latter that became the most developed element of the plan.

COPAMI members believed the solution to preserving the "authentic" character of art and craft production in the colony lay in the workshop. This interest in artist workshops was in part the result of their familiarity with the workshop model used by medieval European artists, and the assumption that traditional models derived from the European past were easily applied to an African present. As such, these workshops were essentially conceived of as spaces for controlled evolutionary transformation based upon deeply racialized conceptions of African "backwardness."[4]

The *politique esthétique* and the planned workshops need to be seen in the context of the rise of colonial and other "development" and "modernization" plans. Indeed, these cultural measures were shaped by the planned interventions into the whole of the colonial economy. The agricultural sector, in particular, was subject to large-scale planning initiatives, and the planning of the large hydroelectric dams on the Congo River, while initially suggested during the 1930s, also reemerged in the 1950s (although they were not executed until 1972).[5] The *politique esthétique* envisaged a modern rationalization of and increased control over the artistic production of the colony, but the Commission members also romanticized pre-industrial, pre-colonial life as a stable and static world that produced objects of lasting value which could be used in the "development" of the colony. Creating economic opportunities in the art and craft industry would keep people local—i.e., within reach of traditional authorities, in "stable master-apprentice relationships", and outside of the cities, which were seen as increasingly politicized—and use past artistic practices in the creation of a modern, yet "authentic," future. The Commission members thus believed the workshops could produce a craft economy rooted in the past that was nonetheless a functional part of a modern export economy—as spaces where Congolese artisans could train in the necessary skills, obtain materials, and sell products in a system that guaranteed quality control and economic development, all under the guidance of colonial officials.

The Execution of the Plan

The Ateliers Sociaux d'Art Indigène (ASAI) were established in 1956 as part of the execution of the *politique esthétique*. Two networks of workshops were created: one in the northeastern part of Congo, led by Belgian painter Lina Praet, and another in the Kasai region, led by painter Robert Verly. Very little information remains about the workshops in the north, but the network of workshops in the Kasai region expanded steadily, and it is the remnants of these that museum employee Seeuws encountered on his trip in 1974. The choice of location was no coincidence. Verly had established a network of workshops centered roughly around the town of Thsikapa in Pende country, but they reached into the regions populated by the Chokwe, Luba, and Lulua, all of which possessed desired and respected sculpture traditions. By 1957 the ASAI network consisted of 12 separate workshops in the Thsikapa region, and in the year before Congo's independence (on June 30, 1960), the network had expanded greatly, to 80 locations.[6] Verly, who returned to Belgium because of health reasons, was replaced by a high school teacher, Paul Timmermans, who was less focused on artistic merit than Verly, but who greatly expanded the sale of the production from the workshops. ASAI insisted on the use of so-called traditional tools (and particularly adzes), believing tools influenced the process of creation. All the workshops had apprentices who were encouraged to focus on the acquisition of skills and to follow the examples of the master artists.[7]

The art trade and accompanying export economy, however, very much pre-dated the *politique esthétique* and its planned workshops. In fact, its existence enabled the plan, not the other way around. Congolese artists (sculptors, in particular) had been producing for expeditions, visitors, and colonial officials since the earliest days of European contact. Moreover, the circulation of objects (both in their materialities and as concepts, genres, or styles) was not exclusive to those contacts; as objects traveled, genres went in and out of fashion, and objects and practices were transformed through cultural exchange across the region in the pre-colonial era.[8] Both Verly and Timmermans relied heavily on the existing cultural economy for determining the location of workshops and the recruitment of artists. When one looks closely at a map of the ASAI workshops, it is clear that many locations had in fact simply absorbed existing artist's ateliers as workshops. This was certainly the case with Kaseya Tambwe's atelier. The support of the sales network of the ASAI, however, helped him expand his operation.

Verly played a crucial role in the implementation of the plan. He functioned as intermediary between the artists and artisans the plan targeted and the colonial government and the COPAMI commission in Brussels. He was also the sole colonial representative (until 1959) involved in implementing the plan on the ground, which meant he had a significant impact on the resulting network and production. Colonial media depicted him as a teacher and the artists and artisans he recruited as his pupils, despite the fact that senior artists and artisans were put in charge of the day-to-day operation of the workshops.[9] Verly saw the latter as artists and as repositories of knowledge and skill, while viewing himself as their protector; counteracting the devastating effects of colonialism and Christianization on what he imagined as "ancient" Congolese cultures. He romanticized his mission as "discover[ing] and contact[ing] the last valid artists, spread out in the distant bush, many of whom no longer practice, in order to try to revive their self-confidence and conscience of their artistic mission."[10] Contrary to the preservation principles that dominated the *politique esthétique* and his own rhetoric, Verly nonetheless encouraged (guided) experimentation for the purpose of commercial sales, such as the practice of using the heads of existing sculptural genres and developing them as busts.[11] He encourage Kaseya Tambwe to develop kishikishis with shortened legs, sculpted on to small shelves, in order to make them more marketable to a Western audience.

A sole surviving report on the workshops from 1959–1960 teaches us about the operation of the system. Then supervisor Timmermans traveled around the Kasai, visiting the various workshop locations. During the year of record, he paid 216 visits to a total of 80 workshops. His visits had a number of purposes: to exercise quality control, learn about pieces and their significance, give advice on materials and execution, and buy objects. He only gathered objects to sell if the artisans were not part of local sales cooperatives (the majority were not) and in turn distributed the material to sales counters in the cities, or used it to fulfill orders from museums, galleries, and foreign companies that specialized in craft sales. Even on the eve of independence, he was still expanding the system, foreseeing no real changes in the colonial relationships that undergirded the system.[12]

The Aftermath of the Plan

The workshop network, along with its sales counters and cooperatives, had some longevity, likely because it was created around on a preexisting economy. Almost 15 years after Congolese independence in 1960, several of the ASAI ateliers were still in operation as studios or workshops although the central coordination no longer existed. This postcolonial afterlife of the workshops supports the interpretation of the network as an intervention, appropriation, and expansion of existing patterns of production and trade in arts and crafts.

Seeuws visited one of the former ASAI workshops in 1974 when he went in search of Kaseya, whose preexisting reputation had only expanded through his participation in the ASAI network. Rather than being a timeless sculpture form that was deeply authentic in the sense that it did not reflect any "outside" influence, art historian Zoë Strother thinks Kaseya's kishikishis were likely inspired in part by Catholic statues of the Madonna and child and were characteristic of his push toward innovation, introducing a more naturalist style in Pende sculpture.[13] The lasting popularity of exactly this type of object, as characteristic of the long-term impact of the colonial workshop system, raises questions about the unintended consequences of the original colonial plan. This demonstrates the limitations and superficiality of the colonial commitment to preservationist ideals and highlights the economic motivation for the workshops: the desire to exploit the trade in arts and crafts marketed as "authentic." This history renders visible the reality of the plan as one for the extraction of art as an economic resource, first and foremost, and as a means for exerting social control over existing modes of production and rural populations. Nonetheless, long-standing local dynamics of change and innovation in artistic and cultural traditions continued within the context of workshops.

Whether we could call the *politique esthétique* and its workshops a successful plan is another matter and perhaps depends on one's perspective. Despite their relatively brief existence as a colonial institution, they did have an impact, but it was one that ran counter to the professed preservationist goals of the planners. The implementation of the plan was deeply dependent not only on its "target population", namely the sculptors, but also on its intermediaries, such as Verly, who encouraged particular commercial innovations in Kaseya Tambwe's work.

Today, the kishikishi that Seeuws bought from Kaseya Tambwe in 1974 is still part of the collection of the Institute for National Museums in the capital of Kinshasa; although it is not currently on display there. Its counterpart at the AfricaMuseum in Belgium is today located in the museum's gallery, "Unrivalled Art." Although the workshop may no longer exist as a colonial institution, the preservationist ideals of the *politique esthétique* still carry weight in postcolonial museums.

Notes

1. Zoë S. Strother, *Pende: Visions of Africa* (Milan: 5 Continents Editions, 2008), 33.

2. Olbrechts F. to Vanhamme, J., July 29, 1950, dossier AA.2-D.2.1950.276, Cultural Anthropology and History, Archives, Royal Museum for Central Africa, Tervuren, Belgium.

3. For more on the history and activities of COPAMI, see Sarah Van Beurden, *Authentically African: Arts and the Transnational Politics of Congolese Culture* (Athens, OH: Ohio University Press, 2015), 61–99.

4. Johannes Fabian, *Time and the Other: How Anthropology Makes it Subject* (New York: Columbia University Press, 2002).

5. Bogumil Jewsiewicki, "Rural Society and the Belgian Colonial Economy," in *History of Central Africa*, vol. 2, eds. David Birmingham and Phyllis M. Martin (London: Longman, 1983), 123–24; Jacques Depelchin, *From the Congo Free State to Zaire (1885–1974): Towards a Demystification of Economic and Political History* (Dakar: Codeseria, 1992), 191.

6. "Rapport de la délégation envoyée au Congo Belge et Ruanda," 1957, portefeuille 4797, lias 5/2, Classement Provisoire, Archive, Ministry of Foreign Affairs (Ministerie van Buitenlandse Zaken—BuZa), The Hague, Netherlands; Paul Timmermans, "Rapport ASAI," 1960, box 1470, COPAMI Files, Colonial Archive, BuZa.

7. Robert Verly, "L'art africain et son devenir," *Problèmes d'Afrique Centrale*, 13, no. 44 (1959): 145–51.

8. Zoë S. Strother, *Inventing Masks: Agency and History in the Art of the Central Pende* (Chicago: University of Chicago Press, 1998), 171–27; Cecile Fromont, *The Art of Conversion: Christian Visual Culture in the Kingdom of Kongo* (Raleigh: University of North Carolina Press, 2014); Enid Schildkrout and Curtis A. Keim, *The Scramble for Art in Central Africa* (Cambridge: Cambridge University Press, 1998); Mary Nooter Roberts and Allen F. Roberts, "Audacities of Memory," in *Memory: Luba Art and the Making of History*, eds. Mary Nooter Roberts and Allen F. Roberts (Munich: Prestel, 1996), 17–48; and Allen F. Roberts, "Peripheral Visions," in *Memory*, 22–45.

9. J. Collard and C. Lamote, "Les ateliers d'art du Kasaï/Ba ateliers Ya Art na Kasai," *Nos Images* 10, no. 169 (1957).

10. Verly, "L'art africain," 149.

11. Verly, "L'art africain," 150.

12. Timmermans, "Rapport ASAI," 1960, Colonial Archive, BuZa.

13. Zoë S. Strother, "A Terrifying Mimesis: Problems of Portraiture and Representation in African Sculpture (Congo-Kinshasa)," *Res: Journal of Anthropology and Aesthetics* 65–66 (2014/2015):141–42; Léon de Sousberghe, *L'art pende* (Brussels: Palais des Académies, 1959), 124–25. Kaseya Tambwe in all likelihood first experimented with this new style in the late 1940s, but its (local and international) popularization dates to the 1950s.

Land Parcel: Lebanon, 1990
Mona Fawaz and Nada Moumtaz

The cadastral map is a common language for planners' representations of space. They privilege representation of the area under study as a net of contiguous lots, recognized as clearly delimited own-able land parcels. Each parcel is assigned a specific tag—either a number or a few letters, or a combination of both. Each of these tags links the delineated sub-area to a "property title," a legal document that assigns this area to a claimant who can be an individual, a group of individuals, a company, a religious entity, and/or a public agency. This "owner" holds the right to exclude, exploit, and transmit. This is what is termed the "ownership model" of the landscape. Most but not all roads have a different form of tagging that identifies them as public domain, areas to be managed in the name of a collective outside of the frameworks of market exchange. The lines and tags are not necessarily reflected in the material reality of the landscape. They nonetheless impose on the terrain a specific set of relations and constraints that largely dictate possible social practices (and consequently spatial futures) by allowing certain circulations and encounters while precluding others.

Each of the maps shown here forms the starting point or "base map" through which commissioned urban planners involved in the two post-war reconstruction projects we examine here (Solidere and Waad) were first introduced to their respective "intervention areas." In both cases, the "cadastral map" is the foundation of the urban planning toolkit, the basis of post-disaster reconstruction projects. It reflects a reading of the landscape as "owned" and translates in both cases, and numerous others, as a strategy of identifying "stakeholders" or "project claimants," individuals and/or agencies that will need to be accounted for by each of the two projects. Given the different

LAND PARCEL: LEBANON, 1990   153

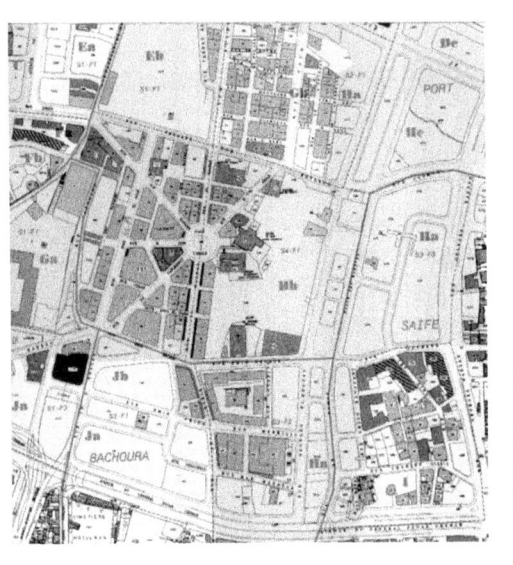

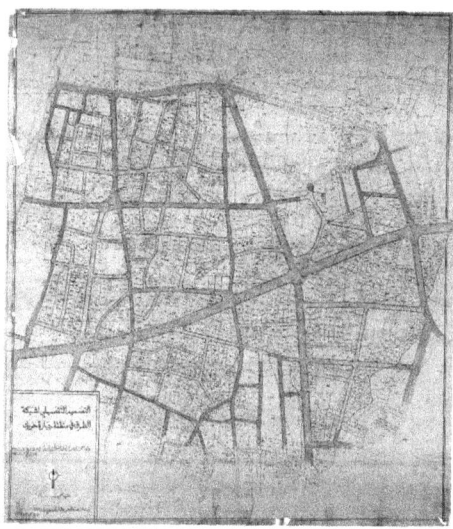

Figure 6. Cadastral map of Beirut's city center with zoning, 2004 (left) and Haret Hreik, proposed road development over cadastral map, ca. 2005 (right)

timeframes and contexts in which the two above documents were deployed, however, a comparative analysis is insightful in showing both similarities and differences in terms of what the map dictates and, perhaps more importantly, what it conceals and forecloses.

Locating each of the projects in its historical and spatial context reveals important differences between them. The "Solidere" base map is the starting point of the post "civil war" (1975–1990) reconstruction of the city's historic core. Once Beirut's dense commercial and business heart, the area had become the setting of a war economy populated by, among others, militia members, sex workers, drug dealers, and arms brokers. In the aftermath, Beirut's historic core was to be rebuilt by a real-estate holding company named Solidere, whose main founder and stakeholder was the Lebanese billionaire and future prime minister, Rafic Hariri.[1] Conversely, the "Waad" base-map of Haret Hreik is the starting point of reconstruction after the 2006 Israel war on Lebanon, a thirty-three day war that ravaged areas tagged by Israeli air forces as "military strongholds"—in this case the high-density, vibrant residential and commercial neighborhood where Hezbollah had established its main headquarters. In the aftermath of the war, Haret Hreik was to be rebuilt by Waad, a nonprofit company set up by Hezbollah.[2] Solidere introduced a new vision for its intervention area through an urban renewal project that would radically modify the geometry and sizes of the parcels and shift the nature of the area from popular, affordable souks to high-end real estate and business offices. Conversely, Waad only introduced minor urban design interventions and aimed to recreate the demolished spaces as they were, replicating as faithfully as possible building height, volumes, sizes and locations. Solidere expropriated all parcels in the downtown area and all parcel owners and rights-holders (usually meaning tenants) received shares in the company's stock at the estimated value of their rights. Waad delved in the property records of its reconstruction area, identified all those who had owned stores, apartments and/or shares and sought to restitute the demolished buildings in the same volume and location while assigning to each unit a clear set of claimants. In sum, reconstruction in these cases involved two opposite strategies: one completely expropriated buildings and turned every right into a fungible asset; the other began by accounting for/measuring actual volumes and spaces and looked to restitute each to a clear set of claimants. Nonetheless, in both these strategies, the unit of preservation or expropriation was the parcel, and the "stakeholders" were the property owners and rights-holders who could claim these parcels within the prevailing property framework.

Solidere's strategy of expropriation, with remuneration in the form of shares, elicited critiques and counter-proposals by urban planners and architects, as well as legal battles involving various rights-holders. To no avail. A less known voice of opposition to expropriation came from parts of the Beiruti Muslim Sunni community who had knowledge of and were involved in the administration of Islamic charitable endowments, known as *waqfs*, particularly the Directorate General of Islamic Waqfs, an agency loosely attached to the Prime Ministry but whose budget mostly comes from waqf revenues.[3] Besides mosques and other religious buildings considered to be heritage buildings, the city center included forty-eight Islamic charitable endowments: land, apartments, offices, rooms, or shares thereof, whose owners had surrendered the right to alienate (to sell, gift, or mortgage) them while dedicating their use and yields to a charitable cause of the owner's choosing. These charitable purposes varied from supplying spaces for prayer in mosques, salaries for teachers and students, bread for the poor, and income for one's family. To the donor of this waqf, the founding of such an endowment was, among other things like estate planning, an act of charity that brought him or her closer to God. These endowments, defined in Islamic law and mostly founded before the twentieth century, were inalienable possessions, each tied to a particular locality and purpose, and connecting current users to a particular history. In exceptional circumstances, like destruction or flooding, Islamic legal scholars allowed for substituting the asset of the waqf for another. Despite their inalienability, these waqfs very much depended on the market for their survival, as their income derived from rents. Various state legislation since the early twentieth century, especially during the French Mandate (1920–1943) have eroded this inalienability, normalizing and encouraging substitutions and even the reversion of waqfs to private property.

While the expropriation of waqfs against company stock was possible in current legislation, various Muslim Sunni groups mobilized the inalienability of waqf as grounded in the Islamic legal tradition to challenge the expropriation, publishing statements in newspapers. For instance, the Association for the Preservation of the Quran "unanimously decided to work to preserve the Islamic endowments in downtown Beirut totally, in terms of their limits and location, without any change or exchange. It will not accept any attempts at harming, decreasing, or changing the waqfs or their locations." The Association of al-Azhar Graduates also requested the preservation of Islamic charitable endowments in downtown Beirut and other Lebanese regions, and their development "based on existing regulations while taking into account the stipulations of the founder [a concept operative in Islamic law but rendered irrelevant in contemporary law]." The mobilization of the Muslim Sunni community around the inalienability of the waqf framed as requirement of Islamic law put a halt on the plan of expropriation of waqfs for shares, and resulted in the promise to return the waqfs as parcels or parts thereof.

Ultimately, the recuperation of waqfs was not a straightforward and simple opposite to the exchanges for shares model, based on a possible recognition of the inalienability of all waqfs—big or small, religious or commercial—and their individual and communal significance. Instead with the use of planning and accounting tools, waqfs were disciplined to a different conception of property, fungible and reduced to its dollar value, where rights of use, usufruct, and alienation were consolidated with one "owner." The recuperation followed Solidere's strict bylaws, which only allowed the recuperation of empty parcels or standing and usable buildings, and prohibited the recuperation of buildings and parcels touched by urban renewal according to the new master plan. Recuperation also required the payment of large dues for the cost of infrastructure and renovation, so the cash-strapped Directorate settled these costs by a debt swap, surrendering small waqfs. Only seven of the twenty-one waqf parcels were recuperated. The value of the recuperated waqfs depended on their assessed price; the most "expensive" waqfs were retained. Their values as acts of charity or as a connection to a family or a charitable foundation were rendered completely irrelevant. Nonetheless, it was the transformation of smaller waqfs into a dollar value that allowed the perpetuation of the larger waqfs. Furthermore, as an owner of real estate assets (and not Solidere shares), the Directorate was an essential beneficiary of Solidere's project, as it gained tremendously from the increased rents that the project generated. Therefore, while challenging the model of fungible, exchangeable parcels, the waqfs were also very much entwined in it.

In Haret Hreik, Hezbollah used the cadastral map to reduce the pool of claimants to the sum of property owners who could claim an "asset" in the area based on the property title. The "loss" to be compensated was translated into a square meter construction surface that needed to be rebuilt and restituted in the exact same volume (i.e., reproducing buildings and apartments within them). Excluded from these calculations were however the multiple other forms in which people claim and own places: as visitors, users, tenants, shoppers, workers, and resident property owners whose experience of Haret Hreik consisted of more than the lost financial asset and/or the function of "shelter" that was to be compensated. Responding to a questionnaire in the immediate aftermath of the war, numerous residents described a neighborhood bustling with activities from shopping to working, but they also invoked the neighborhood as a space of remembrance and/or sociability. It was there, in the 1980s, recalled residents, that the southern suburbs of Beirut had seen its first (now closed) cinema and its first office building—but also, its first important religious congregations. These were communal, shared functions that went well beyond the individual definition of lost property and extended to a multiplicity of shared—yet sometimes conflicting—commons. To many users and inhabitants, the "pledge" made by Hezbollah's General Secretary Hassan Nasrallah to rebuild Haret Hreik "even more beautiful than it was" translated into lots of greenery, less traffic congestion, and large sidewalks, where one could push a stroller and walk easily. These communal or shared dimensions were however rapidly dismissed with the redefinition of the "project's tasks" and with them, possible claimants: stakeholders invited to the participatory meetings were property owners and/or long-term residents with valid property claims, and the task of rebuilding the neighborhood "more beautiful than it was" applied only to the individual *homes* that would be equipped with modern sanitation, elevators, and stone cladding.

The pool of claimants defined as the sum of property owners was part and parcel of the making of the planning project, rather than a reality to which the planners adapted. The project designers indeed had to confront the reality of the "messy" forms of owning through which people typically claim space: shares, accommodations, rentals, and numerous other forms of arrangements to secure shelter. These arrangements hardly fit the imagined representation of a clear property title assumed when tagging a parcel. Instead ownership had to be reassessed through negotiation to eventually streamline and redefine it to fit the extant ownership model. In that sense, the project not only reduced neighborhood claimants to property owners, but also created those very property owners.[4]

Along with surveying, scaling, and other forms of interventions, the practice of territorial planning, we propose, should be considered as one of the technologies that continuously reproduce and consolidate the dominant private property regime. This property regime has been in the making since the Ottoman reforms of the nineteenth century, especially with the promulgation of the Land Code of 1858 and the conducting of cadastral surveys that sought to implement new forms of taxation based on fixed rates on property values and revenues.[5] Indeed, the code furthered the idea of land as an individually owned and alienable thing as it required individual registry, did not allow for collective ownership registration, and extended inheritance rights to usufruct right-holders on state lands.[6] The new code transformed the practice of claiming the same parcel for multiple functions, with rights of use, revenues, and alienation (e.g., right to cultivate, right to a portion of the harvest, right to taxes, right to sell) and replaced it with individual and absolute ownership.[7] These reforms radically transformed the relations among people in relation to land introducing a whole new form of seeing and understanding natural landscapes as objects. The "administrative constitution of the individual ownership right . . . entailed the separating of the object of property from its subject or the legal owner from his right of ownership."[8] Such attempts were hotly contested,[9] and it was only with the French colonial powers and the comprehensive cadastral mapping they introduced in Lebanon and Syria in the 1930s that the new property regime became generalized and dominant. In fact, the French director of the cadaster in modern Lebanon and Syria headed to Istanbul to recruit trained soldiers from the dismantled tsarist Russian army in order to implement this "science of mapping,"[10] because in Ottoman mapping efforts the engineers were Europeans, even if local subjects were in charge of the assessment of property values and revenues.[11]

Transforming the landscape into a series of tagged lots constitutes the first yet essential step in the creation of these parcels as fungible, exchangeable, assets, whose ultimate value lies in their dollar value.[12] Thus, the commodification of land, and consequently a private property regime, are often essential for the accumulation of capital.[13] Over the past decades, urban renewal has been a particularly widespread planning approach used to enhance the circulation of capital in the built environment and to maximize profit from its exchange. While the Solidere project may fall more directly under the label of urban renewal, both Solidere and Waad illustrate clearly how the planning intervention eliminated conflicting property claims, considerably facilitating processes of property exchange.[14] This private property regime is similarly critical for the governing of people and their places[15] as private property constitutes a measurable and easily definable unit that makes it possible to deploy tools such as land-use planning[16] and to allocate specific building rights and regulations to organize the built environment.[17]

At least in the Arab Levant, the historical record points to the connection, during the colonial period, between the introduction of cadastral maps and modern urban planning. In Lebanon and Syria, the French colonial archive (particularly the archives of the head of the Land Registry, Camille Duraffourd) indicates that both were seen as interconnected pieces of the modernizing apparatus of the occupiers, essential tools for the organization of land and the government of people. The correspondence between the director of the land registry under mandate authority and the French planners to be commissioned from Paris to develop the first master plans for several cities of the region shows the planners arguing that cadastral plans were necessary because zoning assignments and building factors, essential elements of modern land-use planning, were to be assigned through the tags of individual lots.[18] Therefore, the development of cadastral maps went hand-in-hand with the introduction of modern land-use planning in the region.

While helping produce the world in the ownership model, the cadastral plan's reach is however not total—as it is, in and of itself, based on measurements and particularities (within the limits of the technology, the tools, and human medium) that are then made objective and universal. As evidenced by the persistence of waqfs and by the residents of Haret Hreik, there remains—outside of the cadastral plan—ways of thinking of the environment that are not foreclosed by the plan. However, these ways of conceiving the environment are not simply "outside" the plan; their reproduction actually feeds off the private property regime ushered by the cadastral map. The map and its outside are deeply mutually intertwined.

Notes

1. Saree Makdisi, "Laying Claim to Beirut: Urban Narrative and Spatial Identity in the Age of Solidere," *Critical Inquiry* 23, no. 3 (1997): 661–705, https://doi.org/10.1086/448848.

2. Mona Fawaz, "Hezbollah as Urban Planner? Questions to and from Planning Theory," *Planning Theory* 8, no. 4 (2009): 323–34, https://doi.org/10.1177/1473095209341327.

3. Nada Moumtaz, "Gucci and the Waqf: Inalienability in Beirut's Postwar Reconstruction," *American Anthropologist* (forthcoming).

4. Mona Fawaz, "The Politics of Property in Planning," *IJURR* 38, no. 3 (2014): 922–34.

5. Alp Yücel Kaya, "Les Villes Ottomanes Sous Tension Fiscale: Les Enjeux de l'Évaluation Cadastrale Au XIXe Siècle," in *La Mesure Cadastrale: Estimer La Valeur Du Foncier En Europe Aux XIXe et XXe Siècles*, eds. Florence Bourillon and Nadine Vivier (Rennes: Presses universitaires de Rennes, 2012).

6. Huri İslamoğlu, "Property as a Contested Domain: A Reevaluation of the Ottoman Land Code of 1858," in *New Perspectives on Property and Land in the Middle East*, ed. Roger Owen (Cambridge, MA: Harvard Center for Middle Eastern Studies, 2000), 3–62; Attila E. Aytekin, "Agrarian Relations, Property and Law: An Analysis of the Land Code of 1858 in the Ottoman Empire," *Middle Eastern Studies* 45, no.6 (2009): 935–51, https://doi.org/10.1080/00263200903268694.

7. İslamoğlu, "Property as a Contested Domain."

8. İslamoğlu, "Property as a Contested Domain," 25.

9. As detailed in İslamoğlu, "Property as a Contested Domain," 35–39.

10. Ministry of Foreign Affairs (MAE), Fonds Camille Duraffourd, AE118, cartons 21–22, Nantes, France.

11. Kaya, "Les Villes Ottomanes."

12. This responds to the argument advanced by the Peruvian economist Hernando de Soto for the regularization of informal property rights as a strategy to build wealth. Hernando de Soto, *The Mystery of Capital: Why Capitalism Triumphs in the West and Fails Everywhere Else* (New York: Basic Books, 2000).

13. Karl Polanyi, *The Great Transformation : The Political and Economic Origins of Our Time* (Boston, MA: Beacon Press, 2001).

14. Douglass C. North, *Institutions, Institutional Change, and Economic Performance* (Cambridge: Cambridge University Press, 1990).

15. Michel Foucault, "Governmentality," in *The Foucault Effect: Studies in Governmentality: With Two Lectures by and an Interview with Michel Foucault*, eds. Michel Foucault, Graham Burchell, Colin Gordon, and Peter Miller (Chicago: University of Chicago Press, 1991), 67–104.

16. Nicholas Blomley, "Land Use, Planning, and the 'Difficult Character of Property,'" *Planning Theory & Practice* 18, no. 3 (2017): 351–64, https://doi.org/10.1080/14649357.2016.1179336.

17. James C. Scott, *Seeing Like a State : How Certain Schemes to Improve the Human Condition Have Failed* (New Haven, Conn.: Yale University Press, 1998).

18. MAE, Fonds Camille Duraffourd, AE118, cartons 21–22, Nantes, France.

National Budget: Sudan, 1946
Alden Young

Google claims that "If you live next to the cemetery, you cannot cry for everyone" is an old African proverb, though my father often retold it with a Russian lineage. Living next to the cemetery is an apt metaphor for the experience of being a planner in Sudan. Decolonization, like any transition, meant that something would have to die in order to for something else to live. The generation that came to power with independence celebrated the dying of colonialism, but the uncertainties that accompanied the passing of the colonial era also generated emotions of fear, anxiety, and hope.[1] One of the ways in which the unknown was tackled by elites across the postcolonial world was to imagine that the formulas of political economy contained answers about the future. Consequentially, in Sudan as in states across the world, planning was embraced as a new means of taming uncertainty.[2] Colonial and postcolonial officials hoped that planning could give order to the process of decolonization, which at least in the administrative sphere you can hear them defining silently as: *killing the old and embracing the new*.

But what to let wither and what to cultivate? For the first half of the twentieth century, Sudan referred to a vaguely defined political and economic region more than a clearly defined territorial entity. The political history of the territory known as Sudan after the 1899 creation of the Anglo-Egyptian Condominium rendered the question of what constituted the Sudanese economy particularly acute, but also open-ended. After all, before the outbreak of the Second World War, colonial officials stationed in Khartoum had considered the territory alternatively a province of Egypt or a part of the imperial trading and war-fighting machine, but rarely as an independent economic entity.[3]

After independence, episodically in Sudan as in many postcolonial states, the object of planning appeared self-evidently to be the *national economy*. Yet, recent research has pointed out how fleeting the national moment often proved. Equatorial Guinea's allegiance to the national form manifested itself only for a decade here or there, simply to vanish again; while Southern Rhodesia/Zimbabwe constructed a very explicitly national economy during the 1960s, only to have the boundaries of the economy increasingly obscured as efforts to define the nation have become more contentious.[4] Almost tautologically, postcolonial officials asserted that strengthening the nation, their patriotic duty, meant strengthening the national economy. But what was the national economy and where did it come from in a new state like Sudan, where the very idea of the nation itself was a work in progress? The political theorist Timothy Mitchell offered a potential answer when he posited that the idea of "the" economy was only possible after the crumbling of the European empires. Therefore, for Timothy Mitchell, decolonization was a necessary precondition for the spread of the economy as a ubiquitous technopolitical entity.[5]

Undoubtedly, there is much truth to Mitchell's insight, but it needs to be prefaced by two caveats, both of which the Sudanese example exemplifies. The first is to remember that thinking about an/the economy did not begin in the era of decolonization; rather the ancient Greeks could speak of the economy of the city, the British in their imperial heyday of the Empire, and the Germans during the late nineteenth century.[6] What was new was the notion that in order to be a good financial steward "the forms of trade, investment, currency, power, and knowledge that might be constituted as an economy" should privilege a particular and clearly defined territory.[7] The view that the world should be divided into units that are structurally alike, states and economies, dates to the era of decolonization and was an innovative feature of the intellectual life of that moment.[8] Yet, a large body of work now calls into question an earlier assumption that "nationalists leaders" were the principal proponents for the nation-states' global spread.[9] Instead, we are left with a much messier picture, one that prominently features development. While "development" refers to a concept that almost no one agrees upon, it became a challenge signaling the end of the European empires.

When budding development planners in Khartoum reflected on the task ahead of them, they could be forgiven for being frightened by the ill-defined magnitude of the challenges they confronted. Finance officials tried to tame enormous and open-ended questions about the nature of the state by reformulating these questions as problems of political arithmetic. Three images of the structure of the Sudanese economy were competing with one another. One was the economy of livestock, which could be marshalled from the western and southern regions of the country and would tie Sudan firmly into a regional network of trade. This economy would be grounded in private ownership. The second economy was an economy of foodstuffs, one which would tie Sudan ever more into union with Egypt as grains traveled north along the Nile to feed the teaming cities of Cairo and the Delta. The third option, the one the first generation of Sudanese politicians and finance officials eventually choose was to reinforce the cotton complex, explicitly orienting Sudan towards the international market. While Sudanese planners understood the dangers that the vagaries of the market posed, the market was also seen as the guarantor of Sudan's independence. Yet, the cotton export strategy demanded huge sums of resources that crowded out the possibility of alternatives.

The international market played a contradictory role in Sudan's state formation process. On the one hand, it allowed Sudan to imagine economic and political independence. In 1954, just as after Sudan achieved self-government, the senior British civil servant John Carmichael reflected on why Sudan was choosing to bet its future on the cotton complex, when he said:

> It is hardly true to say that the Sudan is dependent on cotton. A more correct statement would be that, without cotton, the Sudan could be more or less self-sufficient and self-supporting: with cotton, the Sudan can continue to improve its general standards of living and to lay aside funds for future development.[10]

Yet on the other hand, the international market restricted planners' ability to direct the development of the state and the economy on which it depended. Sudanese policymakers were well aware of the vast inequalities that plagued their country; yet their choice to invest in the cotton complex helped them to justify ignoring those inequalities in favor of the goal of catching up economically with the states that they aspired to have as peers. The central triangle, from Omdurman to Kosti and Sennar, not only dominated the nation's politics, but it was also significantly richer than the rest of the country, with a per capita income more than three times that of parts of southern and western Sudan.[11]

The central region of the country received most of the development funding allocated by the Sudanese government in Khartoum. This inevitably led to unrest: in 1955 at Torit in Equatoria province of southern Sudan, where a rebellion broke out over the failure of the central government to include southerners in the civil service list, and in the early 1960s as rebellions broke out throughout the southern provinces protesting neglect and exclusion from development decision making and a general unwillingness on the part of the central government to interact with southerners outside of militarized counter-insurgency.¹²

The assumption that development could be centrally planned had come under unrelenting attack by the mid-1960s in Sudan. The criticism of Aggrey Jaden, the president of the Sudan African National Union during the 1965 Khartoum Conference on the Southern Sudan, attacked the Achilles' heel of the development project in Sudan, essentially questioning whether there was a Sudan to be developed at all. Aggrey argued that:

> The Sudan falls sharply into two distinct areas, both in geographical areas, ethnic groups, and cultural systems. The Northern Sudan is occupied by a hybrid Arab race who are united by their common language, common culture, and common religion; and they look to the Arab world for their cultural and political inspiration. The people of the Southern Sudan, on the other hand, belong to the African ethnic group of East Africa . . . There is nothing in common between the various sections of the community; no body of shared beliefs, no identity of interests, no local signs of unity and above all, the Sudan has failed to compose a single community.¹³

When Aggrey made this statement in the mid 1960s, it was clear to many northern and southern Sudanese people that the attempt to suppress issues of regional and local identity in the name of national economic development had hopelessly failed. In fact, national planning could not be justified if the Sudanese did not believe in the nation. This dynamic eventually culminated in the partition of the Sudanese state in 2011 into Sudan and South Sudan.

I have just outlined the macro-story. Yet, I want to argue for a minute about how we reached a point where the cotton complex could stand in for the entirety of the Sudanese economy measured according to the per capita growth rate of the gross domestic product. The story begins in 1946 with a rather modest object, a thirteen-page aggregate of Sudan's development priorities, which was written by the finance secretary, Sir Eddington Miller; his deputy, Arthur L. Chick; and the deputy assistant finance secretary, John Carmichael.¹⁴ This document, entitled *The Five Year Plan for Postwar Development in Sudan: 1946–1951* was a hodgepodge of projects, reflecting a multitude of different desires and priorities.¹⁵ The Development Priorities Committee, which was composed of the financial secretary, the civil secretary, the comptroller-general of war supply, and the general manager of the Sudan Railways, was charged with prioritizing what would be included in the plan. The Five Year Plan allocated LE 11,480,470 to a wide variety of schemes suggested by the governors of various provinces and the heads of departments.¹⁶ The first page of the plan explicitly stated that the goal was to create "a store of development plans out of which projects could be selected from time to time," provided that the government possessed the resources and inclination necessary to invest in these capital improvements.[17]

Planning involved a process of interdepartmental meetings, budgeting, the oversight of specific project plans, project modifications, and re-budgeting.[18] Budgeting and re-budgeting are central to this story, because it was in the process of allocating and managing funds that Sudan was eventually defined as the primary economic unit. Because the plan gathered up projects promoted by imperial and Sudanese bureaucrats, it attempted to blend local development projects, aimed at regional self-sufficiency and increasing the governability of local populations, with territory-wide programs, and supra-territorial programs, notably the further expansion of major irrigation schemes, which were tied to the management of the entire Nile Basin. Nonetheless, the first five-year plan did not primarily pursue development goals that were focused on the territorial unit of the Anglo-Egyptian Sudan, for the planners did not regard Sudan as the primary unit of economic development. Instead, the plan developed economic units that were both much larger and smaller than the administrative unit of the Anglo-Egyptian Sudan.

The planning process was centralized through the creation of the General Development Account. Those managing this account found themselves in a position to coordinate the funding of new capital investments. In the budgets of the Finance Department, capital expenditures were kept distinct from recurring expenses covered by allocations from the ordinary budget. Deputy Financial Secretary A. L. Chick and his assistant John Carmichael were constantly being asked to supervise and comment on the priority of particular schemes and projects and whether they were worthy of being funded. Altering the purpose of particular projects was not within the original mandate of the Department of Finance. However, finance officials soon found themselves involved in the details of designing projects and setting actual development policy. Once Chick and Carmichael became involved in development policy, because of their advocacy of certain positions in the midst of bureaucratic competition, they were inexorably led to promoting a national territorial perspective. As the officials in the Finance Department reviewed development projects and plans, they saw their influence grow, and they increasingly based their funding choices on whether or not individual projects contributed to the territory's overall development.[19] By the 1950s, the "national" territorial perspective had won out over the concepts of imperial and local economic management that had had the upper hand in the 1940s.[20]

Getting to the point where a colonial or postcolonial official in Sudan could ask what was good for the Sudanese economy was a long process. The blueprint for the Sudanese national economy was created in the Department of Finance during routine and repeated discussions about prioritizing and implementing various projects included in the centralized financial decision-making system created as a result of the planning process. The need to centralize the funding for individual projects and to develop a means of prioritizing them followed directly from the logic of the national planning process itself. Planning, as practiced in Khartoum from 1946 until 1951, demanded a centralized system of budgeting, even as it encouraged various parts of the government to design development projects in a dispersed and decentralized manner. Reconciling these two processes prompted finance officials to create an evaluative framework. As a framework, finance officials developed a territorial perspective in which those projects which benefited the Sudanese economy as a whole—in theory those projects which made the largest contribution to the central budget—were prioritized over projects that focused on local development or imperial strategic interests. Thus, as Sudanese nationalists gradually assumed power between 1945 and 1966, governing Sudan was transformed from the management of a collection of distinct populations, each with its own attributes, to the management of a national economy made up of equal individuals, whose preferences policymakers assumed could be aggregated and even maximized.[21]

Notes

1. Alden Young, *Transforming Sudan: Decolonization, Economic Development, and State Formation* (Cambridge: Cambridge University Press, 2018).

2. Mahmood Mamdani, "Beyond Settler and Native as Political Identities: Overcoming the Political Legacy of Colonialism," *Comparative Studies in Society and History* 43, no. 4 (2001): 651–64.

3. M. W. Daly, "The Development of the Governor-Generalship of the Sudan, 1899–1934," *The Journal of African History* 24, no. 1 (1983): 77–96.

4. Hannah Appel, "Toward an Ethnography of the National Economy," *Cultural Anthropology* 32, no. 2 (2017): 294–322 and Tinashe Nyamunda, "Money, Banking and Rhodesia's Unilateral Declaration of Independence," in *The Journal of Commonwealth and Imperial History* 45, no. 5 (2017): 746–76.

5. Timothy Mitchell, *Rule of Experts: Egypt, Techno-Politics, Modernity* (Berkeley, CA: University of California Press, 2001), 6.

6. Quinn Slobodian, "How to See the World Economy: Statistics, Maps, and Schumpeter's Camera in the First Age of Globalization," *Journal of Global History* 10, no. 2 (2015): 307–32.

7. Mitchell, *Rule of Experts*, 6.

8. For an encapsulation of the American idea of international relations, see the works of Kenneth Waltz, *Man, the State, and War: A Theoretical Analysis* (New York, NY: Columbia University Press, 1959); and *Theory of International Politics* (Reading, MA: Addison-Wesley, 1979).

9. Adom Getachew, *Worldmaking After Empire: The Rise and Fall of Self-Determination* (Princeton, NJ: Princeton University Press, 2019); Frederick Cooper, *Citizenship between Empire and Nation: Remaking France and French Africa, 1945–1960* (Princeton, NJ: Princeton University Press, 2014).

10. John Carmichael, "Notes for the Minister of Finance and Economics for General Consideration in the Financial and Economic Field," November 24, 1958, SAD.993/1/132–150, The Sudan Archive at Durham University

11. M. W. Daly, *Darfur's Sorrow: A History of Destruction and Genocide* (Cambridge: Cambridge University Press, 2007), 185–89.

12. Oystein Rolandsen and Cherry Leonardi, "Discourses of Violence in the Transition from Colonialism to Independence in Southern Sudan, 1955–1960," *Journal of Eastern African Studies* 8, no. 4 (2014): 609–25. For a discussion of the extent to which the Government of Sudan in Khartoum was part of a wider mid-century tradition of combining counter-insurgency and the discourse of development, which continues until today, see Mortiz Feichtinger and Stephan Malinowski, "Transformative Invasions: Western Post-9/11 Counterinsurgency and the Lessons of Colonialism," *Humanity* 3, no. 1 (2014): 35–63.

13. Khartoum Conference on the South, March 1965 documents; speech by Aggrey Jaden (mimeo.), 4, quoted in George W. Shepherd Jr., "National Integration and the Southern Sudan," *The Journal of Modern African Studies* 4, no. 2 (1966): 193–212, on 195.

14. J. W. E. Miller, "1945/46 Development Budget," letter to Secretary General's Council, SAD.636/1/1–42, The Sudan Archive at Durham University. Miller had arrived in Sudan in 1920; Chick and Carmichael came in the 1930s. The Sudan Political Service, created in 1901, operated independently of the other British civil services, such as the Colonial Service, the Home Service, or the Indian Civil Service. The vast majority of its members were recruited from a rather narrow demographic base of the lower gentry, trained at public schools and had their undergraduate degrees from Oxford and Cambridge. The narrowness of the social base from which these officials were recruited, combined with the Sudan Political Service's autonomy from other administrative orders, created a buffer between administrative practices undertaken in Sudan and in the rest of the empire. See Robert O. Collins, "The Sudan Political Service: A Portrait of the 'Imperialists,'" *African Affairs* 71, no. 284 (1972): 293–303. See also M. W. Daly, *Imperial Sudan: The Anglo-Egyptian Condominium 1934–1956* (Cambridge: Cambridge University Press, 1991), 26. Daly suggests that even during the 1930s Sudanese officials were divorced from larger debates about colonial governance, despite struggling with similar challenges.

15. Sudan Government, *Five Year Plan for Post-War Development* (Khartoum: Department of Finance, 1946).

16. LE stands for Egyptian Pound. Until 1957 the Egyptian pound was the legal tender in Sudan. From World War I until 1962 the Egyptian pound was pegged to the British Pound at almost 1=1 parity or 0.975 Egyptian pounds to 1 British pound. The exchange rate with the US dollar was 0.25 Egyptian pounds to 1 US dollar. In today's US dollars the funds allocated by the Government of Sudan for development LE 11,480,470 equals $526,257,670 (2010).

17. Sudan Government, *Five Year Plan*, 1.

18. Mary S. Morgan, "'On a Mission' with Mutable Mobiles," (working paper, in *The Nature of Evidence: How Well do 'Facts' Travel?*, no. 34/08, August, 200): 6–7, http://dx.doi.org/10.2139/ssrn.1497107.

19. Helen Tilley, *African as a Living Laboratory: Empire, Development and the Problem of Scientific Knowledge, 1870–1950* (Chicago: University of Chicago Press, 2011), 6. See also the frequent use of the terms territorial and territory in D. A. Low and J. M. Lonsdale, "Introduction: Towards the New Order 1945–1963," in *History of East Africa*, eds. D. A. Low and Alison Smith (Oxford: Clarendon Press, 1976), 1–64.

20. The idea that different groups of officials and departments within the same government can simultaneously be working towards divergent conceptions of the state and even mutually competing policies within the same bureaucracy has been explored by Boaventura de Sousa Santos, "The Heterogeneous State and Legal Pluralism in Mozambique," *Law and Society Review* 40, no. 1 (2006):42–44; Boaventura de Sousa Santos, *Toward a New Common Sense: Law, Science and Politics in the Paradigmatic Transition* (New York: Routledge, 1995).

21. Alden Young, "African Bureaucrats and the Exhaustion of the Developmental State: Lessons from the Pages of the Sudan Economist," *Humanity* 8, no. 1 (2017): 49–75.

Orangutans: Borneo, 1962
Juno Salazar Parreñas

In 1962, colonial administrative officers traveled throughout the British Crown Colony of Sarawak, located in the northwestern part of the island of Borneo, to survey fellow Sarawakians about their opinions concerning decolonization.[1] At the same time, another uncertain investigation into the question of independence after the end of empire was also occurring within the boundaries of Sarawak's newly founded Bako National Park. For three years, 1962 to 1965, Bako's dipterocarpus forest, encased by mangrove shorelines and accessible to people only by motorboat, hosted an experiment. The question guiding the experiment was, could humans rehabilitate orangutans and make them wild, despite transformative contact with people and the impacts of modernity?

Modernity for orangutans meant deforestation and urbanization. Newly built logging roads and loggers' motorized chainsaws shrank their habitats and made them vulnerable to the lucrative and illegal pet trade. On behalf of the colonial state, the Sarawak Museum of Natural History and Ethnology cared for displaced infants that came into the colonial state's custody. Modernity for infant and juvenile orangutans who were captured by humans and then taken into the custody of the state meant learning how to climb trees in the household garden of Barbara and Tom Harrisson's home in the city of Kuching. Orangutans, as observed in the wild by later primatologists, spend an extraordinary amount of time with their mothers learning skills, often up to seven years.[2]

The tentative plan for orangutan rehabilitation that was provisionally tested by Barbara Harrisson of the Sarawak Museum inside the home she shared with her husband the curator, a Malay housekeeper, and a young Selako man suggests that the global political project of decolonization was not just intersubjectively generated and planned between people in particular settings. Rather, the efforts taking place in a home and a newly formed national park in Sarawak suggest that their attempt at decolonization, like others' attempts elsewhere, was an experiment: Whose vision of independence and freedom would guide the experiment? What alternatives to colonial governance could be implemented? What future outcomes might unfold? These questions by no means were unique to Sarawak, but rather confronted and continue to confront all colonies and former colonies of the world. The answers to these questions were uncertain and tentative. Orangutan rehabilitation's nascent history shows how political questions of decolonization transform the lives and deaths of orangutans.

This essay suggests that decolonization was an unplanned plan, one that tentatively attempted new possibilities by trial and error. In the contemporary moment, decolonization has renewed salience both politically and intellectually in an ongoing colonial present that is framed by frustration concerning the unfulfilled potentials of mid-twentieth-century decolonization. Yet it is important to not romanticize decolonization as the remedy to colonial violence, madness, and cruelty.[3] Idealization only results in disappointment. Instead, the experimentation of decolonization works as a reminder that decolonization can never be about following an established plan, but about exploring unplanned and uncontrolled possibilities.

Experiments of decolonization differed from colonial experimentation. Colonial experiments were (or are, in the case of ongoing colonialism) intended to engineer imperial control over people and land, even as contingencies and uncontrollable agents evade attempts at being harnessed;[4] whereas experiments of decolonization sought liberation from colonial order. Likewise, the experimentation of decolonization was markedly different from imperial scientific exploration. From the late fifteenth century throughout the nineteenth, projects of exploration and the collection of curiosities were bankrolled by imperial wealth and meant to potentially harness biological diversity into familiar taxonomies and potential commodities for the benefit of imperial economies, as Tahani Nadim in this volume describes in the case of sisal ("Seeds: German East Africa, 1892"). The differences between colonial experiment and experimental decolonization are not because of historical eras and periodization. Rather, the differences lie in the experimental senses of uncertainty when unyoked from colonial domination. The experiments that interest me here took place without control samples, without the expectation of duplicability, without demanding an audience of modest witnesses.[5] These actors experimented at the peripheries of science, even as they attempted to usurp scientific authority.

Experiments involving the rearing of endangered Bornean orangutans help convey the idea of experimental decolonization. We get there by comparing mid-twentieth-century orangutan rehabilitation efforts to colonial experimentation in rearing an orangutan. As a site of these different engagements, the island of Borneo is not merely backdrop or object of their knowledge making, but a place that structures their efforts.

This entry begins with Alfred Russel Wallace's colonial-era exploration of Sarawak from 1854 to 1856 at the behest of the Rajah of Sarawak, which included Wallace's experiment in raising an infant orangutan. It then walks readers through Barbara Harrisson's tentative planning, or better yet non-planning, of instilling independence on orangutans in the context of Sarawak's own uncertain independence from 1956 to 1965. The essay then follows Barbara Harrisson in the months after Sarawak's and Sabah's official decolonization in 1963 as they became states in the federal nation-state of Malaysia. This was when Harrisson helped found a rehabilitation center in Sabah under the leadership of Gananath Stanley de Silva involving the transfer of the orangutans Arthur and Cynthia in 1965. Both Sarawak's and Sabah's mid-twentieth-century efforts at experimentation with orangutan independence offers a window into understanding the plurality of tentative and conflicting visions of decolonization.

The natural historian Alfred Russel Wallace was a younger contemporary of Charles Darwin. The second White Rajah, Charles Brooke, invited Wallace to explore newly expanded and pacified territories of the Raj of Sarawak. Brooke inherited Sarawak from his uncle, who had usurped a claim on Sarawak by squelching an uprising at the mouth of the Sarawak River on behalf of the Sultan of Brunei. The Brookes were British citizens and subjects but gave themselves the title of Rajah once James Brooke, with a gunboat, forced the Sultan to concede his claim of Sarawak. Brookes' territorialization differed from previous forms of rule in which sovereignties overlapped: places typically paid tribute to one or more sovereigns. Unlike coastal sultanates, the White Rajahs made territorial claims to the interior. Hence, the Rajah was eager to host Wallace and other explorers/hunters, like the Italian Oduardo Beccari, and the American William Hornaday. However, Wallace was exceptional among the others. He not only "discovered" new species, naming the Rajah Brooke butterfly after his host and came up with the Sarawak Law (a theory of evolution contemporaneous to Darwin's theory of natural selection), but he also succeeded in convincing the Rajah to found the Sarawak Museum of Natural History and Ethnology—the first museum of its kind in the tropics.[6] Additionally, unlike other explorers, his orangutan hunt lent itself to the possibility of rearing an infant orangutan.[7]

Wallace's orangutan hunting was spent in the company of Dayaks, a term that references a person indigenous to Borneo who is not Muslim, which conveyed nineteenth-century European concerns about colonial subjects and competing empires (of Christendom against otherwise). The region of Sarawak in which Wallace shot orangutans is considered Iban country.[8] Among Ibans in an interior district decades later, Charles Hose found that they revered different kinds of animals as *nyarong* or spirits of their ancestors for whom it was forbidden to hunt, lest misfortune fall upon them.[9] Among the Ibans with whom I lived in the 2000s, the common feeling about seeing an orangutan is the word *geli* or creepiness: one is seeing somebody they should not be seeing and touching somebody they should not be touching.[10] Perhaps the Dayaks in Wallace's presence similarly felt transgressive discomfort.

Wallace had shot and killed an adult female orangutan in the month of May, a time when the abundant wet season peters out and the dry season portends hunger for Bornean orangutans. Nearby, they found an infant orangutan face down in the mud. Wallace could not find milk among the Dayaks with whom he lived, explaining that they do not consume the substance. He seemed to not explicitly ask for human breast milk or perhaps he hesitated to write about their refusal. Wallace resorted to feeding the infant a diet of sugared rice-water and occasional coconut milk. His description of the orangutan's behavior suggests starvation: "when I put my finger in its mouth it sucked with great vigour, drawing in its cheeks with all its might in the vain effort to extract some milk . . ."[11] It also suggests depravation, with its "desperate" and "constant" clinging of its opposite shoulder.[12] The intention guiding Wallace's experimentation, which he declares after the infant dies within three months, is colonial in its extraction: "I much regretted the loss of my little pet, which I had at one time looked forward to bringing up to years of maturity, and taking home to England."[13] Wallace uses the language of pet ownership with its implicit sense of domination.[14] He explicitly wishes to extract the orangutan from Borneo. His plans for transplantation ended with the orangutan's unexpected death.

Wallace lived with Dayaks during his field research. What unspoken, mortal dangers might this contact have entailed for those in his surroundings? What might this mean for someone's *nyarong*, their ancestors, or their loved ones? The experimentation is not about manly rigor that we know from laboratory scientists who experimented on themselves.[15] Rather, this experimentation falls into the familiar traps of colonial science, in which ignorance of what something is ultimately results in its killing, collection, and eventual display.

Almost a century later, Wallace's natural history haunts Barbara Harrisson's own efforts as a fieldworker observing orangutans in the very place where

Wallace had hunted and killed them. While Wallace had stayed on the ground, she climbs up and cinches herself to a tree in order to facilitate an orangutan's perspective of their forest habitat.

Barbara Harrisson carried out the world's first orangutan rehabilitation site a year following the Asian-African Conference in neighboring Indonesia, which articulated the Third World stance against "colonialism in all of its manifestations."[16] This was fifteen years after the last White Rajah promised Sarawak's eventual self-governance in 1941, ten years after he reneged on that promise and transferred Sarawak to Britain in 1946, and seven years after an anti-colonial assassination of the first British appointed governor of Sarawak in 1949. As a form of punishment, the colonial state forbade the participation of Malays in political parties.[17] The problem of how to instill independence was as much a political dilemma for Sarawak's colonial society as it was a programmatic one in the Harrisson household. Her husband Tom was the curator of the Sarawak Museum since the end of World War II and her capacity as a German-born university-educated volunteer and then wife of the curator made her an ambiguous representative of British colonial governance.

How was Barbara Harrisson in the position of testing freedom for semi-wild orangutans in an era of decolonization? Her being German and taking up decolonizing efforts has some valence in other moments of decolonization elsewhere. Hannah Reitsch, an aviatrix and National Socialist Iron Cross recipient who was an associate of Adolf Hitler, became close to Third World leader Kwame Nkrumah. Indeed, she helped establish Ghana's air force following decolonization.[18] However, unlike Reitsch's motivations, confident futurism of a new nation-state did not direct Harrisson's efforts. Rather, Barbara Harrisson's efforts were guided by experimentation, especially in respect to trial and error. Her qualifications were on par with her husband's. In the moment of transition between colonial and postcolonial governance, many of the same figureheads assumed their old jobs. Her husband Mr. Harrisson, however, was not one, but she was able to continue finishing projects at the Sarawak Museum after his departure, including the project of caring for displaced young orangutans brought into the custody of the state and sent to the museum.

Barbara Harrisson felt that the orangutans could not be sent back to the hilly forests of their ancestors and earlier infancy unable to fend for themselves. Neither could they be kept as pets, fated to acquire fatal anthroponotic illness and a diet that almost guaranteed malnutrition. Sending them to the concrete confines of London Zoo was something Barbara Harrisson was loath to initiate. She instead considered another plan. It would be, in her words, an "experiment."[19] The driving question of this experiment was, how might humans teach orangutans to live independently after contact with modernity?

By 1962, the Harrissons scrambled to solidify material orangutan protections before Sarawak would cease its status as a British Crown Colony the next year, and while they still had the ear of the Chief Secretary and Conservator of Forests. Tom Harrisson's letter to him explains that the Harrisson's "experiments of letting young orphaned Maias grow up half-wild has now proved that it is possible to educate them back to wild living."[20] He granted the museum permission to continue the experiment at Bako National Park. The first, only, and last orangutans at Bako were Arthur and Cynthia.

Arthur and Cynthia had been transferred earlier to the custody of the Sarawak Museum, which meant in practice Barbara and Tom Harrisson's home. Their Conservator wanted them to undergo the experiment of rehabilitation instead of sending them to an overseas zoo. Arthur's and Cynthia's stories reveal how decolonization entails uncertainty and experimentation.

Barbara Harrisson personally returned the two orangutans to Sabah for what she planned to be a period of four to six weeks, where the experiment of orangutan rehabilitation was to continue at Sepilok Orangutan Rehabilitation Center after Bako National Park ultimately proved to be too small and too busy with beach-going visitors.[21] A plaque at Sepilok Orang Utan publicly thanks Barbara Harrisson for her instrumental service in its creation. The game warden in charge of Sepilok, Stanley de Silva, described the site and their efforts as an "experiment" that started in 1964 in the immediate aftermath of Sabah's decolonization as a member of the Malaysia Agreement, which was signed on September 16, 1963.[22] While Barbara Harrisson's efforts in Sarawak show a colonial perspective of decolonization, efforts at Sepilok show otherwise.

In a mere matter of days, Barbara Harrisson was called away one afternoon. During her brief absence, one of Sepilok's two workers shot and killed Arthur. Both workers reported to De Silva and said that Arthur "ran amok." Having pushed down a water tank, he cornered the workers. They felt their lives were endangered, so they killed him. De Silva did not believe them because the ground was dry. Yet he felt that he could not punish them: he needed them to keep working at a moment when they could have easily gotten jobs as loggers. Barbara Harrisson, when speaking about it with me decades later, was convinced that: "De Silva could not punish the civil servant. He was Ceylonese, just like de Silva. De Silva could not punish his own countryman on the command of a British colonial." De Silva figured that it was his Malay worker, not the Ceylonese. Years later, De Silva was forced to retire and a Malay officer took his place. To what extent might postcolonial racial politics inform De Silva's memory in the way they did Barbara Harrisson's? We cannot know.

Cynthia died in a different set of circumstances. Her death is illustrative of the open possibilities offered through liberation following decolonization. She was seen examining a clay mound with a hole in it. Her hand became swollen and blue, then her entire arm. She soon died, most likely because a cobra inhabited the hole in the clay mound. If we think the cause of her death was a lack of planning, of having too much freedom when she was insufficiently trained for the responsibility such freedom would entail, then Cynthia was a victim of neglect. But if we recognize Cynthia's pursuit of curiosity as her own experimentation, we see her as a subject experiencing liberation and freedom of movement. I am inclined to see her actions as experimentation. At the very least, she pursued her curiosity. Experimentation, whether with orangutans or with decolonizing, entails trying things out and building ad hoc plans. It is not about resolutions. Nor can it end future inquiries into what other plans might arise. Indeed, experiments open other possibilities: some hopeful, some fatal, and oftentimes combinations of both.

Notes

1. James Chin and Jayl Langub, "The Sarawak Administrative Service: Its Origins," in *Reminiscences: Recollections of Sarawak Administrative Service Officers*, ed. James U. H. Chin and Jayl Langub (Subung Jaya: Pelanduk, 2007), 1–20, on 15.

2. Birute Galdikas, "Orangutan Reproduction in the Wild," in *Reproductive Biology of the Great Apes: Comparative and Biomedical Perspectives*, ed. C.E. Graham (New York: Academic Press, 1981), 281–300.

3. Frantz Fanon, *Black Skin, White Masks* (1952; repr., New York: Grove Press, 2008); Frantz Fanon, *A Dying Colonialism* (New York: Grove Press, 1967); Juno Salazar Parreñas, *Decolonializing Extinction: The Work of Care in Orangutan Rehabilitation* (Durham: Duke University Press, 2018).

4. Timothy Mitchell, *Rule of Experts: Egypt, Techno-Politics, Modernity* (Berkeley: University of California Press, 2002); Helen Tilley, *Africa as A Living Laboratory: Empire, Development, and the Problem of Scientific Knowledge, 1870–1950* (Chicago: University of Chicago Press, 2011).

5. Steven Shapin, Simon Schaffer, and Thomas Hobbes, *Leviathan and the Air-Pump: Hobbes, Boyle, and the Experimental life: Including a Translation of Thomas Hobbes, Dialogus physicus de natura aeris by Simon Schaffer* (Princeton, NJ: Princeton University Press, 1985); Donna Jeanne Haraway, *Modest-Witness@Second-Millennium.FemaleMan-Meets-OncoMouse: Feminism and Technoscience* (New York: Routledge, 1997); Christine von Oertzen, Maria Rentetzi, and Elizabeth S. Watkins, "Finding Science in Surprising Places: Gender and the Geography of Scientific Knowledge. Introduction to 'Beyond the Academy: Histories of Gender and Knowledge,'" *Centaurus* 55, no.2 (2013): 73–80, https://doi-org/10.1111/1600-0498.12018.

6. I am inferring this based on the founding date of other museums housed in older structures but refurbished as museums in the twentieth century. See Franciza Toledo and Clifford Price, "A Note on Tropical, Hot, and Humid Museums," *JCMS Journal of Conservation and Museum Studies* 4 (1998): 11. Sarawak Museum's old building has served as a museum throughout the history of the structure.

7. Wallace described his collecting as "hunting." Alfred Russel Wallace, *The Malay Archipelago: The Land of the Orang-utan and the Bird of Paradise* (London: Macmillan, 1890), 31.

8. Simanjon in *Malay Archipelago* and Simanjun in today's parlance.

9. Charles Hose and W. McDougall, *The Relations Between Men and Animals in Sarawak* (London: Anthropological Institute of Great Britain and Ireland, 1901).

10. Juno S. Parreñas, *Decolonizing Extinction: The Work of Care in Orangutan Rehabilitation* (Durham: Duke University Press, 2018).

11. Wallace, *The Malay Archipelago*, 33.

12. Wallace, *The Malay Archipelago*, 33.

13. Wallace, *The Malay Archipelago*, 35.

14. Yi-fu Tuan, *Dominance and Affection: The Making of Pets* (New Haven: Yale University Press, 1984).

15. Rebecca M. Herzig, *Suffering for Science: Reason and Sacrifice in Modern America* (Piscataway: Rutgers University Press, 2005).

16. Asian-African Conference, *Asia-Africa Speaks from Bandung* (Djakarta: Ministry of Foreign Affairs, Republic of Indonesia, 1955).

17. Michael B. Leigh, *The Rising Moon: Political Change in Sarawak* (Sydney: Sydney University Press, 1974).

18. Jean Allman, "Phantoms of the Archive: Kwame Nkrumah, a Nazi Pilot Named Hanna, and the Contingencies of Postcolonial History-Writing," *The American Historical Review* 118, no. 1 (2013): 104–29.

19. Barbara Harrisson, *Orang-utan* (London: Collins, 1987).

20. Tom Harrisson, Letter to Chief Secretary and Conservator of the Forest, 1962, MU/444/9-12, 27-2-62, Sarawak Museum, Kuching Sarawak.

21. Years later, primatologists have come to agree that orangutans are the least social of all hominids and a typical habitat for a single female orangutan is ideally about 7 km^2, see Serge A. Wich, S. Suci Utami Atmoko, Tatang Mitra Setia, and Carel P. van Schaik, eds., *Orangutans: Geographic Variation in Behavioral Ecology and Conservation* (Oxford: Oxford University Press, 2009).

22. G. S. de Silva, "The East Coast Experiment," Conference on Conservation of Nature and Natural Resources in Tropical Southeast Asia, Bangkok, Thailand, 1965.

Parasite: Liberia, 1926
Gregg Mitman

In the West African republic of Liberia, parasites have mobilized immense amounts of capital, people, and expertise. Research invested by Firestone Tire & Rubber Company on tropical diseases aimed to ensure that its plan to build what would become the world's largest contiguous rubber plantation met with success. Firestone regarded parasites, in particular, as a drain on the productive efficiency of laborers, capable of eating up the company's valuable investment capital.

The term parasite comes from the ancient Greek, παράσιτος a person who eats at the table of another. In Liberian English usage "eating" is characterized as "using up," the act of making disappear what is there to be shared. Through its establishment of a plantation economy, dependent upon large-scale land dispossession and the appropriation of vast amounts of indigenous labor, Firestone itself might be seen as a macroparasite eating up lives and livelihoods in Liberia, a parasite that left the country in a chronic state of underdevelopment.

"When microparasites and macroparasites exist," the medical anthropologist Peter Brown has observed, "there are three mouths to feed, and the host nearly always gets fed last."[1] Such has been the case with respect to Liberia's rural indigenous population, whose land and bodies served as the table upon which microparasites and macroparasites feasted. In 1929, a tiny flavovirus—yellow fever—had fixed the attention of all the world's nations on the small republic of Liberia. William T. Francis, American Minister and Consul General to Liberia, succumbed to the yellow fever outbreak that summer. So, too, did American missionary Maryland Nichols, American educational adviser

James Sibley, a French foreign affairs officer, and a British trader, as the death toll in the capital city of Monrovia climbed.

The US grew increasingly alarmed by the "menace" that threatened the lives of American citizens and subjects residing in Liberia. France and Great Britain also expressed dismay at the threat posed to their neighboring colonies in Guinea and Sierra Leone. But it was the Firestone Plantations Company that most feared the outbreak of yellow fever. Just three years earlier, Harvey Firestone, chief executive of the US rubber and tire manufacturing conglomerate that bore his name, had negotiated with the Liberian government for a ninety-nine-year lease on up to one million acres of land. America sorely needed a source of rubber free from British control, and Firestone intended to grow that rubber in Liberia. When yellow fever erupted in Monrovia in 1929, the Firestone Plantations Company had already mobilized a transient workforce of more than 10,000 laborers, coerced at least in part by a Liberian government quota system, to cut down and clear the rainforest. In place of rainforest, Firestone workers planted thousands of seedlings of the Pará rubber tree, *Hevea brasiliensis*, a species that originated in the forests of Brazil but was bred and cultivated for industrial rubber production on the colonial plantations of Southeast Asia. The combination of men and trees on such a scale posed considerable risks to Firestone's capital investment. Labor was in short supply. Conflict and strife could erupt at any moment. And disease could easily destroy this newly built plantation world. Yellow fever, in particular, was a disease notorious for bringing commercial trade to a grinding halt in port cities like Monrovia, should the need for quarantine arise.[2]

To combat the parasites that threatened his livelihood, Firestone had at his disposal some of the greatest experts in tropical medicine and biology that money could buy. Notable among them was Richard P. Strong, who had been hired by Harvard University to found a new Department of Tropical Medicine in 1913 and whose career was intimately tied to the expansion of American empire. In December 1925, as Harvey Firestone pursued his plans for a rubber plantation in Liberia, Strong met with him to offer Harvard's services in conducting an extensive biological and medical survey of the country's interior. Tropical diseases such as malaria, yellow fever, and onchocerciasis (river blindness) were endemic to Liberia and were a specialty of Strong and his colleagues at Harvard. These diseases had the potential to suck the profits out of American companies trying to gain a foothold in the tropical world and Firestone was happy to accept Strong's offer.

Within six months of the meeting, Strong had put together an eight-member team that included some of the best minds in medical entomology, tropical medicine, botany, mammalogy, and parasitology, and had gathered a supply of the latest experimental drugs for treating tropical diseases. The Boston press touted the expedition as an epic journey that could save "millions of lives."[3] Neither the press nor Strong and his colleagues expressed any apprehension about testing drug compounds, whose harmful side effects at the time were uncertain, on local populations largely unfamiliar with Western medical treatments and practices. Nor did the Harvard doctors question their resolve to harvest tumors, blood, and parasites from people along the expedition's path, despite the objections of indigenous Liberians to such encounters. As Loring Whitman, a first-year Harvard medical student and the expedition's official photographer, observed, when a town chief refused to give up his "charming little tumor . . . we cajoled, we threatened, we vowed he would die . . . and still he coyly refused to part with that most . . . cherished treasure."[4] Such acts of resistance were, in the eyes of Harvard physicians, superstitious, irrational impediments to progress. If only the country would embrace the vision of Harvey Firestone and the promise of American science and medicine, Strong believed, could "a new era of prosperity in the development of the country and the welfare of its people as a whole" arrive in Liberia.[5]

While Western researchers like Strong looked to the endemicity of parasitic diseases such as malaria, schistosomiasis, and trypanosomiasis as an explanation for underdevelopment in tropical countries like Liberia, local populations and the Liberian government often looked upon Western biomedical research and intervention with suspicion and resistance. They had good reason. Time and again, microparasites proved effective allies in efforts by foreign companies like Firestone and Western imperial nations like the United States in legitimating their presence in Liberia and dictating their influence in its local affairs.

The 1929 yellow fever outbreak in Liberia is one such example. The deaths of prominent foreigners prompted the United States, along with France, Germany, and Great Britain, to put heavy diplomatic pressure on the Liberian government to address sanitary conditions in Monrovia that enabled the human-loving mosquito, *Aedes aegypti*, known to carry the yellow fever virus, to thrive. At the same time, the country found itself the subject of a brewing League of Nations investigation into slavery and forced labor, prompted by the Liberian government's use of corvée labor in road building and in providing contract workers to the large cocoa plantations on Fernando Po, a colonial holding of Spain in the Gulf of Guinea. Firestone, having overestimated the ease with which he could recruit a labor force to work his rubber plantations, had a vested interest in the League's investigation.[6]

With its political sovereignty threatened, the Liberian government played into fears of West Africa as the white man's grave dating back to the eighteenth century and frustrated the planning efforts of Howard F. Smith, a surgeon from the US Public Health Service.[7] Smith, sent as Chief Medical Advisor to Liberia by the American government in January 1930 to oversee the work of sanitation and yellow fever control in Monrovia, expressed dismay at the Liberian government's attitude to the outbreak and its alleged belief that the "sanitary work was only for the protection of foreign residents."[8] But the Liberian government was right to be suspicious. When Liberian President C. Dunbar King failed to dispense the entire $18,000 his administration had pledged for Smith's sanitary work, the financial adviser, appointed by the American government under the terms of the Firestone agreement, pressured King to agree to a set of conditions that gave Smith as chief medical adviser significantly greater police and judicial authority in Liberia. When King refused to abide by this new agreement, Smith left Liberia in a huff, citing a "definite spirit of antagonism and opposition."[9] Meanwhile, Justus B. Rice, medical director of the Firestone Plantations Company, anxiously expressed concern that the Liberian government's defiance in the face of an outbreak that threatened Firestone's investment would force the company to build a private port and remove its personnel from the city of Monrovia. Parasites could be used to the political advantage of both empire and resistance.[10]

On the roughly 200 square-mile enclave in Liberia that the Firestone Plantations Company did eventually control, parasites became an excuse and opportunity for experimentation on and surveillance of workers' bodies that proved a boon to American biomedical research throughout much of the twentieth century.[11] In 1931, Justus Rice, Marshall Barber, a malariologist in the International Health Division of the Rockefeller Foundation, and James Brown of Nigeria's Health Department conducted the first experimental trial on the plantation, testing the I. G. Farben anti-malaria drug, plasmoquine, on men, women, and children living in five Firestone camps. Despite its widely assumed highly toxic side effects, the medical team reported no harmful results. Their study was just the beginning, as the plantations became a productive site of what STS scholar Michelle Murphy brilliantly terms experimental exuberance, denoting the ways experimentality "relentlessly produces evidence that then legitimates continued interventions as a self-perpetuating relation of rescue."[12] On the Firestone plantations, experimentation and medical humanitarianism went hand-in-hand. In its planning efforts, the company welcomed and supported experimental research on, and treatment of, diseases such as malaria, onchocerciasis, schistosomiasis, and smallpox that animated company fears and threatened its labor force. While Barber, Rice, and Brown regarded Liberia's "malaria-salted labor" pool beneficial to Firestone, since many workers had over the course of their lives acquired partial immunity to the parasite, the fact that malaria lowered the "general efficiency of labor" posed a threat to the company should a worker shortage arise.[13]

But it was the "protection of the white population, if considered only from the purely economic point of view" that most concerned American doctors.[14] A Firestone operational management plan from the early 1940s for white staff and their families makes this abundantly clear. The manual opens with detailed instructions on the medical regime to be followed by white staff and administered to Black servants to ensure white employees and their families stayed healthy. Firestone dispensed one antimalarial drug for whites and a different one for their Black servants. Servants were subject to plasmoquine, which at the time was believed to have much more harmful side effects than atrabine, the antimalarial given to white employees. But plasmoquine was also known to be more effective in killing gametocytes in *Plasmodium*'s life cycle and thus preventing transmission of the parasite from the blood stream of one host to that of another via mosquitoes. Since Black laborers were widely regarded as carriers of the malaria parasite, cleansing their bodies of gametocytes would eliminate the perceived threats they posed to whites living in close quarters to them on the Firestone plantation.

Out on the divisions of the plantation, where most laborers lived, far removed from the housing quarters of white management, a different set of medical protocols applied. Black laborers seeking employment were subject to medical inspections, and detailed medical records were kept on all Liberian overseers and headman. Dressers visited daily and treated all laborers who reported sick or failed to show up to work, administered quinine twice each week and hookworm treatments once per month, regularly inspected laborers for yaws, and conducted a monthly census of men, women, and children and a sanitary inspection of wells, houses, and latrines in each division.[15]

Firestone promoted its medical care of workers and support of American biomedical research as evidence of how the company benefitted the health and welfare of the Liberian population. But in its singular focus on tropical parasites that most threatened its white staff and hindered the efficiency of labor, Firestone cast health and welfare within a quite narrow frame in its planning efforts. The control of parasites on the Firestone plantations operated largely in the service of industrial efficiency and production. It was a plan instrumental in its focus, guided by the premise and question: What was the minimum needed to guarantee a viable labor force necessary to build and maintain a world that saw profits flow largely from Liberia to Akron and New York City, the centers of Firestone production and finance?

Richard Strong spent his career researching tropical diseases and advising multinational firms like Firestone on the biological threats that hindered the planning and operations of industrial plantations. Reflecting on the parasitic threats that jeopardized company profits, Strong wrote: "the ideal balance," found among organisms living in a mutually beneficial relationship, "even if attained, becomes frequently disturbed, and in some group of animals sooner or later one organism becomes more dependent upon the other and the symbiotic relationship passes into one of parasitism."[16] Immersed in the microscopic world, attentive to the biological relationships structuring the economy of nature, Strong either ignored or failed to see another form of parasitism in his midst that proved far more destructive to the development of Liberia and its people than a host of tropical diseases with which he and other American researchers occupied themselves.

Firestone publicly promoted its relationship with Liberia as a mutually beneficial, symbiotic one, but the terms under which it negotiated its land concession with the Liberian government hardly represented an ideal balance. In exchange for doing business in Liberia, Firestone insisted that the Liberian government take out a $5 million loan from the Finance Corporation of America payable at an interest rate of 7 percent over forty years to pay off its foreign debts and thereby ensure adequate protection from outside creditors of Firestone's "large capital investment" in a "far-off country."[17] The Finance Corporation of America was later publicly revealed to be a subsidiary of the Firestone Tire and Rubber Company. More onerous than the interest payments were the attached terms, which significantly constrained the sovereignty of the Liberian government. Liberian officials entered the negotiations with Firestone, knowing full well, as statesman C. L. Simpson observed, that the loan was a form of "economic domination by a company belonging to a traditionally friendly country."[18] But it was regarded by some Liberians as the lesser of two evils, since American investment would help stave off the very real threat to Liberia from encroaching European countries, notably France and Great Britain.

In 1935, George Brown, an African-American doctoral student working at the London School of Economics, traveled to Liberia where he spent months investigating the different economies at work among Liberia's settler society, foreign business, and rural indigenous population. In *The Economic History of Liberia*, published in 1941, Brown astutely observed an emerging parasitic relationship that no biologist or doctor had seen. In a brilliant ethnography and history of Liberia's indigenous economy, Brown identified the manifold ways in which communal ownership of land structured the activities, culture, values, and exchange relations of Liberia's sixteen indigenous ethnic groups. Land dispossession, Brown argued, was the greatest threat to Liberia's development. The first major land rupture in Liberian history came when free Black people from America settled the West African coast in the 1820s and brought with them a Western system of private property alien to the customary practices and cultural beliefs of indigenous groups already living there. The second major land rupture came with the arrival of Firestone, first through large-scale clear-cutting and then through road building, which ignored the customary rights of indigenous people and made their communal lands vulnerable to enclosure. Loans and concessions brokered by Liberian elites, foreign governments, and multinational firms, drove, Brown asserted, "the parasitic capitalism of the ruling class of Liberia and the financial exploitation of the American and European industrialist." In Brown's view, parasitic capitalism and financial exploitation both rested and fed upon Liberia's indigenous people and communal economy.[19] Land was as equally susceptible to the parasites of capital, as human bodies were to microbial parasites in their midst.

By the late 1940s, workers increasingly began to agitate against this parasitic relationship that Brown had noted. It was an extractive and exploitative relationship far more damaging than the sporozites of *Plasmodium* feeding on red blood cells in their bodies. Labor unrest culminated in a massive 1963 strike in which twenty thousand Firestone laborers brought a halt to all operations on all forty-five divisions of the plantation. Demanding higher wages, improved housing, shorter work hours, and better working conditions, laborers saw their livelihoods threatened by a parasitic relationship that had emerged with the arrival of Firestone in Liberia, one that fed off both their bodies and their land. In 1951 more than 79 million pounds of raw latex, valued at more than $48 million, flowed out of rubber plantations in Liberia. Firestone owned 94 percent of that latex and paid the Liberian government $3.8 million plus six cents per acre for the pleasure of doing business in Liberia; that year, Firestone rubber accounted for 91 percent of Liberian exports. Meanwhile, the average pay for a Firestone worker from Liberia amounted to thirty-eight cents per day.

The promise of planning that Firestone projected onto Liberia is on full display in the 1948 educational film, *Medicine in the Tropics*. The beating of African drums that open the film conjures stereotypical images of West Africa as a place of superstition and darkness. But scenes of neatly planted rows of rubber trees, lines of workers carrying pails of freshly tapped latex, people queued up in rural villages for smallpox vaccinations, and orderly lines of beds filled with patients at the Firestone plantation hospital make clear that the darkness of the dense tropical rainforest is yielding to the progress and light of science, medicine, and industry. Industrial planning had subdued tropical nature, transforming it into a landscape of health and prosperity. When a patient is discharged from the Firestone hospital, seemingly well and in good spirits, he becomes an ambassador of goodwill for modern medicine (and Firestone) in his local village.

But the razor wire atop the hospital fence suggests that the Firestone labor force was not made up entirely of willing and happy subjects. By the 1950s, growing labor unrest on the Firestone plantations made the US State Department fearful of a "situation ripe for Communist activity."[20] Western medicine might readily control diseases such as malaria and smallpox that threatened labor productivity, animated white fears, and helped legitimate Firestone's presence in Liberia in the name of humanitarianism. But what Firestone most needed to fear was the organizational power of labor, an immune response to the parasitic relationships of capital, fostered by Firestone and Liberian elites, that ate at the land and bodies of Liberia's indigenous people. Throughout the history of Firestone in Liberia, parasites built and sustained relationships between empire and resistance; they are an integral part of plantation worlds.

Notes

1. Peter Brown, "Microparasites and Macroparasites," *Cultural Anthropology* 2, no. 1 (1987): 155–71, on 161.

2. On the history of Firestone in Liberia, see Gregg Mitman, *Empire of Rubber: Firestone's Scramble for Land and Power in Liberia* (New York: The New Press, 2021).

3. "Harvard Expedition Off to Africa with Cures for Tropical Diseases," *Boston Traveler*, May 15, 1926.

4. Loring Whitman, Loring Whitman Diary, October 26,1926, Indiana University Liberian Collections, Loring Whitman Collection, Indiana University Libraries, Bloomington.

5. Richard Pearson, Richard Pearson Strong Diary (hereafter RPSD), 4, typescript, 1926–1927, GA82.4, Center for History of Medicine, Francis A. Countway Library of Medicine, Harvard University, Boston.

6. On the Fernando Po controversy, see Ibrahim Sundiata, *Brothers and Strangers: Black Zion, Black Slavery, 1914–1940* (Durham: Duke University Press, 2003).

7. The recalcitrance of the Liberian government to acquiesce to the sanitary imperialism of Western nations may well have been a ploy, as historian Adell Patton argues, to "maintain an image of the nation as undesirable to white settlement" until its "colonial takeover was no longer a threat." Adell Patton, "Liberia and

Containment Policy Against Colonial Take-Over: Public Health and Sanitation Reform," *Liberian Studies Journal* 30, no. 2 (2005): 40–65, on 41.

8. H. F. Smith, "A Resume of the Efforts Towards Sanitation and Yellow Fever Control in Liberia," p. 2, folder 25572: Execution of the Recommendations of the International Commission of Enquiry in Liberia (1931–1932), UNOG Registry, Records and Archives, League of Nations Secretariat, Geneva, Switzerland.

9. H. F. Smith, "A Resume of the Efforts Towards Sanitation and Yellow Fever Control in Liberia," 15.

10. According to American interim chargé d'affaires Clifford Wharton, one Liberian government official allegedly stated that "Germans had their submarines, Great Britain has her navy, and the United States, money with which to turn out war supplies, but Liberia has her mosquitoes." Wharton to Secretary of State, 19 August 1929, p. 5, RG 59, 882.12 Decimal Files, 1930–1939 continued, Box 7008, State Department, National Archives. In fact, white foreigners unexposed to yellow fever were more likely to die of the disease than the local population, given the fact that individuals acquired natural immunity over their lifetime if they had been exposed to yellow fever in childhood and survived. On Rice's concerns, see Henry Beeuwkes, October 1929 entries, "West African Yellow Fever Commission Diary, Vol. 5, 1929," folder 31, box 5, subseries 495, West African Region: O. Yellow Fever, RG 1.1, Rockefeller Foundation Projects, Rockefeller Archive Center, Sleepy Hollow, NY.

11. The intertwined relationships of American empire, capital, and biomedicine reached far and wide across the globe. See, for example, Warwick Anderson, *Colonial Pathologies: American Tropical Medicine, Race, and Hygiene in the Philippines* (Durham: Duke University Press, 2006); Laura Briggs, *Reproducing Empire: Race, Sex, Science, and U.S. Imperialism in Puerto Rico* (Berkeley: University of California Press, 2002); John Soluri, *Banana Cultures: Agriculture, Consumption, and Environmental Change in Honduras and the United States* (Austin: University of Texas Press, 2006).

12. Michelle Murphy, *The Economization of Life* (Durham: Duke University Press, 2017), 79.

13. Marshall A. Barber, Justus B. Rice, and James Y. Brown, "Malaria Studies on the Firestone Rubber Plantation in Liberia, West Africa, *The American Journal of Hygiene* 15, no. 3 (May 1932): 601–33, on 602, 623.

14. Marshall A. Barber et al. "Malaria Studies," 624.

15. General Information for the Information and Guidance of Staff and Families of Firestone Plantations Company and Affiliated Companies in Liberia, 15 June 1941, African Studies Collections, Special and Area Studies Collections, George A. Smathers Libraries, University of Florida, Gainesville.

16. Richard P. Strong, "The Relationship of Certain 'Free-Living' and Saprophytic Microorganisms to Disease," *Science* 61, no. 1570 (1925): 97–107, on 101.

17. "Statement of Harvey S. Firestone," in *Crude Rubber, Coffee, etc., Hearings before the Committee on Interstate and Foreign Commerce House of Representatives, 69th Congress, First Session on H. Res. 59* (Washington: Government Publishing

Office, 1926), 247–272, on 254. On the Firestone loan, see Frank Chalk, "The Anatomy of an Investment: Firestone's 1927 Loan to Liberia," *Canadian Journal of African Studies* 1, no. 1 (1967): 12–32; David Kilroy, "Extending the American Sphere to West Africa: Dollar Diplomacy in Liberia, 1908–1926" (PhD diss., University of Iowa, 1995).

18. C. L. Simpson, *The Memoirs of C. L. Simpson: The Symbol of Liberia* (London: Diplomatic Press & Publishing, 1961), 141.

19. George W. Brown, *The Economic History of Liberia* (Washington, DC: Associated Publishers, 1941), 231.

20. ANE – Mr. Berry, Mr. Sims, "Strikes of Firestone Plantations, Liberia," January 10, 1950, Liberia 1950, 4, Firestone, Box 1, Lot 56D 418, Records of the Office of African Affairs Subject File, 1943-1955, RG59, US National Archives, Washington, DC.

Riverbed: South Korea, 2008
Chihyung Jeon

"Restoration or Devastation?" A 2010 news report in the journal *Science* asked this question about the Four Rivers Restoration Project of South Korea, which was then underway at full speed.[1] The Four Rivers Restoration Project (2008–2012) was one of the largest engineering projects in Korean history (around US$ 20 billion). The government and those who supported the plan claim that it "revived the rivers" by giving them due maintenance and creating spaces for leisure and sports, while the critics say that the project was simply an "environmental disaster" that "killed the rivers" by digging up the riverbeds and blocking water flow with numerous weirs. The idea of "renovating" or "transforming" the nation's land at an unprecedented scale excited President Lee Myung-bak and other supporters of the project. The prospect that its ecological and social consequences would be indelible and irreversible enraged the opponents, including environmental activists, academics from all disciplines, religious groups, and concerned citizens. For both groups, there was no question that South Korea—its nature, economy, and politics—would never be the same after the Four Rivers Project.[2]

The most important and visible technical component of this contested plan was "dredging." The *New Oxford American Dictionary* defines the verb *dredge* as "clean out the bed of (a harbor, river, or other area of water) by scooping out mud, weeds, and rubbish with a dredge." Indeed, the Four Rivers Restoration Project was a gigantic dredging operation that removed 506 million cubic meters of sand and mud from the rivers. Throughout the project, images of excavators, bulldozers, dump trucks, and dredging vessels for on-the-surface and underwater dredging work were widely circulated in the media as evidence of restoration or destruction, depending on how the journalist viewed the project. In terms of budget, dredging operations took up the largest proportion (more than 20 percent) of the project, requiring more money than the construction of sixteen weirs or the management of water quality.[3]

Another definition of the word *dredge* in the same dictionary is "bring to people's attention an unpleasant or embarrassing fact or incident that had been forgotten." In this sense, the Four Rivers Project dredged up the historical memories and legacies of state-led development in South Korea, especially its ruthless and undemocratic aspects. For the critics of the project, dredging on the four rivers represented the worst part of the modern history of planning in South Korea.[4] The dredging scene gave them a visual and physical example of planning without participation (of people), planning without sympathy (toward nature), and planning without foresight (economic, technical, and ecological). In other words, hasty dredging best illustrated the Korean state's habit of executing a plan before deliberating on it—if they deliberated at all. Therefore, those who opposed dredging on the four rivers were also opposing the ways in which the state had drawn up plans and implemented them on its territory under the banner of development.

Dredging for Economic Growth

Since the 1960s, dredging has been an essential part of the South Korean state's plan for land and water development. After acquiring power through a military coup in 1961, the authoritarian government of Park Chung-hee initiated a strong drive for fast economic development, which involved fullfledged plans for infrastructure building and territorial development. Many of these development projects started with digging. Dredging was needed for port development, seaside industrial site development, coastline management, and river management. Both physically and psychologically, dredging *made room* for economic growth, allowing large ships, factories, resources, and products to move in and out and removing whatever blocked the nation's path to development. The more you dredge, it was believed, the faster the economy would grow.

Recognizing the importance of dredging for the nationwide efforts for economic development, the Korean government established Korea Dredging Corporation in 1967. Dredging work was not always visible to the public, but the state planners were well aware that dredging was the groundwork, albeit done underwater, for development. In a congratulatory remark written for *A Seven-Year History of Dredging* (1975)—an interesting time frame for writing a corporate history—the Minister of Construction praised the work of Korea Dredging Corporation as "absolutely indispensable." According to the minister, the corporation's dredging work produced "splendid achievement in expanding the nation's territory by conducting a heated war of 'piercing the sea, moving the land.'" The phrase "piercing the sea, moving the land" was featured in the seven-year history as a slogan that characterized the entire work of the corporation.[5] The "heated war" of dredging in Korea happened to overlap with another heated war in Vietnam where Korea deployed tens of thousands of troops. Fearless dredging around the peninsula manifested, and was justified by, the dominant spirit of "we fight as we construct, we construct as we fight"—another widespread slogan of the era.

Whenever and wherever possible, one could argue, the nation's land and water should be subject to dredging. If you had a dredger—a powerful one issued by the state—everywhere looked like mud and rubbish to be dredged up. In the photo section of the seven-year history, readers encountered an almost poetic description of the Corporation's mission: "The sea is silent, waiting only for development. This means prosperity, and it is the route to the expansion of the nation's territory."[6] The land and the water not just deserved, but *wanted* to be dredged and developed. It would be negligent and irresponsible for the state not to find more sites for dredging and do it.

This historically forged connection between dredging, construction, and development was manifested once again in President Lee's idea for the Four Rivers Restoration Project. It was no coincidence that Lee had started his career at a construction company working on a highway construction project in the late 1960s and had become known for his "bulldozer style" leadership as a CEO, as mayor of Seoul, and then as president. Lee's biggest achievement as mayor was the Cheonggyecheon "restoration" project (2003–2005), which involved demolishing the overpasses built during the 1960s in the central part of Seoul, peeling back the pavement along the path, and then re-creating the old urban stream underneath it. Running through the center of Seoul, the Cheonggyecheon had been a site for many dredging works historically, but it was covered with concrete structure during the development era to make room for the rapidly industrializing city. Three decades later, as he was elected as mayor in 2002, Lee successfully appropriated the "restoration" of Cheonggyecheon as a proof of his new commitment to the environment and quality of life as well as his old experience in construction and economic development. He had no problem in mixing the two potentially competing values.[7]

The political refashioning, however, soon turned out to be superficial. From this large urban development project branded as restoration, Mayor Lee acquired enough political capital to run for president in 2007. He first proposed as an election pledge the Grand Korea Waterway, a network of canals connecting Korea's major rivers for transportation. Lee won the election, but soon faced severe criticism against his Waterway plan from civil and environmental groups. In an attempt to undermine the critique of potential environmental damage, Lee's government revised the Grand Korea Waterway into the Four Rivers Restoration Project in 2008, and proceeded quickly by starting to dig up the rivers. By every measure, the project was rushed. "Rush construction," a term used to describe how the first major highway had been built in the 1960s, was invoked by the critics and the media to characterize the dredging operation at the rivers.[8] Scenes of day-and-night dredging operation on the four rivers reminded Koreans, materially and discursively, of the earlier nationwide development-cum-construction projects. A familiar pattern of militaristic, single-minded, no-questions-allowed construction work was repeated, even though, or precisely because, the project was advertised as a case of Lee administration's "green growth" agenda. Bulldozers still dominated the rivers restoration project.

Dredging as Curing or Killing

Despite the unmistakable continuity with the history of state-driven economic and territorial development, the planners of the Four Rivers Project made efforts to distinguish dredging work at the four rivers in the twenty-first century from dredging operations of the earlier decades. Whereas earlier dredging had been a part of the "river management policies in which economic development was a priority," the Four Rivers Project masterplan document noted, dredging at the four rivers signified a "new concept in the defense against flooding." The idea was that lowering water levels by up to 3.9 meters through dredging would "block flooding disasters fundamentally." Some of the dredged soil would be used to "remodel" the farming fields next to the river, raising the land out of flooding risk.[9] In these plans, dredging was framed as an active, forward-looking practice of preventing risks throughout the rivers.

The proponents of the Four Rivers Project also promoted dredging as treating or curing the rivers that had long been suffering from severe diseases. They liked to compare the rivers to a patient with arteriosclerosis, for which dredging was undoubtedly needed. A promotional video for the project made a strong analogy between the river and the body: "Our rivers have arteriosclerosis. If we dig out sediment at the bottom, just as we remove waste from our body, the water will flow well, curing chronic diseases like flood and drought." A professor of environmental engineering also commented in the video that the Four Rivers Project was "not to destroy, but to cure the dying rivers."[10] In order to be saved by dredging, the rivers had first to be declared as "dying," if not already dead, which became an important point of contention between the project's supporters and critics.

The critics of President Lee and his Four Rivers Project never accepted the assumption that the rivers were like a dying patient. There appeared other forms of personification of the rivers that were in clear opposition to the metaphor of a dying patient. In one newspaper cartoon with a caption that read "Mother, Mother, Our Mother," the four rivers were depicted as an innocent, old mother in humble traditional costume. A Korean woman is lying down with her eyes wide open and tears rolling down her cheek. An excavator marked with the president's initials (MB for Myung-bak) is mercilessly dredging her flesh and blood out of her body. As if not to allow for any misinterpretation, the cartoonist put "four rivers dredging" right above the excavator. Dredging at the four rivers was understood as bloodily gendered depravity against Mother Nature.[11]

Another cartoon critical of the project portrayed the rivers as a man lying on an operating table. The scene implies an organ transplant operation, but the man on the table does not look like a dying patient in desperate need of

Figure 7. Cartoon by Kyung-Soo Kim (April 27, 2010), used with permission from Naeil Shinmun and redrawn by the original artist for this chapter.

Figure 8. Cartoon by Yong-Min Kim (June 30, 2009), used with permission from Kyunghyang Shinmun.

the operation. He rather appears to have been brought to the operating room against his will for a probably illegal organ harvest. The doctors with a spade and an excavator in their hands say to him: "We will neatly remove unnecessary organs, thoroughly dredge at the bottom, pour plenty of cement into the body, but will not connect the remaining organs within the (presidential) term . . . We promise!"[12] The last part was a reference to the government's claim that the project did not aim to connect the rivers and create a nationwide canal system, which had been the original conceptualization of the project before it was refashioned into a river management project due to severe criticism of the idea. The man personifying the four rivers looks stunned at what the doctors say, but is unable to escape as he is tightly buckled on the operating table. With his organs declared as unnecessary, he has no choice but to get dredged to the bottom. This cartoon and other critiques claimed dredging to be an unethical, if not outright illegal, practice against the rivers.

The tools or machinery of dredging provided vivid imagery with which to critique the government's execution of the Four Rivers Project. The cartoonists' favorite was the excavator, which represented everything wrong with the project and the government: brutal, careless, and undemocratic. A less spectacular but more intuitive icon of dredging was the spade. The critics often called the project "spadework." It could be a literal description of the manual work performed as a part of dredging and other operations, but more importantly "spadework" referred to the shoddy, unthinking, and shortsighted nature of the project. With a spade in hand, the government was digging up the entire nation without thinking or caring about the consequences. For anyone who bothered to see, the critics would have said, the consequences were clear. One satirical artwork featured a "spade-fish": a spade's red blade formed the fish's head, and the long handle was the bone and the tail. A couple of curved lines on the sides made the spade look like an angry fish, which said, "I am a spade-fish . . . made by your blind conscience, your nonchalance, your selfishness."[13] Dredging left no real fish, no life, in the rivers.

A Dredged Nation

The Four Rivers Project has had consequences that are still to be grasped, debated, and, if possible, overcome since it was completed in 2012. After the Lee administration ended in early 2013, many reports from the government, media, and environmental organizations have documented damage to the rivers' ecosystem, including water quality degradation, accusations of corruption within the construction circle, and the poor state of maintenance for riverside facilities. The shift of political power in 2017 to the opposition party helped uncover, or dredge up, many more hidden problems.

Once again, dredging offered some of the most striking images of the project's aftermath. Contrary to the initial plan to sell all the dredged materials, huge piles of dredged sand and gravel occupied riverbanks for several years. According to one report, only 35 percent of the materials were sold, and it would take twenty years to sell them all. The "sand mountains" created by dredging work at the Nam-Han River since 2009 occupied an area as large as 224 soccer fields. Residents in the area complained about the impact on draining and farming; some cattle became infertile and even died, suffering from the noise and vibration from trucks. What was once called "golden eggs" has turned into a "white elephant," which no one can get rid of.[14]

Another symbolic scene was that of dredging vessels abandoned after the project's end. Activists and reporters pointed out that these vessels were not properly monitored by the authorities and thus posed serious risk of physical collision, oil spill, and running aground.[15] These ravaged vessels were also a reminder of the just-do-it and take-it-and-run attitudes of the Korean state and businesses for the past half century. The Four Rivers Project, of which the dredging work was such a big part, was the culmination of the developmental drive that transformed South Korea from a poor, agrarian nation to an industrialized one with a turbulent political history.

Both physically and discursively, the Four Rivers Project turned Korea into a dredged nation, so to speak. The project ended up dredging not just the rivers but the nation itself, bringing out mud and rubbish as well as conflicting visions of development, conservation, and restoration. Dredging represented everything that the critics deemed problematic about the project: violence against nature, unchecked state power, lack of foresight and planning, and a greedy clique of construction industry. Would it be possible to undo what dredging did to the rivers and to the nation?

Dredging work is not easily reversible. Undoing the consequences of dredging would require not only technical ingenuity but also political will. In 2017, the new government decided to open the weirs built during the project to see if more water flow through the open weirs would lessen the degradation of water quality and other ecological problems. "Renaturalization" of the four rivers was put on the new government's agenda as well.[16] The most difficult task, however, would be to create an alternative vision of rivers, environment, engineering, and politics. It will take a great deal of planning that is both scientific and democratic.

Notes

1. Dennis Normile, "Restoration or Devastation?," *Science* 327, no. 5973 (2010): 1568–70.

2. Heejin Han, "Authoritarian Environmentalism under Democracy: Korea's River Restoration Project," *Environmental Politics* 24, no. 5 (2015): 810–29; T. J. Lah, Yeoul Park, and Yoon Jik Cho, "The Four Major Rivers Restoration Project of South Korea: An Assessment of Its Process, Program, and Political Dimensions," *The Journal of Environment & Development* 24, no. 4 (2015): 375–94.

3. Ministry of Land and Ocean, *The Masterplan for Saving the Four Rivers*, July 2009, 86, 363.

4. Jin-Tae Hwang, "A Study of State–Nature Relations in a Developmental State: The Water Resource Policy of the Park Jung-Hee Regime, 1961–79," *Environment and Planning A* 47, no. 9 (2015): 1926–43. See also Juyoung Lee, "한국의 제 1차 국토종합개발계획 수립을 통해서 본 발전국가론 '계획 합리성' 비판 [Making of the Developmental State's 'Plan Rational': The Case of South Korea's First

Comprehensive National Physical Development Plan, 1963–1971]," 공간과 사회 [*Space and Society*] 25, no. 3 (2015): 11–53.

5. Korea Dredging Corporation, *A Seven-Year History of Dredging* (Seoul: Korea Dredging Corporation, 1975).

6. Korea Dredging Corporation, *A Seven-Year History of Dredging*, 42.

7. Myung-Rae Cho, "The Politics of Urban Nature Restoration: The Case of Cheonggyecheon Restoration in Seoul, Korea," *International Development Planning Review* 32, no. 2 (2010): 145–65; Chihyung Jeon and Yeonsil Kang, "Restoring and Re-Restoring the Cheonggyecheon: Nature, Technology, and History in Seoul, South Korea," *Environmental History* 24, no. 4 (2019): 736–65.

8. Chihyung Jeon, "A Road to Modernization and Unification: The Construction of the Gyeongbu Highway in South Korea," *Technology and Culture* 51, no. 1 (2010): 55–79.

9. Ministry of Land and Ocean, *Saving the Four Rivers*, 73, 88.

10. Ministry of the Interior and Safety, "To Save the Four Rivers Is to Save Life," June 18, 2010, video, accessed October 15, 2023, https://goo.gl/c7XmrH.

11. Kyung-Soo Kim, "어머니, 어머니, 우리 어머니 [Mother, Mother, Our Mother]," *Naeil Shinmun*, April 27, 2010, http://www.naeil.com/upload/cartoon/n100427.jpg.

12. Yong-Min Kim, "6월 30일 [June 30]," *Kyunghyang Shinmun*, June 30, 2009, https://www.khan.co.kr/cartoon/grim-madang/article/200906292054192.

13. Kyu-Jung Kim, "나는 삽어입니다 [I Am a Spade-Fish]" The artwork is freely available online, for instance, at http://www.ohmynews.com/NWS_Web/View/at_pg.aspx?CNTN_CD=A0001371874.

14. "Once called 'golden eggs,' dredged sands from four rivers became a white elephant," *JTBC News*, February 14, 2017, http://news.jtbc.joins.com/article/article.aspx?news_id=NB11422526.

15. Dae-Sung Sohn, "33 Dredging Vessels Abandoned on the Nakdong River," *Yonhap News*, June 17, 2015, http://www.yonhapnews.co.kr/bulletin/2015/06/17/0200000000AKR20150617045900053.HTML; Seong-Hyo Yoon, "Where Are the Abandoned Dredging Vessels?" *Ohmynews*, October 18, 2012, https://www.ohmynews.com/NWS_Web/View/at_pg.aspx?CNTN_CD=A0001790987.

16. Advisory Committee on National Agenda Planning, "The Five-Year Plan of the Moon Jae-in Government" (Released in July 2017).

Seeds: German East Africa, 1892
Tahani Nadim

Flat, black teardrops the size of a fingernail. That's how the seeds of the sisal agave plant look. They arrived in a small, padded envelope from an online seed seller based in the Kowloon district of Hong Kong. Their seeds are grown in the Gansu province in northwestern China near Jiuquan, one of the world's largest seed production areas and also home to China's satellite launch base. Like satellites, seeds avail themselves of circular, or perhaps elliptical moves that can be obscure (and obscuring) as well as thrifty. They have a talent for returning. In concert with many others, they grow into seedlings and plants and return, after flowering and fruiting, to seeds. This talent, I suggest, is more than a return to form. Instead, it's a sort of counter-clock in the face of universal linear time, the time of progress and conclusions, the kind of time we find with most plans.

Seeds and plants have been the object and tool of some of the most brutal as well as sophisticated planning. Sisal seeds, among other kinds of seeds used for cash crops, have undergirded the violence of colonial plantation economies and enslavement spanning continents and centuries. In a different register, we can think of monocultures, terminator seeds, or the Svalbard Global Seed Vault (the so-called Doomsday Vault), a storage facility in the Arctic Circle supposedly safeguarding the world's plant diversity from present and future disaster. Many of these efforts manage returns and returnings: curtailing the seeds' capacity to spring back to life or keeping it on the cusp indefinitely, all the while eyeing the accrual of yield and profit. Seeds are good to think with, then, about the odd loopings of time and the recursive and enduring fantasies of imperial control which continue to configure and disfigure relations between people and environments and between worlds and the stories we tell about them. They make for good companions when it comes to figuring beginnings and endings, especially because seeds never settle for long on either. The following text traces moments of colonial planning and planting in order to recover their contiguities across different times and spaces. It focuses on three sites—the sisal agave plant, the museum, and the colony—that are both products and agents of planning. The seed acts as method and companion (or protagonist) in unravelling and crossing their plans.

I began thinking with seeds when I found the so-called Mai Collection, a collection of extant plant seeds kept amidst the palaeobotanical collection at the Natural History Museum in Berlin (*Museum für Naturkunde*, MfN).[1] There, kept in small glass vials neatly tucked into 288 drawers, rest thousands of seeds from around the world for the purpose of comparison with fossilized specimens. It's a reference collection, not to be displayed but to be worked with by palaeobotanists and others like an instrument or archive. I came across the collection together with the artist Åsa Sonjasdotter during research for a collaborative exhibition project in 2014.[2] The Mai Collection caught our attention because the labels that accompanied the seeds suggested a vast network of global seed travels. Many came from botanical gardens, some from private seed growers and exchanges, still others from experiment stations and agricultural institutes. Seed exchange continues to be a central aspect of the work of botanic gardens, which regularly publish seed catalogues, like the Linnaean Gardens' *Semina Selecta*. The Natural History Museum in Berlin, like others of its kind, is an imperial institution deeply implicated with colonial histories and presents.[3] These histories are reflected, most ostensibly, in the makeup of collections, with many of the specimens deriving from colonial exploits.[4] Less obvious are the ways in which practices in and of the museum partake in and draw on colonial histories that are ongoing; that is, histories beholden to the "supposed singularity" of Western modernity that endure in the concepts and stories and representational conventions of museums and collection-based research more generally.[5]

The Mai Collection and its seeds prompted Sonjasdotter and I to look into the entanglements of plants and the German Empire, particularly their manifestations in what was then German East Africa (1885–1919). Plants have been integral to imperial formations, providing a resource for relentless extractions to service and secure markets, fortify dominion, and control bodies.[6] Not surprisingly then, Linnaeus regarded botany, the science of plant life, and global trade as mutually dependent: "*Botanico necessaria sunt commercia per totum orbem*" (The botanist needs worldwide trade relations), he wrote in the 1737 *Hortus Cliffortianus*, the catalogue in which he classified and listed the plants growing in the garden of George Clifford, director of the Dutch

East India Company.[7] By then, a considerable network of plant and seed exchange involving botanical gardens and herbaria was already well established. As scholars have noted, the link with global trade secured the material conditions for the doing of botany as well as its epistemological foundations: Only in a heterodox economy comprised of botany, libraries, gardens, books and the idea of globally distributed families could plants become knowable as the kinds of *natural units* required by systematics and its order of life.[8] Thus, taxonomy—the discovery, description, and naming of (new) species—establishes a plan, a tenacious one at that. The aim remains to collect and index all species on earth (or rather, representatives thereof) and deposit them in museum collections for future reference. Yet, as the following story will make evident, the schedule of the plan and its intended futurity do not always, well, stick to the plan.

In 1891 the German planter and agronomist Dr. Richard Hindorf (1863–1954) arrived in Tanga contracted by the *Deutsch Ostafrikanische Gesellschaft* (German East Africa Company). The DOAG was founded in 1885 by the *Gesellschaft für deutsche Kolonisation* for the purpose of seizing, colonizing and administering mainland Tanzania. Hindorf experimented with planting a range of cash crops including coffee, cocoa, tea, and pepper in Derema, about 50 kilometers inland from Tanga in the eastern highlands of the Usambara Mountains. He came across an article in the *Bulletin of Miscellaneous Information* (No. 62, 1892), a newsletter published by the Royal Botanic Gardens, Kew that described the production of sisal hemp in Mexico and its potentials for cultivation in the Bahamas, other parts of the West Indies, and Pacific colonies.[9] Sisal hemp is derived from the sisal agave plant and is a valuable source for cordage and binder twine, at the time extensively used in shipping as well as industrial and agricultural settings, notably for wheat sheaves. Sisal fiber can also be found in carpets, musical instruments, tea bags, paper pulp, and alcohol. Cultivated in Central America since Mayan times, Spanish colonizers had established the hacienda system, "vast agricultural factories" on the basis of peonage, forced servitude.[10] Hence, by the mid-nineteenth century agave fibers could be found across the globe, weaving together the colonial empires of the British, the Dutch, and the Germans.

Yucatán province had what amounted to a monopoly on the export of sisal, and its agricultural practices as well as plant knowledge were guarded secrets. Similarly, the export of plant stock was forbidden. Against the odds and laws, Hindorf managed to smuggle bulbils out of Mexico either in the belly of a stuffed crocodile[11] or in the folds of a large umbrella.[12] However, another story would have it that Hindorf contacted Reasoner Brothers Royal Plant Nursery, plant dealers in Onceco, southwestern Florida, requesting a batch of sisal plants. Their nursery acted "as a portal through which many nonnative horticultural plants were first introduced to Florida," including "some of the most invasive plants in Florida" such as the broad-leaved paperbark tree and the Brazilian peppertree.[13] These doubtful origin stories illustrate the accidental and opportunist groundwork which fueled colonial planning and at the same time point to the haphazard nature of collecting that undergirds the museum's pretense of completeness.

The sisal plant had been introduced to Florida about 40 years earlier by Dr. Henry Perrine, physician and former consul in Campeche, Mexico. He had been granted land by the US government in 1836 on Biscayne Bay in the Florida Keys for establishing a botanic garden to acclimatize economically viable tropical plants, land which was home to the Seminole, a heterogeneous group comprising tribal peoples, escaped slaves, and free people of color. Their resistance to forced removals and ecological devastations became known as the Seminole Wars, and in the course of the wars Perrine was killed but his plants, including sisal agave, flourished unhindered. Hindorf was supposedly sent 1,000 bulbils or "pups," the plantlets that grow at the top of the flower stalk, which arrived in Hamburg in 1892. Eighty percent of the shipment had perished but Hindorf is said to have repackaged the remaining 200 bulbils and sent them on to Tanga where the 62 (or 66 or 72) that survived the trip were planted.

After its introduction to German East Africa, the 62 (or 66 or 72) sisal plantlets became the foundation for large-scale sisal production in East Africa, which at its height in 1961 made the newly independent mainland Tanganyika the world's largest producer, with 200,000 tons per year. Setting up a sisal industry was as much about strengthening the economy as it was about making empire through science and control over people, land, and resources: "in the systematic colonization of German East Africa, the first major institution which was established was the plantation."[14] With the 1895 Imperial Land Decree ownership of most lands was transferred to the German empire (crown land). In addition to dispossessions, the colonial authorities made use of a range of coercive methods in securing the workforce which was made up of migrant labor, local wage labor, and enslaved people.[15] The transformation of land into labor-intensive, agro-industrial plantations in German East Africa was supported by a network of botanic experimental stations, such as the *Biologisch-Landwirtschaftliche Versuchsanstalt* (biological-agricultural experiment station) in Amani in the Eastern Usambara Highlands, coordinated through the *Botanische Zentralstelle für die Deutschen Kolonien* (Botanic Central Office for the German Colonies).[16] The Amani Institute was tasked with providing practical support for plantations and settlers as well as with the mobilization of and instructions for the introduction of commercially valuable plants.[17] It carried out farming experiments to test the suitability of agricultural crops in the regional climates and soils, including investigations into pest control and fertilizer use.

The sisal agave (*Agave sisalana*), a member of the Agavoideae subfamily, is a sturdy plant with little frills. For the exhibition, Sonjadotter and I acquired a sisal plant from *Der Palmenmann*, a large online seller of Mediterranean and topical plants based in Dortmund, western Germany. Now traded as an ornamental plant, it survived eight weeks without sunlight or water in the exhibition and now grows in my office, having developed nine new leaves since its arrival there. The plant's thick leaves are the source for the sisal fiber and can reach a length of almost two meters, terminating in an extremely sharp spine.

By the beginning of the twentieth century, German East Africa had become the world's third-largest producer of sisal, premised on ongoing forced removals, the destruction of subsistence economies and attendant social structures as well as a system of forced labor. Sisal production needs large swathes of land because 100 tons of leaves only yields about 3 tons of marketable fiber and production is year-round, making it a labor-intensive crop. After World War I the British colonial state took possession of most of the territories that once comprised German East Africa. Kew took over the Amani Institute (East African Agricultural Research Station) and its new director noted that "not much of the equipment taken over was serviceable, but we inherited a valuable library, a considerable herbarium, and plantations stocked with introduced economic plants."[18] Regarding sisal agave, Nowell complained that there was little to no information on record for planters and that the existent libraries lacked any records on scientific research into the plants.

During our research, Sonjasdotter and I located the papers and notes of Carl Philipp Johann Georg Braun (1870–1935), a botanist at the Amani Institute, in the Botanical Museum, Berlin. Most of it consisted of reams and reams of paper in all sizes bearing numbers and figures for plant sizes, yield, and other variables. His only published scientific text on agave plants appeared in *Der Pflanzer* (*The Planter*) in 1906.[19] Nowell surmised that the dearth of documented research was due to the plant's vast success which rendered any structured inquiry an unnecessary expense. What he did however find hidden underneath shrubs and trees in an abandoned section of the Institute was a new species of agave plant, a species which proved to be far superior in yield and fiber quality, the *Agave amaniensis*.[20]

Agaves, like ferns, are ingenious at reproduction. They have, as far as we know, three ways of doing it: So-called *hijuelos*, little clones, spring from the plant's root system and grow at the base of the original plant, gradually supplanting it. This can happen throughout the plant's life cycle, unlike the two other techniques of reproduction. Between four to eight years after planting, a central flower stalk will appear which also initiates the plant's demise, as it only flowers once in its lifetime. The stalk can reach a height of six meters and bears yellow flowers emitting an unpleasant odor. It is the source for another clonal method, through the bulbils that grow off the top and fall to the ground, rooting themselves wherever they fall. The third technique of reproduction works through the seeds sprouted by the flowers and constitutes the only way to produce genetically distinct plants, thereby ensuring plant diversity. So while the origin of A. *amaniensis* "seems likely to remain obscure," it was probably through seeds, carried by winds or birds or others, that it made its way to Amani.[21]

There are many different ways to conclude this story: detailing the aftermath of World War I on German colonial botany, its endings and continuations in the devastated ecologies of Eastern Africa; or noting the rise of synthetic fibers and the subsequent demise of the sisal industry; or describing the installation Sonjasdotter and I built showing the sisal plant's implications in German colonialism. But rather than conclusions, I am interested in returns and returnings and, of course, seeds. Seeds are the plant's dispersal unit. They have to be canny travelers, "windborne, animalborne, waterborne," and at times carried by human hands.[22] Seeds can sustain long, some of them extremely long, periods of unfavorable conditions in a state of suspended animation or dormancy, something they can enter into even well into maturation. They spring back to life when the time and place are just right. In fact, seeds in repositories like the Svalbard's Doomsday Vault last an average of five years before having to be renewed to ensure germination viability. And then there are so-called recalcitrant seeds which do not survive the drying and freezing of ex-situ conservation in seed collections.

These latent possibilities contain many moves and turns and returns and suggest how seeds can demand as well as defy planning. While the colonial state apparatus was primarily interested in cultivating cash crops, many of the botanists were more concerned with new species of plants, inundating, as Kaiser noted, the Botanical Museum with unknown specimens.[23] This excess is similarly evident in the overstuffed collections of the major natural history museums, leading some commentators to theorize that a considerable amount of the world's "undiscovered" flora and fauna may actually be found in museum collections. In this, the museum itself returns somewhat to its roots as the central holding place for the stupendous ignorance of imperial formations that stuffed anything it could not make sense of into the shelves and drawers of their many archives. The idea that these materials would quietly submit to such enforced stillness adds another dimension to this ignorance. Although claims for repatriations/rematriations, restitutions, and returns of artifacts and specimens are proliferating, and are, gradually, being honored, the belief that proper preservation and conservation would arrest any unwanted moves remains more intact than ever.

As a kind of counter-clock, the seed makes for a shifty relic that slips in and out of histories, presents, and futures. Rather than a flawless circular move, it might better be described as drawing (on) an elliptical trajectory that sometimes just falls short of its intended objective or that complicates the plan by rendering visible omitted relations and uninvited protagonists. "Elliptical" here refers to both ellipse and ellipsis (. . .), the elision or lapse of words or sentences indicating silence, anticipation, uncertainty or an unspeakable interjection. In reconstructing the doubtful travels of the sisal plant, in and out of planned and unplanned times and spaces, we have to come to terms with their . . . well, seedy temporalities which exceed the sensible possibilities of planning as well as narrative. One important response to this is not letting go (into history) the injuries and inconsistencies, appropriations, and exterminations that have given rise and continue to take effect on historical connections shaping our institutions, disciplines, and futures. Examining how plans and plants coincide lets us draw together the plantation and the museum, botany and empire, and colonial and post-colonial (after-) lives as mutually constitutive orders. In this, shifty relics like seeds afford *"re*turns [that] are products of repetition, of coming back to persistent troublings; they are turnings over. In such re-turnings, there is no singular or unified progressive history or approach to discover."[24] I am now in the process of getting the sisal agave seeds, which were displayed in the exhibition, accessed into the Mai Collection and the museum's catalogue, including the elliptical narrative of their historical travels told above. Planting seeds and their stories might be one way of un-finishing the plan.

Notes

1. For more details, see Tahani Nadim, "Haunting Seedy Connections," in *Routledge Handbook of Interdisciplinary Research Methods* ed. Celia Lury et al. (New York: Routledge, 2018).

2. *Tote Wespen fliegen länger* [Dead wasps fly further], March 1–April 2, 2015, Museum für Naturkunde Berlin.

3. Opened in 1889 by the Emperor Wilhelm II, the Museum united three distinct museums that had been established alongside the Frederick William University (now Humboldt University of Berlin) in 1810: the Museum of Anatomy and Zootomy, the Museum of Mineralogy, and the Museum of Zoology.

4. These exploits comprise expeditions prior to the German Reich's acquisition of colonies (protectorates) such as Friedrich Sellow's work in Brazil 1815–1831, its undertakings in its own protectorates as well as collection activities during the GDR in the colonial empire of the Soviet Union.

5. Amitav Ghosh, *The Great Derangement: Climate Change and the Unthinkable, The Randy L. and Melvin R. Berlin Family Lectures* (Chicago: University of Chicago Press, 2016).

6. Londa L. Schiebinger and Claudia Swan, eds., *Colonial Botany: Science, Commerce, and Politics in the Early Modern World* (Philadelphia: University of Pennsylvania Press, 2005). Londa L. Schiebinger, *Plants and Empire: Colonial Bioprospecting in the Atlantic World* (Cambridge, MA.: Harvard University Press, 2004).

7. Carl von Linné, *Hortus Cliffortianus* (Amsterdam: [publisher not identified], 1737), quoted in Staffan Müller-Wille, *Botanik und weltweiter Handel: Zur Begründung eines natürlichen Systems der Pflanzen durch Carl von Linné (1707–78)* (Berlin: Verlag für Wissenschaft und Bildung, 1999).

8. Müller-Wille, *Botanik und weltweiter Handel*, 316.

9. Adolfo Mascarenhas, "Resistance and Change in the Sisal Plantation System of Tanzania" (PhD diss., University of California, 1970).

10. Jack Ralph Kloppenburg, *Seeds and Sovereignty: The Use and Control of Plant Genetic Resources* (Durham, NC: Duke University Press, 1988), 59.

11. Edward A. Gargan, "International Report; Tanzania's 'Green Gold' Woes," *New York Times*, June 23, 1986.

12. Gwynneth Latham and Michael C Latham, *Kilimanjaro Tales: The Saga of a Medical Family in Africa* (London: The Radcliffe Press, 1995).

13. Robert W. Pemberton and Hong Liu, "Marketing Time Predicts Naturalization of Horticultural Plants," *Ecology* 90, no. 1 (2009): 69–80, on 70.

14. Walter Rodney, "Migrant Labour and the Colonial Economy," in *Migrant Labour in Tanzania during the Colonial Period: Case Studies of Recruitment and Conditions of Labour in the Sisal Industry*, ed. Walter Rodney, Kapepwa Tambila, and Laurent Sago (Hamburg: Institut für Afrika-Kunde, 1983), 7.

15. Nina Berman, Klaus Mühlhahn, and Patrice Nganang, eds., *German Colonialism Revisited: African, Asian, and Oceanic Experiences* (Ann Arbor: University of Michigan Press, 2014).

16. The Zentralstelle was founded in 1891 and based at the Botanical Gardens in Berlin. It was tasked with researching, obtaining, and distributing plants and plant knowledge valuable for the colonial project. For an excellent account of the Botanic Garden's colonial pursuits, see the work of Katja Kaiser, especially "Exploration and Exploitation: German Colonial Botany at the Botanic Garden and Botanical Museum Berlin," in *Sites of Imperial Memory: Commemorating Colonial Rule in the Nineteenth and Twentieth Centuries*, eds. Dominik Geppert and Frank Lorenz Müller (Manchester: Manchester University Press, 2015), 225–42.

17. Adolf Engler, "Das Biologisch-landwirtschaftliche Institut zu Amani in Ost-Usambara," *Notizblatt des königlichen Botanischen Gartens und Museums zu Berlin* 4, no. 31 (1903): 63–66.

18. William Nowell, "Supplement: The Agricultural Research Station at Amani," *Journal of the Royal African Society* 33, no. 131 (1934): 1–20, no 3.

19. Carl P. J. G. Braun, "Die Agaven, ihre Kultur und Verwendung: Mit besonderer Berücksichtigung von Agave Rigida Var. Sisalana Engelm," *Der Pflanzer* 14, no. 2 (1906): 209–24.

20. William Nowell, "Agave Amaniensis: A New Form of Fibre-Producing Agave from Amani," *Bulletin of Miscellaneous Information, Kew*, no. 10 (1933): 465–67.

21. William Nowell, "Agave Amaniensis," 466.

22. Octavia E. Butler, *Parable of the Sower* (New York: Warner Books, 1995), 68. In this sci-fi book set in a dystopian California of 2024, the narrator and "hyperempath" Lauren Olamina is moved to invent (or discover) a new religion, or rather ethics of becoming, which she names "Earthseed":

> Well, today, I found the name, found it while I was weeding the back garden and thinking about the way plants seed themselves, windborne, animalborne, waterborne, far from their parent plants. They have no ability at all to travel great distances under their own power, and yet, they do travel. When they don't have to just sit in one place and wait to be wiped out. There are islands thousands of miles from anywhere—the Hawaiian Islands, for example, and Easter Island—where plants seeded themselves and grew long before any humans arrived. Earthseed. I am Earthseed. Anyone can be.

23. Katja Kaiser, *Wirtschaft, Wissenschaft und Weltgeltung: Die Botanische Zentralstelle für die deutschen Kolonien am Botanischen Garten und Museum Berlin (1891–1920)* (Vienna: Peter Lang, 2021).

24. Christina Hughes and Celia Lury, "Re-Turning Feminist Methodologies: From a Social to an Ecological Epistemology," *Gender and Education* 25, no. 6 (2013): 786–99, on 787.

Steel Plant: Orissa State, India, 1955
Itty Abraham[1]

"One sees Christianized *Adivasi* (tribal) girls stepping out of German cars and station wagons, dressed in silks and georgettes, with flashing lipsticks and wearing high heeled shoes . . . already quite a number of fair-skinned Christian births have taken place and by the time the plant is completed—if at all is completed in time—Rourkela will have a thriving little community of Indo-German bastards."
— FILM INDIA, JUNE 1958[2]

"Virtually every one of these girls came from surrounding *Adivasi* villages; most of them were Christians and some had attended the nearby Harimpur Mission School. It is not difficult to imagine how it was that many of the girls originally engaged as house servants gradually came to be employed 'for other purposes' by the single German men for whom they worked; . . . in many Adivasi villages and families the girls were the only breadwinners. Subsequently, when there was an opportunity of sending the girls into service for considerably more money, this opportunity was also grasped."
— JAN BODO SPERLING, TECHNICAL ADVISOR TO THE GERMAN MISSION IN ROURKELA[3]

Not for the first time, the bodies of tribal women become the terrain on which larger struggles are waged. In this case, the apparent conflict was between the cultural norms and social mores proclaimed by mainstream Indian bourgeois society, with more than a hint of lasciviousness at the prospect of inter-racial unions, and the "opportunities" provided by single German men employed in the Rourkela steel mill. Beyond the ruptures produced by this intersection of race, geopolitics, and sexual desire, the narrative is complicated further by the travails of impoverished and displaced indigenous families, mediated by the discipline imparted in Christian mission schools, where obedience and compliance are valued above all else.

No one knows exactly how many Adivasis were displaced as a result of the establishment of the Hindustan Steel Plant at Rourkela in 1955, a landmark joint project of the Indian and West German governments during the Cold War. Contemporary official Orissa State Gazettes report that initially fifty-three and later eighty-two villages, covering an area of forty-three thousand acres and nineteen thousand acres, respectively, were handed over to the Hindustan Steel Plant. A much later (non-governmental) report notes that 33 villages comprising 13,000 people were evicted from land holdings of twenty thousand acres in order to make way for the plant.[4] Notwithstanding this numerical uncertainty, it is telling that official state figures do not identify how many people—families and individuals—were removed and displaced, but rather concentrate on the number of settlements destroyed and the extent of land alienated. This omission is not due to bureaucratic carelessness; it is an immanent condition of modern planning.

Planning: Past, Present, Futures

A Plan is a particular kind of text devoted to future-talk. In general, regardless of whether a Plan begins from rupture or is projected as a linear extension of the present, its foundations are a radical simplification of a complex and uneven reality. In the Indian context, wherever the indigenous were to be found, the Plan began from the ready assumption of *terra nullius*. Entirely in consonance with the imaginations of European explorers and conquerors of Australia and the Americas, the actions of the Indian state, colonial and postcolonial, treated the presence of the indigenous as at best a temporary hindrance. The displacement of people and territorialization of indigenous bodies referenced in the epigraph are violent symptoms; the core assumption would prevail across huge swathes of mineral and resource-rich central and northeast India and become the hallmark of the colonization and settling of the Andaman and Nicobar archipelago.[5] Once removed from their ancestral lands, tens of thousands of displaced Adivasis became footloose casual labor, alienated from home and history, available to be employed and exploited in factory and household depending on gender, abilities, and demand. When the Plan imagined the future, it did not—could not—contemplate that this time yet-to-come would erase the enshrined hierarchies of the present. Far from it—as numerous accounts of Rourkela make clear, difference would be further consolidated through the installation of social and spatial hierarchies built around status, professional standing, technical expertise, gender, and, not least, race.[6]

Colonial and Postcolonial Spaces

The segregation and hierarchical ordering of space is a characteristic feature of colonial rule. Especially following the 1857 uprising and the imperial takeover of Indian political life, the spatial separation of "native subjects" and the British became increasingly institutionalized in the interests of racial order and colonial security. Racial difference drew a hard boundary between prevalent legal orders within the colony and between the colony and the metropolis,[7] while language formed its own spaces.[8] Urban built environments were divided between spaces restricted to Europeans, including Civil Lines, social clubs, and access to modern sanitation, and Black Towns, overpopulated by indigenes divided by class, religion, gender, and caste. Biomedical asylums separated the sick and the insane from the ostensibly normal, military cantonments and cemeteries were ordered by rank and race, Inner and Excluded Areas were kept apart through the invocation of ethnicity and civilizational difference.[9] And on.

Colonial efforts to segregate and hierarchically order space are usefully contrasted against projects seeking to produce a distinctly vernacular modern, especially in progressive princely states such as Baroda, Travancore, and especially, Mysore. Janaki Nair's seminal *Mysore Modern* is a case in point, the account of a semi-sovereign state that was the "first princely state to use electric power to illuminate cities, found a state bank, start a university, found a chamber of commerce, initiate a program of reservation for Backward Classes, set up a Serum Institute, found birth control clinics for the general public, send a trade commissioner to London, and run an administrative training institute for Indian princes."[10] These two tendencies—spatial separation and desired technological modernity—would come together to define the postcolonial Indian state's strategy to overcome its foundational crisis of legitimacy.[11]

Technological Modernity

The Industrial Policy Resolution of 1948 is considered the Ur-text of the early Indian postcolonial state's strategy for economic development. The Resolution monopolized the "commanding heights" of the economy, namely, heavy industry, defense, minerals, and core infrastructure sectors for itself, while permitting private investment in small and medium size industries in the services and light manufacturing sectors.[12] In carrying out the vision entailed by the Resolution, a particular expression of postcolonial space was imagined and reproduced across the country. Integrated dam projects, steel mills, iron ore and coal mining enterprises, locomotive and aircraft factories, educational institutions, new urban centers—the list goes on—were set up, more often than not in rural and economically underdeveloped communities and areas dominated by tribal populations. Names of hitherto unknown places such as Rourkela, Bokaro, Bhilai, Durgapur, and Koraput entered the national imaginary, with Bhakra-Nangal, a massive integrated dam, hydro-electric, and irrigation project, and Chandigarh, a new city designed by the Swiss architect and urban planner Le Corbusier, standing in as prime metonyms for these massive state projects of national transformation.

Postcolonial future-making. To enter these urban visions was to enter a future space: an India that was yet-to-be. A road network arranged on a grid, smooth flows of traffic, clean sidewalks, hierarchically ordered housing complexes separated by urban parks, functioning water and electricity infrastructures, and the absence of visible poverty struck early visitors to new townships as contemporary visions of utopia. What middle-class visitors saw was one thing. What they did not see is even more telling. It begins with the missing past and its absent people, but didn't stop there. What was also absent from everyday accounts of these futuristic townships was the close collaboration of foreign partners in almost every one of these hyper-national projects.

Geopolitical Imaginaries

These spaces of future India were also nodal sites in the negotiation of external relations. Both the Soviets and the West saw in India's development plans a chance to instantiate their own claims to global supremacy during the Cold War. Helping build a new India through their patented designs would become concrete proof of the virtues of close association with the Eastern and Western blocs. The Soviet Union would offer its much-vaunted steel technology to Bokaro and Bhilai, the British would help build Durgapur, the West Germans, Rourkela. Even the most original and novel of postcolonial projects—atomic power—would be less than completely national in the sense of being indigenously produced and maintained. This would not have been an issue for most developing countries who would have been delighted to get any aid at all, with or without strings. But for postcolonial India, which had defined its international identity in terms of non-alignment with the major power blocs and insisted on the powerful rhetorical slogan of self-reliance, the presence of foreign support for practically all its flagship endeavors was a mark of ambivalence that could not be so easily overcome.

Visions of Planned Modernity

Private "company towns"—including the steel towns of Jamshedpur and Burnpur (Asansol), tea and coffee plantations in the Niligiris, and the Digboi petroleum complex in upper Assam—had long existed but state planning took the scale and extent of infrastructural investment to new heights. New enterprises were set up based on a calculus that combined economic logic with dominant interests. Once a site had been chosen, entire villages were compulsorily acquired, their inhabitants relocated, space cleared, and the project begun. The larger political significance of these projects was explicit. Srirupa Roy explains, "Nehru wanted the steel plants to be special places, inhabited by special people . . . developing these industrial townships or 'steel towns' offered the nation-state an opportunity to realize its vision from scratch."[13] Missing pasts, places, and communities would be further translated into the idiom of a necessary sacrifice in the interests of a common national future.

Opening the massive Bhakra project, Prime Minister Nehru invoked both sacrifice and progress in his imagination of a national future: "Which place can be greater than this, this Bhakra Nangal, where thousands and lakhs of men have worked, have shed their blood and sweat and laid down their lives as well[?] ... I look far, not only towards Bhakra Nangal but towards this our country, India, whose children we are ... The work of a country is never completed. It goes on and no one can arrest its progress—the progress of a living nation."[14] In retrospect, it is easy to see why these huge integrated projects were called "temples of modernity." Yet sacrifice and progress meant the violent erasure of prior existence—people, livelihoods, lands, and memories—in order to set into motion this national vision of the future. Such a vision was predicated on prior absence, an absence that would have to be created if it did not already exist. *Terra nullius*.

New postcolonial spaces were material negotiations seeking to overcome the ambivalences of past and future, domestic and foreign. Embedded in the linear topographies of the townships was a disavowal of histories of past communities, lives, and livelihoods, rewritten into a geography of unoccupied and pristine lands. Equally denied was the presence of the foreign, whether in the form of personnel, capital, design, or expertise, at best glossed as encounters with a global modern expressed in local idiom. But the spectral presences of the past and the foreign could not be erased so easily. To simulate an originary absence and indigenously defined future, violence became an inescapable and necessary supplement to the territorial logic of power.

Territory and State Power

Political geographers have long argued that territoriality has powerful political effects that are often obscured due to the normalization of the idea of territorial control as a necessary precondition for state existence.[15] Paul Alliès notes that territory offers the state "a physical basis which seems to make it inevitable and eternal." Henri Lefèbvre puts it as follows: "Each State claims to produce a space wherein something is accomplished—a space even, where something is brought to perfection: namely, a unified and hence homogenous society."[16] Bringing Lefebvre and Alliès together in a reading of state formation and consolidation makes it clear that control over space plays a vital role in the development of state legitimacy. The political power of the state begins from the "official fiction" that it always has been and is the only legitimate organizer of political life within a particular place. The state produces this illusion through a spatial erasure of the past, a violence expressed as the "territorial logic of power," in order to establish its preferred and dominant relation to society.[17]

The early postcolonial state suffered a number of foundational handicaps that made the easy production of a "unified and hence homogenous society" next to impossible.[18] First and most obvious was the creation of the new state, unmistakably marked by the moment of sovereign independence. More telling however was another kind of timing. To come into being at a moment when a thick international space defined dominant international norms meant that hegemonic meanings of "something brought to perfection," were already well established, especially in relation to economic development (constituting at the very least capital investment, poverty alleviation, improvement of public health, creation of modern infrastructure and institutions). The Indian state had little flexibility in meeting the standards of perfection now expected of it. Structural conditions of underdevelopment would long outlast the immediate (and serious) political crises facing the newly independent state including severe food shortages, secessionist tendencies, rural uprisings, and the need to give asylum to millions of homeless and forced migrants.

Planning the Future

Seen from this perspective, the Plan was overdetermined. The ideological demands of state legitimacy amidst extraordinary conditions of violence, displacement, and impoverishment made planning a necessary and responsible action of a sovereign government. During the early years of the Cold War, however, sovereignty and development were always going to be incomplete projects; not surprisingly, the Plan was a text that reflected those absences, even if it could not always acknowledge them explicitly. Hence, if the Plan was overdetermined from one standpoint, it was also underspecified in critical ways. These immanent contradictions of the Plan led to a distinct mode of displacement, namely, the substitution of time with space.

Aspirations to modernity—expressed as material embodiments of the future—would be written into the spatial present in order for the state to claim success in its project of postcolonial transformation. The future was manifested and exemplified by hyper-modern places carved into spaces deemed empty, areas of apparent backwardness inhabited primarily by the indigenous, producing the doubled condition that defined urbanity in the enchanted palaces of Durgapur, Bhilai, Bokaro, and Rourkela. The success of these new cities was built on the unequal sacrifices made by those least able to articulate resistance and who suffered violence, immediate and structural, then and now. The greatest irony is that it was only when female bodies bedecked in "georgettes, flashing lipsticks, and high heels" stepped in and out of foreign cars did the absent tribal become visible to mainstream eyes. This was not a humanitarian vision; it sought to arrest escape and to re-territorialize indigenous subjects into more familiar postures of deference and compliance.

Notes

1. Thanks to the organizers of the workshop for their invitation to present my work and to Miriam Jaehn for her help with translation.

2. Quoted in Jan Bodo Sperling, *The Human Dimension of Technical Assistance: The German Experience at Rourkela, India*, trans. Gerald Onn (Ithaca: Cornell University Press, 1969): 92–93.

3. Sperling, *The Human Dimesion*, 87.

4. *Sarini* and Adivasi-Koordination, "Adivasis of Rourkela: Looking Back on 50 Years of Indo-German Cooperation: Documents, Interpretations, International Law," (Bhubaneshwar: CEDEC, 2006). In later documents the number of displaced is estimated to be around 2000 households.

5. Itty Abraham, *How India Became Territorial: Foreign Policy, Diaspora, Geopolitics* (Stanford: Stanford University Press, 2014); Itty Abraham, "The Andamans as a 'Sea of Islands': Reconnecting Old Geographies through Poaching," *Inter-Asia Cultural Studies* 19, no. 1 (2018): 1–20.

6. Corinne M. Unger, "Rourkela, ein 'Stahlwerk im Dschungel': Industrialisierung, Modernisierung und Entwicklungshilfe im Kontext von Dekolonisation und Kaltem Krieg (1950–1970)," *Archiv für Sozialgeschichte* 48 (2008): 367–88.

7. Uday Singh Mehta, *Liberalism and Empire: A Study in Nineteenth-Century British Liberal Thought* (Chicago: University of Chicago Press, 1999); Nasser Hussain, *The Jurisprudence of Emergency: Colonialism and the Rule of Law* (Ann Arbor: University of Michigan Press, 2003).

8. Farina Mir, *The Social Space of Language* (Berkeley: University of California Press, 2010).

9. David Arnold, *Colonizing the Body: State Medicine and Epidemic Disease in Nineteenth-Century India* (Berkeley: University of California Press, 1993); Rajnarayan Chandavarkar, *Imperial Power and Popular Politics: Class, Resistance and the State in India, 1850–1950* (Cambridge: Cambridge University Press, 1998); Prashant Kidambi, "'The Ultimate Masters of the City': Police, Public Order and the Poor in Colonial Bombay, c. 1893–1914," *Crime, Histoire & Sociétés / Crime, History & Societies* 8, no. 1 (2004): 27–47; Duncan McDuie-Ra, "Fifty-Year Disturbance: The Armed Forces Special Powers Act and Exceptionalism in a South Asian Periphery," *Contemporary South Asia* 17, no. 3 (2009): 255–70.

10. Janaki Nair, *Mysore Modern: Rethinking the Region Under Princely Rule* (Hyderabad: Orient Black Swan, 2012), 17.

11. Hamza Alavi, "The State in Post-Colonial Societies: Pakistan and Bangladesh," *New Left Review* 74 (July 1972): 59–81; Pranab K. Bardhan, *The Political Economy of Development in India* (Delhi: Oxford University Press, 1998).

12. Vivek Chibber, *Locked in Place: State-Building and Late Industrialization in India* (Princeton: Princeton University Press, 2006).

13. Srirupa Roy, *Beyond Belief: India and the Politics of Postcolonial Nationalism* (Durham, NC: Duke University Press, 2007), 134.

14. Satish Deshpande, *Contemporary India: A Sociological View* (Delhi: Penguin Books, 2004), 67–68.

15. John A. Agnew and Stuart Corbridge, *Mastering Space: Hegemony, Territory and International Political Economy* (London: Routledge, 1995).

16. Henri Lefèbvre, *The Production of Space*, trans. Donald Nicholson-Smith (New York: Wiley-Blackwell, 1991), 281.

17. David Harvey, *The New Imperialism* (Oxford: Oxford University Press, 2003).

18. Lefèbvre, *The Production of Space*.

Surnames: Brazil, 1979
Ana Carolina Vimieiro Gomes[1]

In a 1979 research report that Eliane Azevêdo prepared for the Organization of American States, an image of a Brazilian Bahian woman, a woman of Afro-Brazilian origins, was captioned: "Baiana (woman from Bahia), with typical costume, begging to celebrate her Saint which she carries with flowers and a white towel. Her ancestors' genetic and culture inheritance merge in her, creating a mixed way of being and believing: in genes forming her body and in religious syncretism feeding her soul."[2] The woman in the image is wearing the costume in which she performs her religious rituals. For the geneticist author of the report, the costume announces her ethnic identity.

The Bahian Brazilian geneticist Eliane Azevêdo was reporting her studies on population genetics that had been financed by the Organization of American States; it offered information on "anthropogenetic characteristics of the Bahian population."[3] The representation encapsulates the method of Azevedo's studies and ties to plans to categorize the genetic features of some populations in Brazil. What mediates in connecting costume and genetics is speculation on the racial and cultural origins of the subject's surname.

Since the first half of the twentieth century, the state of Bahia has been romantically and ideally deemed the cradle of Brazilian national identity and, simultaneously, an embarrassing symbol of the backwardness of Northeast region of Brazil.[4] This image of a Bahian woman represents the creative cultural construction behind the social imaginary around the population of this region of Brazil: biological and cultural African heritage and biological and cultural miscegenation, a construction in which scientific discourses and practices played an important role. So, from the perspective of science, the photo also aims to give evidence to her body, marked by "racial" traits and her ancestors' biological features. Translating this into scientific terms: the "Baiana" hides her genes, i.e., the genetic inheritance from black, white, and indigenous populations merging in her. In a wider understanding, the "Baiana" was a symbol of the cultural and biological miscegenated character of the population of Northeastern part of Brazil. One of these cultural aspects then used by the geneticist Eliane Azevedo was mapping the pattern of surname adoption to understand Northeast (focusing on Bahian) population genetic composition.

Surnames were used by geneticists because, according to their reasoning, in some cultures they were inherited like genes. They were deemed easy to sample and could be collected in both historical and present-day populations. According to the language of population genetics, they could thus serve as a trace for kinship, inbreeding, and relationship among several populations (within and among communities), for instance. Eliana Azevêdo's investigations on genetics and Northeast population started in mid-1960s during her PhD training at the University of Hawaii, supervised by the famous US geneticist Newton Morton. After her return to Brazil in the late 1960s, Eliana Azevêdo became professor at the Federal University of Bahia, where she carried on working on surnames and genes at Laboratory of Medical Genetics, financed by OAS.

Studies on the Bahian population were notably part of a wider agenda on the genetics of the Northeast population headed by the US geneticist Newton Morton in the 1960s. This scientific project was one of the branches of population genetic studies in Brazil, planned by US geneticists and devoted to the study of "primitive" peoples (as a biological and demographic phenomena); such as the one dedicated to Amazonian isolated indigenous populations, which was carried out by another renowned US geneticist, James Neel, as part of the WHO International Biological Program.[5] Brazil was viewed as a central place for global theoretical discussions on the micro-evolutionary process of the human species and for the categorization of populations' genetic composition. Brazil's Northeastern population was also framed as primitive (pre-industrial population) and deemed a valuable object for studying human genetic adaptability and variability.[6] Both Neel and Morton's presence in Brazil was enabled by the collaboration of local geneticists and the support of US public funding agencies, such as the NIH and the US Public Health Service. This scientific agenda could be thus strongly tied to broader US Cold War health and scientific programs to Third World countries.[7]

This essay provides an overview of scientists' plans to categorize Brazilian biological features, by focusing on genetic studies that were carried out on Brazil's Northeastern population by US and Brazilian geneticists from the 1960s to the 1980s. Their research and planning involved practices of quantifying, categorizing, normalizing, correlating data, medicalizing, and, especially, geneticizing population features. Nevertheless, these scientists were not able to focus only on their planned genetic approaches; instead, the scientists' research practices became enmeshed with local contemporary social thinking on race and cultural perceptions that shape representations of Brazilian people. The example of the genetic studies carried out on Brazil's Northeast population will help to shed some light on Brazilian scientists' capacity to creatively invent their own approaches and build narratives and scientific explanations for the heterogeneous Brazilian biological features. An exceptional characteristic of these narratives and scientific explanations is the way in which they took up local perceptions and interpretations of the population formation in order to reinforce an ideal of "Brazilians" as a miscegenated people.

Contrary to what was imagined before, Eliana Azevêdo found that surnames in the Bahian population did not indicate a common ancestor, nor were they related to kinship. Instead, these surnames were correlated with ethnicity. Mapping family names and their correlation to genes and ancestry was not immediately adaptable to local tradition of family name adoption.[8] As a solution, surnames were thus sorted by race. The frequency of peoples' surnames was thus interpreted according to family name adoption throughout Brazilian history, especially in Black populations: after the abolition of slavery, the adoption of a master's family name or devotional names (names of saints, religious ceremonies and festivities). Eighteenth- and nineteenth-century manumission documents, checked by Eliana Azevêdo, revealed the two most common procedures of surname adoption: some freed slaves remained without a surname, and for those who acquired a surname, the preferred method was adopting a devotional name, such as Santana, Jesus, Nascimento, Conceição, etc. Following this procedure, racial genetic markers were correlated to this local culture of surname adoption.[9]

Eliana Azevêdo and collaborators found that devotional surnames increase in frequency with Black admixture, especially in women. People with a white phenotype and a devotional surname were more prone to Black admixture, Azevêdo remarked. A cultural index was later created as another way to give order to their data, by grouping people according to the following criterion: the frequency of the type of surname, i.e., with religious, animal, or plant connotations; resulting in the black or indigenous cultural index. Still grounded in racialists' perspectives despite post-1950s genetic approaches, geneticists considered surnames as racial markers permitting the mapping of the genetic origins and cultural traditions of some population groups of northeast Brazil. Geneticists even concluded that the surname was a more powerful indicator of racial ancestries than ABO blood groups or phenotype classifications.[10] The results of surname research in Bahian populations reinforced the arguments of the mixed character of that population, and reflect the racialist point of view held by these genetic researchers.

I will now offer a broader contextualization of the Brazilian debates on the country's biological identity. Since the nineteenth century, the identity of Brazilian people was intertwined with discussions on human biological diversity, either through medical, racial, or anthropological perspectives.

National identity was a central issue in the thinking of intellectuals, scientists, and institutions, who envisioned the past, present, and future of the nation, having in mind the peculiar formation of its population: composed by the mixture of black, white, and indigenous people. The heterogeneity and plurality of the Brazilian population composition was always posing challenges to the scientists' impetus for categorizing, ordering, and normalizing biological features; all of which have been a central scientific issue for biomedical research on human biological diversity in Brazil until recent years.[11]

While racial miscegenation since the nineteenth century was at first sight an embarrassment to the scientists' efforts to define a homogeneous Brazilian identity, after the 1930s it was positively viewed as one of the most important feature of "Brazilian-ness," as suggested by the contemporary view of anthropological composition, conciliatory racial relations, and cultural mixtures that historically formed the nation; such as in Gilberto Freyre's thinking, and interpretations appeasing Brazil's histories of colonialism and slavery.[12] This positive view of miscegenation (also later called the "myth of racial democracy," including social and cultural mixtures) pervaded the collective imaginary, making Brazil a country that was perceived by the international scientific community as a "racial laboratory." Categorizing surnames in relation to genes and race was one of the ways to make sense of the culturally and biologically mixed composition of the Brazilian population.

Representations and narratives of miscegenation and the formation of the Brazilian were at stake in scientists' interpretations of population genetics. The origins of Brazilian peoples were encompassed by the human contingent formed by the indigenous groups existing before Portuguese colonization, and by the migratory flow not only of Portuguese settlers and African people in the slavery period, but Italians, Germans, Spanish, Turkish, and Arabic people, etc., who had come to work in Brazil since nineteenth century. All of these people, geneticists stated, intercrossed with each other (in variegated intensities) and were viewed as being influenced by different environments, i.e., "from temperate zones of the south to the humid tropical forests and arid regions of the Northeast."[13] Interpretations regarding the history of colonization, settlement, and migratory process, racial anthropological features and racial relations, cultures of weddings, and social development were again used to explain the variability of the genetic formation of Brazilian people. Population genetic data were put into a broad framework related to cultural, anthropologic, and demographic issues of the Brazilian people's development.

The cultural and social realities of Brazil's Northeastern people challenged the international scientific ways of knowing and ways of doing.[14] The main genetic studies were carried out by Newton Morton and his Brazilian collaborators in Northeastern migrants who had come to the city of São Paulo to work in industries or in plantations in the interior of São Paulo state, who, as soon as therein arrived, were accommodated during 24 hours at *Hospedaria de Imigrantes*." The *Hospedaria* was public housing for migrants, created in 1888 and supported by the Department of Immigration and Colonization pertaining to the Secretary of Agriculture, Commerce, and Public Works from the state of São Paulo. It was linked to governmental policies for immigration promotion due to the slave abolition process in late nineteenth century.[15] After the end of big waves of foreign immigration until the mid-twentieth-century, the Brazilian Northeast population then became the main workforce supplying economic activities in São Paulo. Due to the operating difficulties for on-site study of the Northeasterners, the *Hospedaria* offered the unique opportunity to gather a varied and significant number of people to be sampled, and a laboratory was therein installed to carry out the collection of biological material. More than a thousand migrant families, most of them from Bahia and Minas Gerais, were exclusively recruited as "samples" for the proposed genetic studies. They went through racial evaluation (anthropological features) and medical anamnesis. The blood samples collected underwent at least seventeen genetic analyses, investigating blood groups, enzymatic, and proteins electrophoretic analysis. Geneticists created their own methods for racial classification, such as the statistic method to estimate the genetic composition, frequency of genes, of every "racial class" correlated to the use of phenotypic indexes, such as anthropological traits.[16] Eliane Azevêdo's PhD dissertation was then developed based on this data collected in Brazil and later analyzed in the US. The results on kinship were the starting point to her later investigations on surnames and genes.

The genetic composition of the Northeastern population remained in the Brazilian geneticists' scientific agenda until the 1980s and included the consecrated interpretations, in science and in culture, concerning the Northeast identity and its place in national identity: racially mixed, modelled by adverse environment conditions (harsh climate and social underdevelopment), and poverty—which, at the time, referred to rural workers living in preindustrial societies. This perspective was grounded on early twentieth-century considerations on the formation of Northeast population: such as the narratives concerning the *sertanejo* (hinterlands inhabitants), and the history of the colonization and settlement of the *sertão* (hinterlands). Scientists had in mind "sociocultural forces modeling the genetic structure of this population" to understand the "[micro]evolutionary dynamic" and genetic mixture of Northeasterns; what encompassed the history of Northeast colonization, culture of marriages, settlement, migratory patterns, and surname adoption of the different racial groups during this period. Therefore, as we see in the image representing a Bahian woman that opened this essay, genetic markers from population genetics studies were thus correlated to culture and consecrated narratives about the "trihybrid" character of Northeast population.

This essay shows how Brazil's Northeastern population and their regional singularities were represented in science and viewed as a central issue for biomedical studies, as well as regional and nation-building discourses in Brazil. It reveals the multifaceted and contingent efforts involved in the scientific planning of Brazilian bodily identity. It also sheds some light on the biocultural character of miscegenation, by showing its construction as a constituent essence of the country's national identity, i.e., a supposedly "multiracial" way of being a Brazilian. The counterpart of miscegenation, nevertheless, was always a Western (implicit or explicit) plan to reach an ideal of purity, order, and social hierarchy.[17] Emphasis on miscegenation was a local discursive strategy used by local intellectual and scientific elites to give positive meaning to the Brazilian heterogeneous reality and an image of Brazil as a place of racial fraternity—the so-called racial democracy.

The representation of the genetic composition of Brazilians as miscegenated peoples has social implications. At first sight, miscegenation was supposedly an ideological symbol of inclusion, harmony, and diversity, shaping the social imaginary about the country's population, both in Brazil and internationally. However, this representation was also replete with tensions. After the 1950s, this ideal of exceptional "miscegenated nation" and "racial democracy" was discredited by social scientists and social movements as a myth. It was condemned for concealing the social inequalities and racial hierarchies, intolerance, and conflicts in Brazilian society. Even in post-racial and multiculturalist discourses in recent years, the "exceptionality of miscegenation" remained a central issue in biomedical science's explanations of Brazilian body composition. Genomic research on the "genetic origins of Brazilians," i.e., the so-called "molecular picture of Brazil" by mitochondrial DNA and Y chromosome sequencing, repeated the shared perception of admixed biological feature of Brazilian people, thereby reinforcing the enduring plans to fix an image of Brazil as a "multiracial" and miscegenated people.

However, in the social-political arena, the reality is more complex. Those scientific plans to reify the biological miscegenation of Brazilians have been falling apart. On the one hand, in science this genetic approach to admixture was then used to demonstrate the nonexistence of race as a biological reality. In contrast, the same scientific ideal of miscegenation has been condemned as a device that works to conceal racism and its effects in Brazil. As a counterplan, race as a social construction has been mobilized as a political ground for recent social policies and for affirmative action, such as the racial quotas and Afro-Brazilians' claims for identity and right to reparation of historical social inequalities and injustices.[18]

Notes

1. Supported by CNPq, Brazil.
2. Eliana Azevêdo, *Características antropogenéticas da população da Bahia, Brazil*, research report presented to Organization of American States, 1979.
3. Eliana Azevêdo, *Características antropogenéticas*.
4. Anadelia Romo, *Brazil's Living Museum: Race, Reform, and Tradition in Bahia* (Chapel Hill: The University of North Carolina Press, 2010), 1–2.
5. Ricardo Ventura Santos, Susan Lindee, and Vanderlei S. de Souza, "Varieties of the Primitive: Human Biological Diversity Studies in Cold War Brazil (1962–1970)," *American Anthropologist* 116, no. 4 (2014): 723–35, 725.
6. Ricardo Ventura Santos et al. "Varieties of the Primitive," 727–28.
7. Ricardo Ventura Santos et al. "Varieties of the Primitive," 725.
8. José Tavares Netto and Eliana Azevêdo, "Racial Origins and Historical Aspects of Family Names in Bahia, Brazil," *Human Biology* 49, no. 3 (1977): 287–99, 296.
9. José Tavares Netto and Eliana Azevêdo, "Racial Origins and Historical Aspects of Family Names," 298.
10. José Tavares Netto and Eliana Azevêdo, "Family Names and ABO Blood Group Frequencies in a Mixed Population Of Bahia, Brazil," *Human Biology* 50, no. 3 (September 1978): 361–67, on 365.
11. Peter Wade, *Degrees of Mixture, Degrees of Freedom: Genomics, Multiculturalism, and Race in Latin America* (Durham: Duke University Press, 2017), 17.
12. Mônica Grin, "Mito de excepcionalidade? O caso da nação miscigenada brasileira," in *O Brasil em dois tempos: História, pensamento social e tempo presente*, ed. Eliana de Freitas Dutra (Belo Horizonte: Autêntica, 2013), 321–40, 324–30.
13. Francisco Salzano and Newton Freire-Maia, *Populações brasileiras: Aspectos demográficos, genéticos e antropológicos* (São Paulo: Companhia Editora Nacional, 1967), 1.

14. Stanley S. Blake and Stanley E. Blake, "The Medicalization of Nordestinos: Public Health and Regional Identity in Northeastern Brazil, 1889–1930," *The Americas* 60, no. 2 (2003): 217–48.

15. Ellis Island in New York, for instance, was only created four years later, in 1892. The Hospedaria hosted and supported migrants until 1978. See Odair da Cruz Paiva and Soraya Moura, *Hospedaria de Imigrantes de São Paulo* (São Paulo: Paz e Terra, 2008), 14.

16. Newton E. Morton, "Genetic Studies of Northeastern Brazil," *Cold Spring Harbor Symposia on Quantitative Biology* 29 (1964): 69–80.

17. Wade, *Degrees of Mixture*, 2–3.

18. Michael Kent and Peter Wade, "Genetics against Race: Science, Politics and Affirmative Action in Brazil," *Social Studies of Science* 45, no. 6 (2015): 816–38, 822.

Taxonomer: United States of America, 1923
Laura J. Mitchell

He gives the impression of looking right at you, though his eyes are shadowed under his brow ridge so they're actually hard to see. Perhaps it is the stern set of his jaw, or the imposing bulk of his shoulders that makes *The Old Man of Mikeno* appear to surveil the library reading room at the Field Museum in Chicago. He seems out of place, but I found his presence comforting; he looked over my shoulder as I read correspondence from and about the artist who created this bronze bust of an adult male gorilla. The sculpture itself is disquieting, despite its beauty and technical accomplishment. We expect busts to be of humans, and a bust in the study area of a natural history museum to be of Linnaeus, or Darwin, or even of Carl Akeley, the bust's creator, who was also a noted naturalist, taxidermist, and inventor whose work changed museum displays in the early twentieth century. This celebratory art form is typically used to commemorate great men and worthy women, not wild animals. So *The Old Man of Mikeno* disrupts our presumptions about the ordering and representation of knowledge in the Western academy, bringing unspoken assumptions to the surface.[1]

Taxonomy—the science of bringing order—has been central to projects of planning and colonization. It is embedded in modernity, visible as a historical intertwining of taxonomy, ordering, and control exercised on people, nature, and space. Taxonomy is also evident in Akeley's professional practice. Because of how scholars order knowledge today, it matters that we understand how Akeley's career was entangled in African ecosystems at a pivotal period in colonial history. Imperial planning for modernization required ordering messy realities on the ground. His story illuminates intellectual foundations of that project.

Akeley's historical notoriety lies in his drive to preserve archetypal mammal specimens through taxidermy and place these mounts in realistic habitat dioramas. He is credited with revolutionizing taxidermy and museum displays by creating realistic three-dimensional representations of flora and fauna in multimedia tableaux.[2] But this was not Akeley's only intervention in knowledge production and the ordering of power. Alongside this pioneering taxidermic work, Akeley also made bronze sculptures of both humans and animals that attest to his artistic accomplishment. His bust of an individualized, named, and solitary animal, however, defies clean categorization. Unlike other examples of Akeley's sculpture, such as a representation of lion hunters with their quarry or a group of elephants, *The Old Man of Mikeno* bends genre.[3] The detail in the sculpture evokes a species type-specimen, but the body is partial and is not the remnants of a once-living animal preserved for scientific study. The subject is a wild animal, but it is not situated in its habitat so it misses the didactic mark intended by most displays in a natural history museum. The generic form of a bust also seems intentional in its challenge to the boundary between human an animal. The existence of a bust—the iconic invocation of an individual of significance—depicting a gorilla "specimen" challenges the taxonomic ordering on which museums, colonial power, and western knowledge is based. This sculpture, then, speaks directly to the problems of classification in scholarship about the colonial world, and to elaborate plans diverted.

Taxonomy was a mainstay of early natural history and the foundation of today's biological sciences. The classificatory impulse inherent in Western attempts to order the natural world expanded in colonial contexts to include

Figure 9. The Old Man of Mikeno, Carl Akeley, bronze, 1923. Field Museum, Chicago. CSZ6288

explicit social, economic, and political categories of rule, leading to the intellectual and material ordering of human communities alongside ecosystems and resources. This kind of categorization was imperative for colonial planning. But as with the naming of species, colonial socio-political classifications were—and remain—open to reinterpretation, a process of ongoing change that highlights the challenges of attaching the right label in the first place. Colonial taxonomy is further complicated by human behavior that elides or transcends categories. Despite our best efforts at making orderly sense of the world, neither species nor human behavior are neatly categorical. Categories persist, though because we need some organizing principles to structure knowledge production and support planning efforts. Taxonomies are unavoidable, but we need to remember that the categories we create for this work can occlude as much as they reveal.

As an example, consider what's illuminated and what's obscured when we apply categories to the professional life of Carl Akeley. He was an artist, technician, inventor, scientist, and an avid exploiter of resources both natural and human—who ended his life as an effective spokesman for nature conservation. Akeley's work may have served the taxonomic project of early twentieth-century zoology, but he defies easy categorization. Considering him only as an early biological researcher overlooks his many technological innovations (he registered over 30 patents, including a device for spraying concrete, and a motion picture camera designed to capture fast-moving animals). Attempting a singular classification for Akeley points out the pitfalls of taxonomies, ignoring the complex ways in which living beings repeatedly altered or thwarted plans born from popular stereotypes, sponsored research, and colonial bureaucracies.

Granted, Akeley's zoological research and museum displays were predicated on taxonomy. But the hunting, fundraising, and creative work that enabled elaborate taxidermy and dioramas depended on disrupting intellectual and political boundaries. In this, Akeley's career mirrors the ways in which the project of Linnaeus and his disciples to make the natural world orderly worked in concert with imperial efforts to structure messy political relationships and rationalize extraction—justified in part by claims of scientific good.[4]

From the eighteenth century onward, science rationalized the harvesting of specimens from around the globe to be conserved and curated in collections in the global north in the name of ordering—and thus understanding—the natural world. We can see still see their efforts today: a moth pinned in a case, a dried a blossom pressed in a florilegium, or a preserved impala pelt stretched over a mounting base. These artifacts imply a static understanding of the world. In this logic *Adela cuneella, Pelargonium graveolens, or Aepyceros melampus*, whether cataloged in a lab or living in the wild, are always and forever the same.

Scholars and planners extend this taxonomic impulse to people, both individuals and communities, from a need to find (or impose) order in an unruly world. Difficulties arise, though, when people consistently fail to conform or comply in categories assigned to them by outside observers (anthropologists or poll-tax collectors, for instance). Historians' take on Carl Akeley are no exception. He was neither entirely a villain nor a savior, but categorizing him as such helped both Donna Haraway and Penelope Bodry-Sanders construct meaningful stories about humans' fraught, changing relationship to the environment in the twentieth century.[5] The premise that categorizing is central to understanding founders even further when we acknowledge that humas are fickle and can change quickly. A re-examination of Akeley's museum career offers an instructive example of how taxonomic impulses worked—and didn't—in the early twentieth century.

Life and Categorization in Modern Museums

Carl Akeley started his professional life in the 1880s at the Milwaukee Public Museum, where his lifelike diorama of muskrats above and below the waterline is still on exhibit.[6] In 1896 Akeley moved to the Field Museum in Chicago where he and his wife Delia created an early taxidermic masterpiece: *The Four Seasons* depicts white-tailed deer in their natural habitat in four distinct phases of the yearly cycle. Delia painstakingly recreated individual leaves, blades of grass, and other details to capture the seasons while Carl continued to improve techniques for preserving hides, capturing life-like poses, and representing the animals' musculature.

Akeley's position in Chicago also offered his first opportunity to travel to Africa. In 1898 he and a colleague went to northeastern Africa to hunt for Somali wild ass specimens for the Field Museum, an expedition that brought Akeley his first brush with death. Akeley was not an especially experienced or gifted hunter; neither he nor Daniel Giraud Elliot, the museum's zoology curator and expedition leader, had experience in or much knowledge about African environments or communities. But they had money, permission to travel, and scientific specimens to collect. They voyaged naively and lived to tell the tales—of drought, being abandoned by their guides, being attacked by nomads, and the ordinary travails of camping among wild animals. (Reading Akeley's memoirs suggests he and his colleagues could have been much better planners.)

Notably, Akeley fought off a leopard with his bare hands, a story—with accompanying photographs—that helped forge his reputation as a rugged adventurer. His subsequent trips to East Africa are fairly well documented: they were approached as scientific expeditions, sponsored by museums—which included fundraising from donors—and intersected with other well-known hunting parties, including the filmmakers Martin and Osa Johnson and former President Theodore Roosevelt's large-scale slaughter for the Smithsonian in 1909 (that expedition killed over 10,000 animals as specimens to ship back to Washington).

In 1909 Akeley moved to New York with big plans for the American Museum of Natural History (AMNH). He courted sponsors with the same dedication that he hunted elephants or developed technology. The resulting Akeley Hall of African Mammals bears the hallmarks of his work and brass plaques honoring the donors. A set of named vitrines form highly ordered snapshots: the many species at an African watering hole; the large antelope of the Serengeti; the expressive faces of mountain gorillas from Virunga. There are 28 windows in all, each a depiction of a habitat frozen in time.[7]

Akeley's grandiose vision for the AMNH, premised on the identification and differentiation of species and reflecting the hierarchical ordering of a colonial world, found eager supporters among the museum's administrators and patrons. His career coincided with a period of active development in public museums in North America and Western Europe—an acceleration of changes that had been underway since the eighteenth century, when personal curiosity cabinets were gradually transformed into museums for the general public—especially Hans Sloan's notable gift to the English crown in 1753 that formed the basis of the British Museum. The new museums reflected the growing influence of taxonomic thinking in both natural sciences and studies of people as displays transitioned from jumbled objects to categorized displays. As modern museums coalesced in the global north, forces of philanthropy, civic pride, and a "civilizing mission" in poor urban communities combined to spur the proliferation of public libraries, museums, and other sites that supported cultural, research, didactic, and frankly hegemonic goals.[8] The uplifting of colonial subjects had a domestic counterpart. In this context, museums were sites where the lower classes could be both educated and socialized through exhibits that explained heroic national pasts, distant cultures, and exotic landscapes—all intended to help visitors understand their place in the world, even if they missed or rejected these implicit hierarchies and their own supposed place in the pecking order.

Donna Haraway describes this function of museums in public life, showing how the AMNH displays class, masculine, and imperial power relations as inherent and universal, when in fact these dynamics were historically contingent and contested.⁹ Natural history museums had a special function as places dedicated to ongoing scientific work—especially species description and classification. This taxonomic work involved gathering and preserving knowledge about extinct or vanishing species, a project that often took the form of collecting examples of rare or exotic animals in order to save them for future generations. In metropolitan institutions, this work served also to extract samples of nature from their typical habitat and present this vision of nature (along with supposedly more "natural" peoples) to urban audiences increasingly alienated from their original, uncultivated states of being.¹⁰

Akeley's role at the AMNH cast him as a scientist doing this work. Through study and field experience he became an expert on African mammals. He and Delia, who was a significant member of his two expeditions in British East Africa, had good luck in their early hunts, acquiring a wide range of species and some type specimens for the museum. World-wide biological science depends on agreement about which plants and animals—despite individual variation—belong to the same species. Consequently, type specimens, which clearly exhibit a species' distinguishing characteristics, are important tools in a research collection. The Akeley's contributions to both display and research spurred even greater ambitions in Carl. His grandest taxidermic masterpiece, eight once-living elephants, now posed eternally in the center of the Africa Hall as a herd charging a foe, were not enough of a spectacle. His vision of the Hall of African Mammals included the little-studied gorillas; he was determined to see the animals in the wild and to collect a representative sampling in order to reconstruct a family scene in New York.

In the early 1920s experts debated the number of gorilla species and the relationship among gorillas, chimps, and humans. Few researchers or members of the general public in the global north had even seen gorillas in 1921 when Akeley planned a journey to the Kivu region in eastern Congo. This major scientific expedition, sponsored by the AMNH, was intended to generate new knowledge about the natural world—contributing to the correct classification of species—and acquire exemplary animals suitable for taxidermic preparation and display.

Getting to Kivu, receiving permission to hunt, acquiring the necessary export permits, and recruiting local guides required the cooperation of British and Belgian colonial authorities, the help of European missionaries working in the Great Lakes region, and the participation of local communities. The territory where Akeley's team wanted to work was claimed as a Belgian colony, but getting there necessitated transiting British East Africa, so two imperial bureaucracies were involved. The projection of geopolitical order and colonial governance was considerably messier on the ground.

The Belgian state assumed governance of Congo in 1908, after international condemnation ended King Leopold II's especially extractive reign over the territory he had claimed as personal colony since 1885.[11] The White Fathers, a Catholic missionary order, established a presence in Central Africa as early as 1878 but did not sustain permanent settlements or recruit large numbers of followers until after Belgian colonial administration was evident in the region. Gradually the White Fathers supported outposts from Lake Victoria south through the Kivu region, with "intense settlement-activity" in Kivu between 1910 and 1914.[12] The missionaries' presence facilitated travel for colonial visitors, including hunters.[13]

Whether hunting for gorilla in the Belgian Congo or other animals in British East Africa, Akeley and his American expeditions worked in colonial Africa not as colonists themselves, but as privileged residents in a world classified and ordered by empires. In short, these American expeditions functioned in the global system and on the ground as beneficiaries of imperial powers—despite the fact that their country did not make formal colonial claims in the regions from which they extracted resources. They enacted a new form empire that they consciously acknowledged, based not on territorial claims but on economic and cultural expansion.[14]

Expedition planners were animated by an imperial ideology that presumed economic development and knowledge creation to be universal goods, creating benefit for all people, not just those who were citizens or subjects of specific empires. In this context, metropoles and colonial territories had porous qualities: people, goods, and knowledge from other polities passed through, not only in the service of economic transactions but also to benefit administration or knowledge production. This situation created "extra-colonial" relationships, circumstances when institutions or individuals could exercise tools of empire even in contexts where direct imperial relationships did not exist. US citizens, for example, exploited African wildlife resources with the complicity of British and Belgian colonial administrators in the name of western science and without regard to the long-term sustainability of the practices, or the needs of the communities that already lived with those resources. In other words, a quest for knowledge justified extraction. And the categories of empire proved permeable.

The 1921 AMNH expedition to Kivu depended on the presence of the White Fathers for basic infrastructure and information, benefitted from the maps and reports of preceding European hunters, and worked in an existing economy of guides and porters. Although the extent of direct Belgian control in Eastern Congo was arguably thin, material structures of colonialism were instrumental in the expedition's ability to realize its plans.

By the time the entourage finally found a group of gorillas, Akeley was feverish and exhausted, but as determined as ever to bag the animals that would complete his vision of the Hall. He killed an adult female, only to have her fall down a steep incline and nearly knock him from his perch. He had to scramble precariously down the ravine to retrieve her body; she was one of the specimens he needed. Later that afternoon he saw a baby from the troupe still squirming on the tip of his guide's spear. He perceived a combination of terror, pleading, and human intelligence on the gorilla's face. Later in the expedition, one of the party shot a large male on Mt. Karisimbi. Upon seeing the corpse, Akeley was moved to write, "It took all one's scientific ardour to not feel like a murderer."[15] His two American biographers treat Akeley's experience in the Congo as an epiphany. He went into the forest on one mission and came out with another, determined to protect gorillas in their habitat.[16] After the 1921 trip, Akeley lobbied for preservation, encouraged Belgian officials to establish a national park in Eastern Congo, and spoke out against hunting, except for scientific purposes. By advocating for a wildlife sanctuary, Akeley was effectively categorizing human interactions with nature, declaring some actions justifiable and others illegal. Indeed, the protection of game animals — regardless of where or when — makes the action of hunting legal or illicit depending only on who kills the quarry, an exquisite classificatory nuance.

Another fine point of categorization — the extra-colonial status of the American expeditions — influenced the AMNH team's access to resources and support in British East Africa, Rwanda, and Congo. Without some colonial structures, Akeley's entourage would not have been able to plan the 1921 trip to Kivu. Planning — whether for colonial administration or scientific research — depends on order, after all. Their travel would likely have been more difficult, though, had this early colonial administration not been quite thin on the ground. So order and classification are neither exclusive nor inviolable markers of administrative or scholarly success.

Conclusion

Taxonomy, then, is both a necessary and fraught element of planning. We rely on categories to make sense of the world and are startled when an object challenges our expectations about where to put it — physically or intellectually. A bust of a gorilla in a library seems misplaced because our anticipation of order is disappointed, because taxonomy — central to natural science, Western education, and governing practices — fails us.

As it confounds the science of categorization, *The Old Man of Mikeno* invokes histories of both ordering and disrupting. Its existence is evidence of a deep change in Akeley's attitude toward nature late his life. The celebratory nature of the statue drives home the transformation of his plans for African wildlife. Akeley's original intention to collect a few individual gorilla specimens for taxidermy became an ambition to conserve the species through habitat preservation. That Carl Akeley—an American taxidermist—had any thoughts at all about the uses or management of African wildlife in the early years of the twentieth century brings into relief the complex relationships among private institutions, states, and colonial infrastructures. The object and the connections that enabled its creation elide straightforward classification. But without attempting the exercise, histories of planning for museums, wildlife conservation, and colonial governance in Africa remain hidden from view.

Notes

1. On the unexamined assumptions of Western epistemology, see Laura J. Mitchell, "Close Encounters of the Methodological Kind: Contending with Enlightenment Legacies in World History," in *Encounters Old and New in World History*, ed. Alan Karras and Laura J. Mitchell (Honolulu: University of Hawaii Press, 2017), 165–80. For an overview of the ordering of the natural world as a colonial project, see Laura J. Mitchell, "The Natural World," in *A Cultural History of Western Empires: Volume 4; The Age of Enlightenment (1650–1800)*, ed. Ian Coller (London: Bloomsbury Press, 2018), 69–91.

2. Mark Alvey, "The Cinema as Taxidermy: Carl Akeley and the Preservative Obsession," *Framework: The Journal of Cinema and Media* 48, no. 1 (2007): 23–45.

3. Carl Akeley, "Lion Spearing in Africa," Bronze sculpture, CSZ59153, The Field Museum, Chicago; Akeley, "The Wounded Comrade" [a group of three elephants], Bronze sculpture, Z93872c, The Field Museum, Chicago.

4. Harold J. Cook, *Matters of Exchange: Commerce, Medicine, and Science in the Dutch Golden Age* (New Haven, CT: Yale University Press, 2008); Richard H. Grove, *Green Imperialism: Colonial Expansion, Tropical Island Edens and the Origins of Environmentalism, 1600–1860* (New York: Cambridge University Press, 1996); Viveka Hansen and Lars Hansen, *The Linnaeus Apostles: Global Science and Adventure* (Whitby, UK: IK Foundation, 2007).

5. Donna Haraway, "Teddy Bear Patriarchy: Taxidermy in the Garden of Eden, New York City, 1908–1936," *Social Text*, no. 11 (December 1984): 20–64, https://doi:10.2307/466593; Penelope Bodry-Sanders, *African Obsession: The Life and Legacy of Carl Akeley*, rev. 2nd ed. (Jacksonville, FL: Batax Museum, 1998).

6. The two biographies of Akeley hew to similar story lines: Bodry-Sanders, *African Obsession*; and Jay Kirk, *Kingdom under Glass: A Tale of Obsession, Adventure, and One Man's Quest to Preserve the World's Great Animals* (New York: Henry Holt, 2010).

7. Stephen Christopher Quinn, *Windows on Nature: The Great Habitat Dioramas of the American Museum of Natural History* (New York: American Museum of Natural History, 2006).

8. Jillian Carman, *Uplifting the Colonial Philistine: Florence Phillips and the Making of the Johannesburg Art Gallery* (Johannesburg: Witwatersrand University Press, 2006); Cynthia Saltzman, *Old Masters, New World: America's Raid on Europe's Great Pictures* (New York: Viking Adult, 2008).

9. Haraway, "Teddy Bear Patriarchy," 24–25.

10. Michael Schudson, "Cultural Studies and the Social Construction of 'Social Construction:' Notes on 'Teddy Bear Patriarchy,'" in *From Sociology to Cultural Studies: New Perspectives*, ed. Elizabeth Long (Malden, MA: Blackwell, 1997), 379–98; Jennifer Price, *Flight Maps: Adventures with Nature in Modern America* (New York: Basic Books, 1999).

11. David Van Reybrouck, *Congo: The Epic History of a People*, trans. Sam Garrett (New York: Ecco, 2014); Adam Hochschild, *King Leopold's Ghost: A Story of Greed, Terror, and Heroism in Colonial Africa* (Boston: Houghton Mifflin, 1998).

12. William Blondeel, "Settlement-Policy of the Missionaries of Africa (White Fathers) in Kivu, Belgian Congo, Phase 1910–1914," *Belgisch Tijdschrift Voor Nieuwste Geschiedenis* 3/4 (1975): 329–62, on 330.

13. Belgian government correspondence documents the presence of hunters in the Kivu region. See, for example, Letter (copy) from Prince Wilhelm of Sweden to M. May, Minstre de Belgique à Stockholm, January 5, 1923, Royal Palace Archives, Dossier Création Parc Albert; and letter to M. T. Serstevens denying permission to hunt gorillas, October 15, 1923, Dossier Création Parc Albert. Royal Palace Archives, Brussels.

14. Emily S. Rosenberg, *Spreading the American Dream: American Economic and Cultural Expansion, 1890–1945* (New York: Hill and Wang, 1982).

15. Carl Ethan Akeley, *In Brightest Africa* (Garden City, NY: Garden City, 1923), 230.

16. Bodry-Sanders, *African Obsession*; Kirk, *Kingdom Under Glass*.

Treasures: Palestine/Israel, 1979
Tamar Novick

Beehives are treasures. The mobile honey-making machines, inhabited by bees that make honey, are usually placed out in the open, and without the sort of protection afforded to other valuable objects. They are tools of extraction — bees use the hive as a site to manipulate plants' nectar; humans extract the honey that the bees produce; settlers extract resources from the land on which they settle. Mobility is a central feature of modern beehives — all parts of the hive are made to be portable in order to enhance production and optimize extraction across time and space. In spite of the fact that beehives are valuable and expensive, unlike most treasures, they tend to be left exposed, unwatched, and unfixed in fields and groves. That is why people like to steal them.

Beehive theft emerged as a problem in the earliest days of the European settlement in Palestine. This problem began to draw new forms of attention in Israel in the late 1970s, when a group of leading beekeepers established a militia, which they called the Action Committee. Its plan was to trace stolen hives, to punish the thieves, and to bring back the hives and the bees to their lawful owners.[1] Beekeepers explained that beehive theft plagued the country, making the situation unbearable for them and their families; they demanded immediate action. The group then wandered around with weapons in areas suspected of harboring beehive thieves, whether within Israeli territory or while infiltrating into the West Bank or Gaza, where most of their suspects resided, to do the work that, according to them, police and military forces failed to do. They made harsh claims against the government, too, saying that not only were their hives not sufficiently searched for, but that beekeepers were also never compensated for their great financial losses. Their fury grew in the late 1990s when they learned that the Israeli government, together with Palestinian authorities, planned to train Palestinian farmers in beekeeping methods. "They will learn to keep bees in the hives they stole from us,"[2] said the leader of the search group, who was later summoned for a parliamentary discussion about the growing problem of agricultural theft, where he was considered the expert in matters of theft.[3] By establishing the Action Committee, beekeepers decided to take the law into their own hands.

According to these Jewish agriculturists, whose families monopolized beekeeping and honey production in the area from the late nineteenth century, the state continuously failed to do justice. While culminating at of the end of the twentieth century, claims about stolen beehives and the mistreatment of beekeepers' complaints are not new; they are as old, in fact, as the European settlement plan in Palestine, and have been consistently apparent in both professional and popular platforms. From the very early days of settlement and the emergence of colonial beekeeping, hives have been stolen, Palestinian Arabs were held responsible for the theft, and the (changing) authorities blamed for doing too little about it.

The Lerrer family, for example, the earliest Jewish family to use movable frame beehives in Palestine, complained about the inadequacy of state authorities in dealing with the theft of hives and honey by the neighboring villagers of Mr'ar. In a letter sent to Va'ad Hatsirim ("The Zionist Commission"), one of the Zionist settling organizations, in 1918, they describe a reality of lawlessness, fear, and violence experienced during their daily work as beekeepers. In earlier days, they explained, there was an agreement with their Arab neighbors: the Lerrers would leave the hive in proximity to the neighboring agricultural fields so the bees can extract nectar, and in return, would pay their neighbors with honey. This agreement did not last very long, they noted. The neighbors were not satisfied with the amounts of honey they got, and cases of theft of honey and hives became frequent at night and even during the day. When "the matter crossed every boundary," the Lerrers turned to the British government, which promised to send a policeman to protect the hive from theft and settlers from violent acts. The policemen never came, and as a result, thievery grew and plagued their everyday lives. The "government is rather negligent," they concluded, and hence they decided to turn to those that consider "the Jewish individual and his property dear to heart," namely, the group of a dozen Zionist representatives that composed the commission.[4]

Complaints about stolen beehives were among the earliest type of complaints voiced by settlers, but the phenomenon grew to encompass the entire realm of agriculture, as crops, animals, and agricultural technologies became deeply connected to the growth and success of the entire settler economy.

Tracking settlers' continuous complaints throughout the twentieth century reveals something about the nature of colonialism, settlement, and resistance. Through an analysis of such complaints, with a focus on cases related to beehives and livestock in particular—being both transportable property and means of food production—agricultural theft emerges as one of the most consistent and successful means of interfering with the construction of a colonial economy and the changing environmental order. Theft of hives and livestock has always posed a threat to settlers, and the legal authorities have systematically demonstrated an incapacity to control thievery or supply satisfactory solutions to settler-farmers.

Reports on agricultural theft in context of the Jewish settlement in Palestine have been brief. They contain information on the animals or equipment stolen, the place from which they were taken (usually Jewish agricultural farms, but Arab villages, too, in some cases), and often the names of the owner. They contain very little information about the thieves themselves (with occasional names mentioned, especially when they were caught and convicted by the courts for stealing), but almost always note that they are Arabs. Most cases of theft remained unsolved and the thieves unknown.

With the establishment of the State of Israel, reports on theft grew dramatically, but remained rather succinct. In some of the rare cases where additional information appears, reports mention that the thieves were infiltrators, and in a few cases, that they executed their act barefoot.[5] "Infiltration" and "infiltrators" were terms used to describe a phenomenon that became widespread with the establishment of the state in 1948. These were usually Palestinian people who lost their homes and property during the war, were forced to reside beyond the borders of the state, and were now crossing borders—temporarily returning to their land.[6] In official reports, and in the public mind, the animals and agricultural technologies stolen were directly linked to such infiltrators; these were no longer only faceless and barefoot criminals, but now also enemies of the state.

The Action Committee of the 1970s, established as an attempt to track stolen hives and return them to their Jewish Israeli owners, operated under the same assumptions, coupling protection from theft and securing state borders. Retrieving the booty meant crossing borders to enemy territory, except in those unexpected cases where stolen hives were found in backyards of Jewish Israelis.[7] The heroic celebrations with the retrieval of hives stand in contrast to decades of concerns, complaints, and the sense of crisis among beekeepers in Israel.[8] Surprisingly, these heroic voices seem to resemble the voices of Palestinian beekeepers in the Gaza Strip whom, in recent years, managed to send their bees across the border under Israeli siege in their search of nectar. This act, they argue, allows them to retrieve the land of Palestine that was stolen from them. "I moved my hive 200 meters close to the border," noted Abdallah, a Palestinian beekeeper from the town of Beit-Hanun, "so that the bees could search for nectar in the green fields across the Israeli border." In a 2008 interview to the Iranian Press-TV network he explained that "one hundred of my beehives were destroyed four times in the last years, but the Israeli soldiers cannot prevent or fight our bees which bring us honey from our stolen land occupied in 1948."[9]

Such similarities in the understanding of the retrieval of the plunder and the moral standing of theft blur the boundary between victim and thief; who is stealing from whom, we may ask. The inherent mobility of the hive opens up the possibility of moving it across the intended border, but it also exposes the limits of that mobility; what moves and who is constrained from movement, we may also ask. While the thief is unknown, and the act of theft presumably haphazard, there are always holes along the borders of the plan. Theft, in this sense, is intimately tied and relational to the plan; what is planned and what isn't, we may also ask. Due to its consistency over time, furthermore, and in contrast to other criminal acts such as sabotage or arson (which are not directly related to production but assume a level of planning), stealing (and agricultural theft in particular) might be considered a highly efficient tool for reclaiming the land in settler colonial contexts. Its assumed haphazardness (or barefootedness) is that which allows theft to continuously interfere with the banality of the plan.

Notes

1. Aharon Priel, "Cavarot gnuvot yimtaku" כוורות גנובות ימתקו [Stolen hives are sweeter], *Ma'ariv*, January 5, 1979, 30.

2. Amos DeWinter, "Falestinaim yilmedu legadel dvorim bekavarot sheganvu meitanu" פלסטינאים ילמדו לגדל דבורים בכוורות שגנבו מאיתנו [Palestinians will learn to keep bees in hives they stole from us], *Mashov Hakla'I* 55 (2000).

3. Ronnie Feldman, "Legal Actions and Protection of Farmers from Thefts Originating in the (Palestinian) Autonomy, and the Government's Demand of the Palestinian Authority to Cooperate for the Purposes of Preventing Theft and for Law Enforcement," Protocol of the State Controller Committee Meeting, Israeli Parliament, March 7, 2000, https://www.nevo.co.il/law_html/law103/15_ptv_497726.htm (accessed August 19, 2022).

4. Lerrer Family, "Ness-Ziona, to Va'ad Hatsirim," August 2, 1918, folder L4/423-1, 1–4, Central Zionist Archives, Jerusalem, Israel.

5. See, for example, "Nignav eder mekibbutz gaviv banegev" נגנב עדר מקיבוץ גביב בנגב [A herd was stolen from Kibbutz Gaviv in the Negev] *HaTsofeh*, September 2, 1949, 8.

6. Orit Rosin, "Infiltration and the Making of Israel's Emotional Regime in the State's Early Years," *Middle Eastern Studies* 52, no. 3 (2016): 448–72.

7. "'Mistanenim' yehudim chashudim bigneiva" "מסתננים" יהודים חשודים בגניבה [Jewish "infiltrators" are accused of theft], *Ma'ariv*, April 12, 1950, 4; Priel, "Cavarot gnuvot yimtaku."

8. Beekeepers have been concerned with the problem of theft, but also about the decline in open field areas over the years, as well as a variety of financial concerns. See, for example, "Hakavranim dorshim ezrat hamemshala" הכווראנים דורשים עזרת הממשלה [The beekeepers demand the help of the government], *Al-Hamishmar* May 18, 1954, 4.

9. Interview with Gazan Beekeepers, Press TV, 2008, www.youtube.com/watch?v=BM1nj8Wc4AA&feature=player_embedded, accessed June 15, 2017). Similar views were voiced by beekeepers interviewed by Motasem A. Dalloul in Motasem A. Dalloul, "Gaza Honey Production Stung by Israeli Policies," *Middle East Eye*, May 21, 2015, https://www.middleeasteye.net/features/gaza-honey-production-stung-israeli-policies.

Water Samples: Treaty 8 Territory, Canada, 2012

Sarah Blacker

A representative sample is one that has strong external validity in relationship to the target population the sample is meant to represent. As such, the findings from the survey can be generalized with confidence to the population of interest.[1]

The Peace-Athabasca Delta in Treaty 8 Territory, Alberta, Canada, is one of the largest inland freshwater deltas in the world; it is more than 3,900 square kilometers in size. This is a particularly sensitive ecosystem, and one that has long provided food and drinking water for two Indigenous communities that live along the shore of the delta: the Mikisew Cree First Nation and the Athabasca Chipewyan First Nation. The Peace-Athabasca Delta is located downstream of the Athabasca river, and at the other end of the Athabasca river, 223 kilometers to the south, is the Athabasca oil industry (also known as the Oil Sands and the Tar Sands). Having begun operations in 1967, this is currently the largest scale industrial project in the world by land cover. Oil companies in Alberta use two separate processes of extraction: open pit mining, in areas where the oil sands lie closer to the surface, and the in situ Steam Assisted Gravity Drainage method when the oil sands lie too deep to be reached through open pit mining. The oil company Syncrude operates the largest mine in the world here. This mine covers more than 140,000 square kilometers (more than three times the area of the Netherlands). The oil is in the form of bitumen, which is heated and washed out of the sands using steam and chemical solvents. Once the bitumen is separated from the sand, the now-contaminated sand and water are deposited into "tailings ponds," which contain high levels of polycyclic aromatic hydrocarbons (PAHs), which leak into the groundwater and enter the food chain.

One of the most important by-products of the extraction process in northern Alberta's Oil Sands are these tailings ponds, which are engineered dam and dyke systems that contain the waste products of the bitumen extraction process. Occupying more than 200 square kilometers, tailings ponds contain 1.2 trillion liters of water contaminated with substances such as bitumen, naphthenic acids, cyanide, and heavy metals.[2] The effects of this industrial contamination have been difficult to measure because of the lack of an accurate baseline[3] and have been fiercely contested both in the media and in court.[4] The attention paid to tailings ponds reveals a critical point of epistemic contestation in settler colonial Canada: environmental contamination produced by an extraction economy that is disproportionately harming Indigenous health and communities.

While technical manuals written for toxicologists figure environmental contamination as static and observable, obvious even to those visiting the region for the first time, chemical contamination in waterways can be elusive and difficult to capture. After observing a sharp increase in premature deaths of animals and humans in their communities in Treaty 8 Territory, the Athabasca Chipewyan First Nation and the Mikisew Cree First Nation organized a community monitoring program to investigate levels of contaminants that have flowed northwards from the Athabasca oil industry into the Peace-Athabasca Delta.[5]

Facing rapidly rising rates of disease following years of persistent water contamination, the two First Nations communities needed a way to render their observations and knowledge into a form of "evidence" that would be recognized by the federal government of Canada so that the community could obtain much-needed resources, including medical care. The form that this evidence would take needed to bring together the context provided by the First Nations' Traditional Ecological Knowledge with the credibility offered by government-recognized toxicology metrics. The aim was to track the movement of contaminants: from industrial tailings ponds into bodies of water, into bodies of animals, and into the bodies of humans, where the contaminants led to "people dying earlier and from different illnesses than in the past."[6]

Elaborate measurement protocols were devised to enable members of these First Nations to collaborate with non-Indigenous toxicologists who were invited to visit the First Nations to help plan a study that would provide evidence of contamination in a form that could be recognized by Canadian federal government policymakers. This collaborative study was run from 2011 to 2014, and it competed for credibility with two other studies run by Health Canada, following the government's dismissal in 2006 of the First Nations' community physician after he alerted Canadian media to elevated cancer rates in the community. The collaborative study brought together the First Nations' Traditional Ecological Knowledge of the land and waterways, "extend[ing] back thousands of years,"[7] with toxicology measurement practices to provide evidence of extensive contamination of groundwater. The study measured levels of PAHs, as well as arsenic, cadmium, mercury, and selenium in groundwater, plants, and animals in the region.[8]

One of the founding mythologies that has undergirded settler colonial epistemic regimes is the construction of Indigenous communities as characterized by an epistemic "primitiveness." This mythology has had many deleterious effects, including the cultural denigration of Indigenous Traditional Knowledge. It has also been used as an attempted justification for settler colonial researchers to conduct research on Indigenous lands without collaborating with or asking permission from Indigenous communities. The non-Indigenous toxicologist Stéphane McLachlan, who has worked extensively with the Athabasca Chipewyan First Nation and the Mikisew Cree First Nation, maintains that settler colonial scientific practices in Canada have intentionally set up and maintained a binary between Indigenous knowledge practices and the *choice* to participate in science, and, through science, modernity.[9] The Canadian state's long history of colonial violence toward Indigenous Peoples includes the residential school system that operated across Canada between 1883 and 1996, forcibly removing Indigenous children from their families and communities, and punishing children when they spoke their own languages or practiced their culture. Recent scholarship has documented further abuses, such as medical experimentation and intentional starvation in the schools;[10] in 2015 the settler colonial residential school system was found by the Truth and Reconciliation Commission to constitute a form of cultural genocide.[11]

In May 2021, the Tk'emlúps te Secwépemc First Nation announced that ground-penetrating radar had detected the remains of 215 Indigenous children buried in an unmarked mass grave located next to the former Kamloops Indian Residential School.[12] Indigenous communities have long known that these unmarked mass graves exist and that the very fact that they are unmarked and hidden provides evidence that the genocidal actions taken by the settler colonial state and residential school administrators were intentional.[13] Indigenous communities have long known, too, that settler colonial states are particularly adept at hiding evidence of their violence and destructiveness. The onus is consistently placed upon marginalized communities to produce evidence of harm despite often having very limited resources to support this production of evidence; in this instance, ground-penetrating radar produced this evidence of colonial eugenic and genocidal practices. Residential schools and their successor institutions and practices—including forced sterilizations, forcing Indigenous girls as young as 9 years of age to have IUDs inserted by doctors, "birth alerts" and systemic racist practices in foster care[14] that wrench Indigenous children away from their families[15]—exert a biocolonial form of control over Indigenous reproduction.

This biocolonial control also extends into the making of scientific knowledge in settler colonial Canada. However, colonial domination in science does not announce itself as such. Instead, it often operates through technologies of quiescence—of quietness, stillness, inactivity, or dormancy; of making quiet or calm—that are developed and mobilized by settler states in order to minimize or displace public concern and to produce apathy.[16] Such technologies work to conceal and suppress evidence that colonial violence has been done (and is still underway), and thereby enable the settler state's ongoing colonial practices to proceed, often uninterrupted and unrecognized.

According to McLachlan, when it comes to the question of participation in science and the making of evidence of contamination, Indigenous Peoples in Canada "can't have one foot in each canoe."[17] In the settler colonial context, they are forced to "choose one tradition": either to participate in science or to "live in the past" by practicing Traditional Knowledge.[18] The dichotomy between science and Traditional Knowledge is naturalized in settler colonial societies, not necessarily in order to intentionally exclude Indigenous people from participating in science, but because science and Traditional Knowledge are understood as incompatible knowledge systems.

In order to move beyond this dichotomy, and to prove the compatibility of the two knowledge systems, McLachlan designed a "three-track" methodology that aimed to place into dialogue two forms of knowledge that are often held in opposition with one another. The "three-track" methodology presents evidence in three distinct forms: the first form of evidence is an articulation of unmediated and untranslated Indigenous Traditional Knowledge presented in narrative form; the second form of evidence provides numerical measurements of contamination levels taken using current industry standards to render it legible and credible to industry scientists and government policymakers alike; and the third track attempts to integrate the first two knowledge systems, allowing the Traditional Knowledge to guide the interpretation of the data collected.[19] The third track is designed to remedy the tendency for science and Traditional Knowledge to "speak past one another" in a two-track methodology. As McLachlan notes, in a two-track methodology, "the two tracks are now seen as independent and, like any train track, they have been constructed so that they remain parallel such that they rarely, if ever, meaningfully intersect."[20]

The members of the First Nations communities participating in the program reached out to McLachlan and said that they could not endorse a study of regional contamination that did not engage with Traditional Knowledge. In one of my interviews with McLachlan, he emphasized that it was the Traditional Knowledge that directed their study, and not the Western toxicology.[21] The reason for this, he said, is that "the Traditional Knowledge is often richer, more place-sensitive, and longer-term in nature than its scientific counterpart."[22]

The ways in which chemicals flow through river water at uneven concentrations make it impossible to obtain the same measurement twice. And yet, government manuals written for field scientists regarding the measurement of contamination in this region figure chemical contaminants as static, visible, and identifiable, even to the scientists as first-time visitors to the area, lacking the knowledge of the land possessed by the First Nations. In contrast, the community-monitoring program's "three-track" methodology is designed to incorporate situated knowledge, such as the subtle and gradual changes in animals and waterways over decades that are observed by First Nations Elders and other community members.

Another point of divergence between the two knowledge systems is their conceptualization of temporality. Work is currently being done in developing technologies for measurement that can capture change over time, to better incorporate Traditional Knowledge into the act of measurement. The spectrophotometer is designed to take a snapshot of a moment in time, and to translate that measurement into a "representative sample" that is then circulated with an aura of objectivity as a knowledge claim about the state of the delta water. The contingency embedded in the knowledge claim is rendered invisible. When the First Nations monitoring work is carried out through the use of these same tools that are used by the toxicologists, then a great deal of interpretive and contextualizing work needs to be done to flesh out the processes through which each kind of knowledge is produced.

The Canadian settler colonial state has remained focused primarily on control over land and natural resources. Despite modernization and urbanization, Canada's economy continues to depend on the extraction and sale of its natural resources.[23] I emphasize this because it is crucially important to connect the Canadian state's valuation of the land and its natural resources with the state's devaluation of the Indigenous Peoples who live and have lived on these lands.²⁴ The state's continued valuation of the land and the devaluation of Indigenous Peoples have played determinative roles in the shaping of multiple Canadian institutions, including those of science and medicine.

Practices of measurement in science—including the selection of sampling sites—are often seen as uncontroversial and politically impartial, whereas the analysis and interpretation of collected data has been understood as more susceptible to "bias." However, measurement practices are not a neutral form of observation. Calculations and judgements made by scientists, that in turn inform the sites and methods of measurement, are conditioned by the social and political context in which the research is carried out.²⁵ Government metrics of existing levels of contamination regularly serve as arbiters of truth—framed as objective and politically neutral—and remain difficult to challenge, particularly for Indigenous communities who are often most affected by environmental contamination in Canada.²⁶ As multiple scholars have shown, there is much at stake for states that are invested in continued industrial development to produce metrics showing minimal contamination, ideally at levels considered inconsequential for human health.²⁷

Planning processes are multiple—and sutured to the set of epistemic practices that place them in motion—in the context of the knowledge dynamics around measurement of levels of contamination of Indigenous land adjacent to industrial sites in Canada in the latter half of the twentieth century. One way to understand the dynamic at play here would be through a binary: on one hand, the settler colonial complex collusion of state and industrial interests in planning for the suppression of scientific data that reveal the ecological harm done by the oil industry, as well as its costs for human health and especially Indigenous health. The community-based monitoring project that the Indigenous communities are carrying out in collaboration with non-Indigenous scientists, could then be seen as a form of counter-planning, or resistance, as this project aims to create public data on contamination that the government has devoted significant resources to rendering undocumentable.

Colonial technologies such as the spectrophotometer can be subverted and used to support a form of counter-planning, which begins to gain visibility in the tension between the different spatialities and temporalities that different actors in the story inhabit, which shape different forms of planning. The primary distinction is that between the placelessness of the government scientists' use of the spectrophotometer versus the in-place-ness of the Mikisew Cree First Nation and the Athabasca Chipewyan First Nation's selection of sampling sites on the basis of Indigenous Traditional Knowledge.

For the Mikisew Cree First Nation and the Athabasca Chipewyan First Nation, the epistemic approach to planning is the modality through which community members are able to observe and identify change in the environment that may not be visible to field scientists who are unfamiliar with the patterns of the region. Other forms of planning are much more short-sighted and are carried out without an understanding of the context.[28]

In 2012 in the Peace-Athabasca Delta, downstream of the Athabasca river which flows through northeast Alberta, two regimes of measurement practices embedded in contesting knowledge traditions abutted and abraded. Each claimed adequacy as a form of measurement from which valid epistemic claims about the state of chemical contamination could be made. Each claimed to be a valid basis for planning protocols of environmental management that would be imposed on mining operations.

The two measurement regimes both relied on water sampling techniques, and in many cases an observer watching what the experimenters actually did on the ground might perceive little difference. Both plunge spectrophotometers into water and record sets of numbers. There is however significant divergence in the sampling sites selected; that is, the actual places chosen in which to immerse the measuring instrument. The federal government set guidelines for "the design and implementation of an effective monitoring system" based on a particular definition of what constitutes scientific rigor: "a science-based approach that uses robust indicators, consistent methodology and standardized reporting, including peer-review, that will result in independent, objective, complete, reliable, verifiable and replicable data."[29]

The different protocol of timing between the two sampling techniques expresses a differing conceptualizing of temporalities in the "three-track methodology." These different places and differing timing protocols are identified through working in the first track of the epistemic work of the collaborative monitoring project. This involves a focus on community members' observations and insights in order to determine the spatial locations and the temporal intervals that are best suited for measurement. This is the form that Traditional Ecological Knowledge takes for these traditional landowners.[30]

The epistemic practices through which the places were known differed profoundly in each of the contesting regimes. Among other things, the epistemic practices of the Indigenous landowners' "three-track methodology" differed ontologically from the methodology invoked in dominant Western monitoring regimes. Different forms of knowledge and divergently configured knowers were generated in the contesting epistemic regimes. The places Indigenous landowners know differ from the places that environmental scientists trained in conventional Western scientific practices know.[31] These differences in what it is that is known—divergences in the forms that place has in the "three-track methodology" compared to dominant Western scientific practices—makes a difference when it comes to planning which protocols of environmental management are appropriate in the context of the ongoing colonial dispossession of Indigenous communities, lands, and waterways.

Notes

1. Paul J. Lavrakas, *Encyclopedia of Survey Research Methods* (Thousand Oaks, CA: Sage Publications, 2008).
2. Bob Weber, "Showdown Looming for Alberta's Oil Sands over Cleanup of Tailings Ponds: Report," *The Globe and Mail*, March 30, 2017.
3. Matthew S. Ross, et al. "Quantitative and Qualitative Analysis of Naphthenic Acids in Natural Waters Surrounding the Canadian Oil Sands Industry," *Environmental Science and Technology* 46, no. 23 (2012): 12796–805.
4. Meagan Wohlberg, "Court Case Claims Suncor Tailings Pond Leaking into Athabasca," *Northern Journal*, July 8, 2013.
5. This chapter draws on material that was published in the following article: Sarah Blacker, "Strategic Translation: Pollution, Data, and Indigenous Traditional Knowledge," *Journal of the Royal Anthropological Institute* 27, no. S1 (2021): 142–58.
6. Stéphane M. McLachlan, *Water Is a Living Thing: Environmental and Human Health Implications of the Athabasca Oil Sands for the Mikisew Cree First Nation and Athabasca Chipewyan First Nation in Northern Alberta* (Winnipeg: Environmental Conservation Laboratory, University of Manitoba, 2014), 216.
7. Stéphane McLachlan, *Water Is a Living Thing*, 214.
8. Yifeng Zhang et al., "Airborne Petcoke Dust Is a Major Source of Polycyclic Aromatic Hydrocarbons in the Athabasca Oil Sands Region," *Environmental Science & Technology* 50, no. 4 (2016): 1711–20.
9. Stéphane McLachlan, interview by Sarah Blacker, May 7, 2015.
10. Ian Mosby, "Administering Colonial Science: Nutrition Research and Human Biomedical Experimentation in Aboriginal Communities and Residential Schools, 1942–1952," *Histoire Sociale / Social History* 46, no. 1 (2013): 145–72.
11. Truth and Reconciliation Commission of Canada, *Honouring the Truth, Reconciling for the Future: Summary of the Final Report of the Truth and Reconciliation Commission of Canada* (Ottawa: Truth and Reconcilliation Commission of Canada, 2015), 1.
12. Courtney Dickson, and Bridgette Watson, "Remains of 215 Children Found Buried at Former B.C. Residential School, First Nation Says." *CBC News*, May 27, 2021.
13. Mark Gollom, "How Radar Technology Is Used to Discover Unmarked Graves at Former Residential Schools," *CBC News*, June 14, 2021. https://www.cbc.ca/news/canada/ground-radar-technology-residential-school-remains-1.6049776

14. Cindy Blackstock, "67-Million Nights in Foster Care." *Maclean's*, October 16, 2019. https://www.macleans.ca/opinion/67-million-nights-in-foster-care/.

15. In her resignation speech given in the Canadian House of Commons on June 15, 2021, Nunavut MP Mumilaaq Qaqqaq stated: "Colonization is not over. It has a new name. Children are still being separated from their communities. Foster care is the new residential school system. The suicide epidemic is the new form of Indigenous genocide" (Qaqqaq quoted in Teresa Wright, "Foster Care Is Modern-Day Residential School System: Inuk MP Mumilaaq Qaqqaq." CBC News, June 4, 2021. https://www.cbc.ca/news/politics/foster-care-is-modern-day-residential-school-1.6054223). In response, the Minister for Crown-Indigenous Relations, Carolyn Bennett, acknowledged that "There are more kids in care now than there were at the height of residential schools and it's unacceptable and harmful" (Bennett quoted in Wright "Foster Care").

16. Sarah Blacker, "Technologies of Quiescence: Measuring Biodiversity, 'Intactness,' and Extractive Industry in Canada." *Catalyst: Feminism, Theory, Technoscience* 8, no. 2 (2022): 1–26. https://doi.org/10.28968/cftt.v8i2.37828.

17. Stéphane McLachlan, interview by Sarah Blacker, May 7, 2015.

18. Stéphane McLachlan, interview.

19. Stéphane M. McLachlan, *Deaf in One Ear and Blind in the Other: Science, Aboriginal Traditional Knowledge, and Implications of the Keeyask Hydro Dam for the Socio-Environment. A Report for the Manitoba Clean Environment Commission on Behalf of the Concerned Fox Lake Grassroots Citizens.* (Winnipeg: Environmental Conservation Laboratory, University of Manitoba, 2013), 36.

20. Stéphane McLachlan, *Deaf in One Ear*, 36.

21. Stéphane McLachlan, interview by Sarah Blacker, May 7, 2015.

22. Stéphane McLachlan, "Water Is a Living Thing," 191–192.

23. Hayden King, "New Treaties, Same Old Dispossession: A Critical Assessment of Land and Resource Management Regimes in the North" in *Canada: The State of the Federation 2013: Aboriginal Multilevel Governance* (pp. 83–98), eds. M. Papillon and A. Juneau (Montreal: McGill-Queen's University Press, 2016); Shiri Pasternak, Hayden King, and The Yellowhead Institute, "Land Back: A Yellowhead Institute Red Paper." 2019. https://redpaper.yellowheadinstitute.org/wp-content/uploads/2019/10/red-paper-report-final.pdf.

24. Audra Simpson, "The State Is a Man: Theresa Spence, Loretta Saunders and the Gender of Settler Sovereignty," *Theory & Event* 19, no. 4 (2016).

25. Maggie Walter and Stephanie Russo Carroll, "Indigenous Data Sovereignty, Governance and the Link to Indigenous Policy," in *Indigenous Data Sovereignty and Policy*, ed. Maggie Walter et al. (London: Routledge, 2021), 1–20.

26. S. G. Donaldson et al., "Environmental Contaminants and Human Health in the Canadian Arctic," *Science of The Total Environment* 408, no. 22 (2010): 5165–234.

27. Javiera Barandiaran, "Chile's Environmental Assessments: Contested Knowledge in an Emerging Democracy," *Science as Culture* 24, no. 3(2015.): 251–75; Fabiana Li, *Unearthing Conflict: Corporate Mining, Activism, and Expertise in Peru*

(Durham: Duke University Press, 2015); Kathleen H. Pine and Max Liboiron, "The Politics of Measurement and Action," in *CHI 2015—Proceedings of the 33rd Annual CHI Conference on Human Factors in Computing Systems* (New York: Association for Computing Machinery, 2015), 3147–57.

28. Joshua R. Thienpont et al., "Comparative Histories of Polycyclic Aromatic Compound Accumulation in Lake Sediments near Petroleum Operations in Western Canada," *Environmental Pollution* 231 (December 2017): 13–21.

29. L. Dowdeswell et al., *A Foundation for the Future: Building an Environmental Monitoring System for the Oil Sands* (Ottawa: Environment Canada, 2010), 7.

30. Michaela Spencer, Endre Dányi, and Yasunori Hayashi, "Asymmetries and Climate Futures: Working with Waters in an Indigenous Australian Settlement," *Science, Technology, and Human Values* 44, no. 5 (2019): 786–813, https://doi.org/10.1177/0162243919852667.

31. Helen Verran, "A Postcolonial Moment in Science Studies: Alternative Firing Regimes of Environmental Scientists and Aboriginal Landowners," *Social Studies of Science* 32, no. 5/6 (2002): 729–62; Helen Verran, "Transferring Strategies of Land Management: Indigenous Land Owners and Environmental Scientists," in *Research in Science and Technology Studies, Knowledge and Society*, ed. Marianne de Laet (Oxford: Elsevier, 2002), 155–81.

Weeds: Laos, 2006
Karen McAllister

Tall weeds conceal the official government map that depicts the territorial resources belonging to Ban Samsum,[1] a small ethnic minority Khmu village located in the highlands of Northern Laos. Weeds also grow alongside farmers' crops, resiliently colonizing and disrupting neatly planted fields and competing with the desired upland rice and annual commercial crops. They pop up unwanted in carefully planned fields, and farmers spend long hours and many days each year manually uprooting them, a task undertaken mainly by women.

The weed species obscuring the view of Ban Samsum's map—*Chromolaena odorata*, or Siam weed—is an invasive species native to the Americas that has proliferated widely across Southeast Asia. A relic of colonialism, it was originally brought to Asia by the British in the late 1840s as an ornamental plant for a botanical garden in Calcutta, from where it escaped and subsequently crept across the subcontinent. The weed also boarded boats in the West Indies and dispersed itself through the shipping routes of the British India Company.[2]

Chromolaena reached Laos in the 1930s and has since become one of the most abundant weeds in the Lao mountains.[3] It is locally known as *Nya Kiloh* or *Nya Falang*,[4] the latter meaning "French weed" (or "foreign weed") because it began to appear during the period of French colonial rule.[5] Although farmers manually remove *Nya Kiloh* from cultivated fields, locals also use it as an indicator of soil fertility, and unlike other weed species, the uprooted plants are left lying on the soil as fertilizer for the growing rice crop.[6] Villagers in Ban Samsum also use the leaves of *Nya Kiloh* as medicine to stop bleeding and to treat burns, skin, and stomach ailments.[7]

Figure 10. Weeds concealing an official village land use map, Northern Laos. Photo by Karen McAllister, June 2012.

In addition to their various uses and nuisances, weeds are an important consideration in shifting cultivation (also known as swidden cultivation or slash and burn)—an extensive form of agriculture in which farmers cut and burn forest or secondary fallow vegetation and cultivate the land for a limited number of consecutive years before moving to clear and crop another area and allowing the temporarily abandoned land to lie fallow.[8] Vegetation regenerates, and the land can be cleared and burned again after several years, providing natural fertilizer for the crops. Fast-growing weedy species such as *Nya Kiloh* are early colonizers of fallow fields, followed by slower-growing tree species that eventually shade them out and control the spread of their seeds. In Laos, most highland communities traditionally practice rotational swidden cultivation, moving between the same land parcels within a relatively large territory and leaving land fallow for five to fifteen years before returning to clear and cultivate the same area again. Under low population densities, swidden cultivation is ecologically and socially sustainable. However, with increased land pressure, weeds, soil fertility, and erosion become problems, crop yields decline, and farmers are compelled to invest more labor to weed, to apply chemicals and fertilizers, and/or to switch to alternative crops. Since at least the 1990s, many Lao villagers have been facing increased land pressure from growing populations and from state policies that limit villagers' access to land, that concede village lands to companies for plantation crops or other forms of development, and/or that resettle remote mountain communities to areas near roadsides where land is already occupied.

Colonial and post-colonial governments across Southeast Asia have consistently demonized the practice of shifting cultivation as wasteful and destructive and have enacted various policies intended to stop it. The Lao government perceives it as destructive of forests and soils and as "backwards" because it is largely subsistence-oriented, condemning it as a cause of poverty. The government would like highland farmers to replace "traditional" shifting cultivation for upland rice with "modern" permanent intensive cultivation of cash and tree crops intended for the market.[9]

Shifting cultivation creates a dynamic, patchy landscape which is both social and natural, with areas of cropped land interspersed with forest-fallow at different stages of vegetative succession. Areas that are not actively cultivated are used as village commons, for gathering forest products and hunting wild game, and for grazing livestock. Anna Tsing describes the landscapes created and inhabited by shifting cultivators as a "gap" between taken-for-granted categories of natural/wild forest and social/cultivated farmland.[10] She uses "weediness" as a metaphor for this gap—to represent what is at once natural and social—as human activities such as clearing and burning forest and fallow facilitate the "natural" growth of weed species, disrupting easy classifications between what is wild and what is cultivated. Tsing also uses "weediness" to describe landscapes "degraded" by human activity and people perceived as being "left behind" by "modernization," such as shifting cultivators. However, weeds are ambiguous plants—the classification of *Nya Kiloh* as a weed obscures its alter ego as a resource.[11] The classification of "weed" is in part a value judgement about a plant's undesirability in a specific time and place. What is identified as a weed by some people and in certain contexts may be considered a resource in others. Weeds are unplanned plants that are out of place—that literally invade and disrupt the territorial plans of gardeners and farmers. They may also act as a metaphor for the "unplanned" that emerges out of and outside of planned human activities—the resilient, dynamic place-based human and non-human practices and assemblages that entangle, unravel, and obscure easy classification and the authority of even the best laid plan.

Maps are a type of plan because they not only provide abstract representations of social and territorial spaces but provide models to which social and territorial realities are intended to conform.[12] However, just as *Nya Kiloh* conceals the village map of Ban Samsum, the map itself conceals the reality of "weedy" spatial practices in the village. Maps represent and make visible the roads, paths, buildings, zones, and borders selected by their makers, yet omit the fabric of human and non-human assemblages and the practices that flow through and create these territorial spaces, making them meaningful and functional.[13] Like the resilient weeds that both disrupt and are shaped by farmers' agricultural plans and actions, these dynamic practices undermine, transform, and are shaped by territorial plans, producing dynamic realities that are influenced by yet belie abstract and simplified representation.

The map outside Ban Samsum is typical of the official maps that mark the entrances of most Lao villages. Like *Nya Kiloh*, "birds-eye" maps and their implicit ideology of bounded territories that neatly correspond with political authority or "property ownership" are colonial imports. Winichakul describes how the Siamese notion of sovereignty, as authority over people rather than over a bounded area of land, was transformed through encounters with the British who, following their European conception of property, sought to demarcate clear linear borders between Siam and British Burma.[14] Prior to this, borders between the various kingdoms of Thailand, Laos and other parts of mainland Southeast Asia were unproblematically fuzzy, overlapping and flexible, and political authority was defined by social relationships rather than by defined territories. The import of maps and borders representing a political authority that coincided with a defined area of land made visible distinctions between Thai and British territories for colonial authorities.

Contemporary village maps in Laos—hand-painted on large wooden boards and posted outside each village—similarly attempt to transform local conceptions of authority over land and to clarify village boundaries and land use to the Lao government. These maps were erected during the implementation of the Land and Forest Allocation Program (LFAP) in the late 1990s to mid-2000s, a policy that was influenced by the international donor community and that broadly resembled a general model of land-use planning and zoning that is being implemented in many countries around the world. The LFAP was designed to formalize (i.e., to officially codify and document) village land rights to facilitate greater state control of forest resources and their use by villagers by making them "legible"—that is, simplifying, encoding, and mapping land and local practices of tenure to make these transparent to government authorities.[15]

The policy also ecologically zoned and attempted to impose state-approved resource uses within areas classified as "state forests." Forests in Laos are "political forests"[16] defined by administrative maps and associated legal frameworks rather than by actual tree cover or land use.[17] Much of this land is mountainous and is classified as state forest primarily based on slope and remoteness from roads. State forests include the weedy patchy forest-farm landscapes that are inhabited and under customary claims of highland villagers who practice shifting cultivation, such as the Khmu of Ban Samsum. Government maps classify early-succession weedy fallow land as "degraded" state forestland, while villagers simply consider this as another type of forest (*pa*) that is under customary claims and that will eventually regenerate into older tree-covered forest or be cleared and re-cultivated on a later date, accounting for the dynamic nature of weedy fallow and tree-covered forestland.[18] For example, Ban Samsum villagers refer to a field of *Nya Kiloh* as "*Nya Kiloh* forest" (*Pa Nya Kiloh*).[19] In contrast, the wooden village maps represent different forest types as static, bounded, different-colored zones that are fixed in time and space.

One important goal of the LFAP was to demarcate village territories to clarify boundaries between the "state" forestland that villagers would be allowed to use from the forestland that would be managed by the government for conservation, forestry, or agro-industrial plantations. Another goal was to provide farmers with a limited number of private land holdings for cultivation on forestland zoned as "degraded," by formalizing customary claims to only three specific land parcels. In theory, this would both encourage and compel farmers to invest in intensifying their agricultural system by providing tenure security to individually held land parcels and by limiting their land access, making shifting cultivation no longer ecologically nor socially viable. However, instead of providing tenure security, the LFAP created livelihood insecurity as villagers were forced to shorten fallow periods, leading to soil degradation, increased weed growth, greater labor demands, and declining crop yields. Essentially, the proliferation of weeds made the plan untenable for farmers.

Customary property systems governing the use of swidden landscapes in Northern Laos are flexible and complex, and they vary between locations and by ethnic group. They are designed to be responsive to dynamic place-based social and ecological needs, such as weed, pest, and fire management and new economic opportunities, and both create and are influenced by the socio-ecological environments in which they are situated. In Ban Samsum, villagers recognize the pioneer property rights of the person/household who initially cleared a land parcel of forest. However, these rights are not exclusive and other villagers are permitted to cultivate the land if it is not being used by the recognized owner that year. Farmers hold "privileged" rights to many parcels of land scattered across village territory, each with different physical and biological attributes and histories of use, which they cultivate on a rotational basis. This allows land to recover, weeds to be shaded out by trees, and farmers to take advantage of ecological diversity to grow a range of different rice varieties and commercial crops to help manage livelihood risk. Essentially, villagers have "farm" tenure (the right to a place to cultivate as members of a community) rather than land tenure (the right to cultivate a specific piece of land), and farmers choose which plots to clear and cultivate each year based on social, economic, and ecological considerations that are adaptive to changing contexts.

Land formalization programs such as the LFAP and their associated maps, seek to simplify and clarify complex local property regimes that are ecologically, socially, and spatially defined, with the goals of making these "legible" to the state and of bringing land rights under the management of an allegedly impersonal government system. Programs like LFAP try to fix land rights in time and space and associate them with abstract bounded uniform land parcels, but intersect with dynamic socio-ecologies and customary conceptions and practices of property, which like weeds in a rice field, interfere with, transform, and are transformed by their desired goal. In the district where Ban Samsum is situated, the LFAP was first implemented in easily accessible and well-established roadside villages populated primarily by economically better-off Lao and Lue (lowland Lao—*Lao Loum*) ethnic groups, and only later in remote Khmu villages like Ban Samsum.[20] In theory, this created villages with state-legible private property rights adjacent to villages that remained under customary tenure. In reality, weeds and the ongoing weedy practices of local officials and villagers unraveled the desired legibility and goals of state plans, as is illustrated by the following stories.

Bounsavath,[21] a relatively well-off Lue farmer and trader from the roadside village of Houay Loum,[22] owned a small tractor, and regularly purchased crops from Khmu from remote highland villages, which he transported from the hills and resold for profit in the town. Khmu villagers often borrowed money from him and other traders when they ran short of rice or faced a family emergency, promising to repay after harvest. However, when crops failed, some repaid their debts with land. Bounsavath had accumulated many parcels of land in remote Khmu villages as repayment for debts, allowing him to comply with the LFAP within the boundaries of his own village, while alleviating the livelihood problems created by the proliferation of weeds on his fields by extending his land holdings into remote communities where the LFAP had not yet been enforced. In Ban Samsum, Bounmee, a middle-aged Khmu farmer who held pioneer rights to many land parcels, began to deny villagers with customary rights access to his land, claiming it had been allocated to him by the LFAP even though the policy had not yet been enforced in his village. This transformed customary tenure practices. He had also sold some of his land to roadside Lue villagers seeking land to compensate for the livelihood constraints created by the LFAP. None of these transactions were formally documented and de facto land rights remained illegible to the state as the plan was unraveled by local practice. On my visit to district offices in search of land allocation maps and cadastral records, I discovered that these had been misplaced or damaged by rats and water, making these documents literally illegible. It is possible that the lack of care taken to preserve these documents signified local officials' awareness of their irrelevance in reflecting actual land rights and practices in the villages.

The implementation of the LFAP also coincided with Chinese companies seeking land to establish rubber plantations in Laos. In Ban Samsum, district officials misused the LFAP to expropriate village land for a Chinese company by allocating individual household plots within only half of village territory, thus "legally" freeing up the most fertile land to lease to the company.[23] This dispossession was legitimated by representing village land as state forest, as meeting national goals to increase "tree cover" through industrial tree plantations, as promoting cash crops to replace highland rice, and by promises of plantation wage labor opportunities to alleviate poverty. At the time, the government was also promoting rubber tree plantations as a positive alternative to shifting cultivation, perceiving them as more modern, economically lucrative, and environmentally beneficial, regardless of the greater ecological diversity of shifting cultivation systems compared with a monocrop of rubber trees.[24] Villagers of Ban Samsum resisted dispossession by refusing to work for the company, enacting everyday forms of resistance[25] such as surreptitiously trampling and uprooted rubber saplings and "accidentally" setting fire to the trees while they were burning weeds, and by seeking support from different state officials.[26] These weedy acts of disobedience functioned to undermine the planned success of the plantation. *Nya Kiloh* and other weeds also played a role in this resistance, as without farmers manually removing them from the plantation, they grew thick and tall, competing with the young rubber saplings for light and nutrients. Without farmers' labor and in context of their anonymous disruptive actions, the rubber trees withered.

The plan of the LFAP and its associated maps to provide the state with legibility and control of local rights to and uses of state forestlands was uprooted, reshaped, and resisted by the ongoing undocumented weedy practices of villagers and local officials alike, as well as by the proliferation of the weeds themselves, which influenced farmers' actions by making the policy ecologically unviable. The tangle of *Nya Kiloh* obscuring the village map of Ban Samsum is itself a form of weedy disobedience, reflective of villagers' acts of ignoring and adapting to state plans that do not make local sense. While such plans do have an impact, they become entangled in messy place-based socio-ecologies and relationships of power and inequality, germinating unplanned practices and effects that sprout from them like weeds in a farmer's field, reshaping and resisting plans in unexpected ways.

Notes

1. The name of the village has been changed to protect those who participated in this research.

2. K. Biswas, "Some Foreign Weeds and their Distribution in India and Burma," *Indian Forester* 60, no. 12 (1934): 861–65; Xiangqin Yu, Tianhua He, Jianli Zhao, and Qiaoming Li, "Invasion Genetics of *Chromolaena odorata* (Asteraceae): Extremely Low Diversity across Asia," *Biological Invasions* 16, no. 11 (2014): 2351–66.

3. W. Roder, S. Phengchanh, B. Keoboulapha, and S. Maniphone "*Chromolaena odorata* in Slash-and-Burn Rice Systems of Northern Laos," *Agroforestry Systems* 31(1995): 79–92.

4. Marita Ignacio Galinato, Keith Moody, and Colin M. Piggin, *Upland Rice Weeds of South and Southeast Asia* (Los Banos, Philippines: International Rice Research Institute, 1999).

5. Roder et al., "*Chromolaena odorata*." Interestingly, in France *Chromolaena* is popularly known as *l'herbe du Laos* (Lao weed). The local naming of weed species after political events or periods of rule coinciding with their spread has also been noted in Indonesia by Michael R. Dove, "The Practical Reasons of Weeds in Indonesia: Peasant vs. State Views of *Imperata* and *Chromolaena*," *Human Ecology* 14, no. 2 (1986): 163–90; and in Thailand by Tim Forsyth and Andrew Walker, *Forest Guardians, Forest Destroyers: The Politics of Environmental Knowledge in Northern Thailand* (Seattle: University of Washington Press, 2008).

6. This is not true of all types of weeds, and different species are managed differently. See Karen Elisabeth McAllister, "Shifting Rights, Resources and Representations: Agrarian Transformation of Highland Swidden Communities in Northern Laos" (PhD diss., McGill University, 2016), 502; Roder, "*Chromolaena odorata*."

7. McAllister, "Shifting Rights."

8. There are many different systems of swidden cultivation, but all involve sequences of burning, cultivation and fallow.

9. Unlike paddy rice, which is grown on flat flooded fields, upland swidden rice is grown on dry soils and often on relatively steep unterraced hillsides.

10. Anna Lowenhaupt Tsing, *Friction: An Ethnography of Global Connection* (Princeton: Princeton University Press, 2005).

11. See also Dove, "The Practical Reasons," and Roder, "*Chromolaena odorata*."

12. Thongchai Winichakul, *Siam Mapped: A History of the Geo-Body of a Nation* (Honolulu: University of Hawai'i Press, 1994).

13. Henri Lefebvre, *The Production of Space* (1991; repr., Oxford: Blackwell, 2000); Michel de Certeau, *The Practice of Everyday Life* (1984; repr., Berkeley: University of California Press, 2013).

14. Siam was the previous name for Thailand.

15. James C. Scott, *Seeing Like a State: How Certain Schemes to Improve the Human Condition Have Failed* (New Haven: Yale University Press, 1998).

16. Nancy Lee Peluso, and Peter Vandergeest, "Genealogies of the Political Forest and Customary Rights in Indonesia, Malaysia and Thailand," *The Journal of Asian Studies* 60, no. 3 (2001): 761–812.

17. "Forestlands" are legally defined as "all land plots with or without forest cover, which are determined by the State as Forestlands" (Article 2, Lao Forest Law 2007).

18. Keith Barney, "China and the Production of Forestlands in Lao PDR: A Political Ecology of Transnational Enclosure," in *Taking Southeast Asia to Market: Commodities, Nature, and People in the Neo-Liberal Age*, ed. Joseph Nevins and Nancy Lee Peluso (Ithaca: Cornell University Press, 2008), 91–107; McAllister, "Shifting Rights."

19. Ibid.

20. Karen Elisabeth McAllister, "Rubber Rights and Resistance: The Evolution of Local Struggles against a Chinese Rubber Concession in Northern Laos," *Journal of Peasant Studies* 42, no. 3/4 (2015):817–37; McAllister, "Shifting Rights."

21. All personal names have been changed to protect the identity of the individuals.

22. Village name has been changed to protect confidentiality.

23. Village name has been changed.

24. This claim is based on assumed negative effects of swidden landscapes and positive effects of "tree plantations" that are contested by empirical evidence (see Forsyth and Walker, *Forest Guardians*; and Dove, "The Practical Reasons").

25. Scott, *Seeing Like a State*.

26. McAllister, "Rubber Rights and Resistance."

Zoomorphic Wickerwork Figure: Australian Administered British New Guinea, 1908

Helen Verran

The zoomorphic figure of my title is also known as "Object 1985.0339.1215 Papuan Official Collection, National Museum of Australia."[1] In the ethnographic literature it is called the "rattan-cane carcass of a *ruru*" (a place-being).[2] I have never met this object face-to-face so to speak, but I do hope to make its acquaintance in the future. Using the website maintained by the National Museum of Australia I have gazed often at a single photograph of the object, and re-read the rather sparse notes that appear alongside the image on my computer screen. The object is the brownish-greyish color of old cane; 2840 mm in length, 660 mm high, and 900 mm wide. It has an open smiling mouth and a slightly worried look created by a double woven brow. It seems to have an eye painted on it in white, but sometimes when I study the image that seems more like a museum-applied inscription, a sort of index number. When that happens, when I am unable to decide who painted the eye/inscription and when, then the two very different plans in which this object has participated become difficult to disentangle.

Two looped rattan canes have been bound together at their ends, and a third circular cane bound about half way along the loops' length so that the looped canes separate in two dimensions to form the lips of an open-mouthed fish which, seemingly, is smiling. Using these canes as a structure, the surfaces of the fish-shaped head have been created by woven split cane, thus the object's title "zoomorphic wickerwork figure" in the register of items that comprise Australia's Papuan Official Collection. That collection is one legacy of Australia's supportive role in British imperialism, establishing and enforcing the boundaries of some of the territories over which the British empire claimed protected extractive and global trading rights.

Using this object in comparatively inquiring into past contrivings of two quite distinct plans for enacting governance of Papua New Guinea's Elema peoples, whose lands fringe the coastline of what is now Gulf Province, I utilize the "amodern" analytic framing developed by social philosopher Bruno Latour.[3] This will have me offering a novel exegesis of ethnographic reports of Elema life in the early twentieth century. Utilizing a framing which refuses the modern constitution of the social as a separate and distinct domain within a natural cosmos, has me "reading-into" colonial anthropology reports, to generate two novel contrasts between plans.

In Papua New Guinea, modern twentieth-century anthropology saw Papuans as humans arrested at an early stage of development[4] engaging in a primitive social life through "magico-religious impersonation."[5] In that rendition, Elema society is merely expression of fanciful myth and hence irrational. However, using the analytic that Latour has articulated, and reading against the grain of the ethnographic texts, I take the Elema participation in elaborate and splendid ceremonial occasions which are described in full and vivid detail in ethnographic reports, as contributing to the collective enactment of a form of parliament, what in using a very old form of that modern word *parliament*, we might call a "parlement of place." This parlement of place involves immanent "earth-beings"[6] as other-than-human participants.[7]

As Latour elaborated in in his project *Making Things Public*, this involves quite different entities and quite different procedures than modern parliaments.[8] Perhaps the most useful way to read Latour's initiative here is to see parliaments of all sorts, modern and amodern, as concerned with the processes of "making the publics of things". In the case of modern parliaments, making the publics of things is concerned with various means for bringing into being particular publics committed to promoting happenings of particular things, including places. In the case of Indigenous Elema parlements of place, the effect of gatherings for the making of things' publics, was that particular groups of Elema would identify with and promote the interests of particular immanent earth-beings. Thus, Elema parlements of place enacted an ontology of people-place.

The object "zoomorphic wickerwork figure" which colonial anthropologists believed was merely an expression of some far-fetched magical beliefs amongst a primitive people is, in my reading an object active in an early twentieth century Indigenous Elema governance.[9] In working as such, the processes it was involved in were informed by an elaborated tradition of epistemics enacted in cultural forms unfamiliar to the modern Western men of the British empire, who believed absolutely in the workings of science and its naturalistic cosmos.

Articulating contrasting parliaments/parlements is my first reading-into the ethnographic archive; my first eisegetic move. Focusing on the concept of object of governance, which is what I take the zoomorphic wickerwork figure to have been in a dual sense in the early twentieth century, is a second reading-into. I am contrasting its alternative roles as on the one hand, "mediator," which "transform, translate, distort, and modify the meaning or the element they are supposed to carry," with that of "an intermediate" which "transports meaning or force without transformation . . . for all practical purposes an intermediate can be taken . . . as a black box," on the other.[10] Passing from Elema hands and institutions to come into the clutches of Australian institutions, very likely via the interventions of a Motuan Papuan who was an Elema trading partner, this zoomorphic wickerwork figure has performed in both these roles in the early twentieth century.

Although some of the ends of the split cane are now loose and flailing, the surface still holds together well enough for the object to seem quite lively; it makes me smile, when I examine it on the screen. I chose this object to focus my stories through because with some piecing together of archival fragments it is possible to infer a biography for this ethnographic item.

At the beginning of the twentieth century this zoomorphic wickerwork figure played a part in a plan by an Australian administration to generate the information by which administrative policies, to produce "a better brown man,"[11] might be devised. The Australians sought to intervene in and to edit native culture[12] in order to produce compliant native populations, so that restiveness did not disturb establishment of imperial governance designed to facilitate resource extraction. The irony of this policy that saw the Australian state collecting ethnographic objects was that ethnographic objects themselves came to be a resource to be extracted by colonizing entrepreneurs.[13] The colonial Australian state found itself passing legislation to curtail the trade in ethnographic objects. I propose that in participating in the Australians' plan, the ethnographic item was rendered as a modern object to work as an intermediate in modern governance.

Before it became entangled in that elaborate imperial plan which partially backfired, and at best only ever generated equivocal insight into "the native mind," the object carried a different name and came to life as part of a very different plan intervening in a quite alternative governance regime. The reason it became part of the Australians' plan was that it had already played a part in a plan hatched by an unknown young Elema man to recognize and to publicly perform the entanglement of his life with that of a particular friendly immanent being of place. His maternal uncle or other male kin with whom his family was linked in a formal gift-giving relation, would have woven this object out of cane as the beginning in this young man's plan to conjure up a splendid vision of *Mara'ope*, a fish-being of place amongst the Elema people of Gulf Province in Papua New Guinea—a place-being with which his family had special connection. The plan for the evocation of this particular being of place involved participation in what I read as a parlement involving humans and other-than-humans, particularly those other-worldly place-beings.

It is likely that in a highly decorated version, this object had a short but vivid life as an *Eharo* participating in ceremonial proceedings. Later, after being stripped of its covering of colorful bark cloth, and "highly valued feather decorations, and the shell ornaments which hung on [the *Eharo's*] breast... [this] bare body of a mask was stuck up out of the way, on some wall at the rear of the *eravo* [the men's house]."[14] I surmise that this temporarily abandoned carcass of a *ruru* (a place-being), hung there until through several changes of hands, and after being carried on several journeys, it became one of the first objects to enter and be registered as part of Australia's colonial era Papuan Official Collection.

The Elema Plan

The Elema Peoples of Papua are a group of thirty to fifty thousand people whose villages spread along a hundred or so miles of the Papuan coastal deltas and immediate coastal hinterlands, stretching from Cape Possession (named by the British in 1848) to the Purari River Delta. Fast-flowing rivers carry fresh water from the very high mountain ranges to the north. Several decades into the twenty-first century, many Elema are bilingual, spending their time predominantly in Port Moresby while retaining strong connections to their family village. Elema Peoples see themselves as divided into three inter-related language groupings and recognize themselves as having village-level governance where age-grade generational grouping remains significant. Communal farming produces ample supplies of food—pigs, yams, taro, sago, bananas, and coconuts, which is supplemented by beach fishing. Many of the earth-beings who are associated with the one whose likeness has been contrived in the zoomorphic wickerwork figure that is the lens by which I bring Elema governance to light, are still recognized, revered, and feared in the ongoing life of villages, albeit that the *Hevehe* ceremonies, or what I am naming as parlements of place involving other-than-humans, when place beings and village people negotiated life collectively, are no longer held.[15]

Elema oral traditions speak of a migration to the coast they now inhabit from a remote inland region to the northwest, from the high mountains at the headwaters of the Purari and the Tauri rivers. Archaeological investigation at early village sites indicates that this migration took place around three centuries ago. This was an unsettled period in the Papua New Guinea Central Highlands, generating flows of refugees. It seems that the Elema migration was part of this general movement amongst central highland peoples, the choice for the Elema being the down-river coastward journey.

Soon after they arrived to populate the lands on the beach side of the river deltas, as their oral history traditions tell it, they took to building imposing men's houses and to holding elaborate ceremonies as part of which local beings of place were invited to participate in village life for set periods of time. These ceremonial times in village life ended in extensive feasting involving a number of proximate villages. The masked ceremonies became more and more magnificent over several centuries. In the late nineteenth century British visitors recorded the Elema peoples as living in well-ordered village communities, with an identifiable *Pukari* (a chief-like head man) with social organization that featured strong segregation of the sexes, age sets, and age grading, as part of a loose and dispersed clan system. In villages the *Pukari*, the headmen, were responsible for the maintenance of peace and order. What impressed the missionaries greatly were the imposing men's houses (*eravo*), from which elaborate ceremonies were staged and which women did not enter. Central in governance of the complex assemblies of neighboring Elema villages, was the multifaceted *Hevehe* creed.[16]

In beginning to account an Elema plan involving what would later be named as a zoomorphic wickerwork figure, I imagine a time in an Elema village at a specific juncture in the years-long collective enactment of the many stages of the *Hevehe* creed, when members of the other-world, many of whom were and are, not well-disposed to humans, came to stay in the village. In one sense this happened through humans embodied as male adults taking on the alternative exterior forms, but in another sense—in the other-worldly time-place of these other-than-human earth-beings, the occasion happened because these beings had themselves decided to fraternize with their human neighbors. According to the stories[17] those humans needed constant reminding to act in proper ways and to conduct *Hevehe* ceremonies. *Eharo*, is that part of the long ceremony that features friendly place-beings, *Mara'ope* among them.

In beginning his account of a *Hevehe* gathering, Williams, a long-serving colonial government anthropologist, bids us imagine a very large and airy *eravo*, a men's house,

> a huge structure built along the lines of [an everyday dwelling] but enormously magnified...hog backed rising from rear to front, the roof culminating in a high and forward-extending peak...It covers a length of some 110 ft. and at the front reaches a height of over 50 ft.[18]

The planning for the final gathering and mingling of other-than-human place-beings and villagers across a period of three weeks or so involves choreographing daily events associated with presence in the village of various particular beings of place. At the explicit invitation of the older men, various of these earth-beings have come to stay and partake of the spoils of village life, rather than remain in their places in the forests or the sea. Planning begins with discussion of a secret excursion. Agreement is made for a group of particular men to go unnoticed and unannounced beyond the village and its gardens

> to the bush to procure the rattan cane of which skeletons of the masks [that are the feature of the ceremony] are made. As each man cuts his cane, [with great sincerity] he utters the traditional name of his *ruru*, the name which his father, and grandfather used before him, and calls upon it to leave the forest and live for a space [of time] in the village, for the time has come for him [the man] to reveal the mystery [of the beings of place] to his son. *Arulavai!* [the beings of place in our midst!] *Meravakore* [nearby beings of this place uninhabited by humans!], *Lepulela!* [beings living amongst the creepers!] he may cry, 'Come to our village, I have a pig waiting for you'. On the return of the expedition the cane is smuggled into the *eravo* by night, unseen by women or children.[19]

A month or so later a group of younger Elema men would be busy at work fashioning *Eharo* costumes of many different types. The *Eharo* mask is constructed of cane with a covering of bark cloth; but it may assume the most fanciful forms. Some are ornate and lavishly decorated with sprigs of leaves and tufts of feathers. The planned contrivance of *Eharo* costumes including the masks occurs in the liminal space of the men's house the *eravo*, amongst its many supernatural inhabitants, many malign, some benign, hidden from the eyes and experience of female and younger Elema.

The *Eharo* emerge near the end of a *Hevehe* ceremony, having been made in neighboring villages they arrive en masse in the village staging the ceremony. On the occasion that Williams describes, around forty members from nearby villages were present. It is they who have brought the *Eharo* masks. Many such masks returned to their home villages after the ceremony, but a number of exterior forms of *Eharo* entered the *eravo* from which the ceremony was being staged.[20] There they would be disrobed and the rattan-cane carcass of the *ruru* hung up, out of the way, on a back wall.

As I read the ethnographic account Williams has provided us with, in the early twentieth century Elema peoples maintained a governance regime that attested to the power of place-centered non-human forms of other-worldly life.

In a form of parlement of place periodically these beings came to life when male adult humans took on and became their embodied forms. The presence of other-than-humans was evoked in part through the agency of mask skeletons bedecked in colorful finery. The object that has become my familiar by virtue of my studying its form on my computer screen, was part of the means by which the power of place was harnessed in ordering human affairs. In other words, more than a century ago, the object played a minor, but direct role in a plan to re-enact the means of sustaining a particular historical and political moment in an Elema village.

The Australians' Plan

> In Papua there is the opportunity to prove that it is possible to rule a native race without destroying it and that it is left to Australia to make of this splendid dream a glorious reality . . . to serve the dictates of humanity and [Australia's] own best material interests.[21]

These words were written by Hubert Murray (1861–1940), a first-generation, Sydney-educated Irish Australian. Hubert Murray, a judge in the New South Wales State Court, was appointed as Lieutenant Governor of the colonial territory of Papua or British New Guinea in 1908. Here Murray sees himself speaking on behalf of a newly fledged Australian state, formed in 1901 by the union of five British colonies as a federation, governed by Edward, King of the United Kingdom, Emperor of India, and King of the British Dominions Canada, Australia, and New Zealand. Murray's appointment coincided with the Protectorate of British New Guinea being handed over to Australian administration, and in carrying out the tasks of his office he would loudly and often condemn much of what happened in the previous administrative regime as "unscientific bungling."[22] He was determined that, freed of the British Colonial Office, his rule would express "Australian ideals" in its administration.[23]

As Murray saw it, Australian ideals of colonial justice and administration were informed by the science of anthropology. But, he would specify, not the sort of anthropology that depended on "forced inferences," but rather of an economically informed empirical anthropology. Murray's administration would institute a scientific anthropology, "pursued properly on the basis of evidence."[24] Evidence would take the form of material objects that would provide information, on the basis of which knowledge of the primitive societies to be governed would become available. Articulating this knowledge base would be the work of the government anthropologist.

[Murray's] investigations reaped more and more detail of native mentality. ... He ... found amazing ignorance among the New Guinea people, but he ... also found astonishing intelligence and capacity ... His knowledge of classical lore and ancient history reminded him that the most startling and barbaric practices among the native of New Guinea had their exact counterpart in the habits and ceremonies of the older races of Europe ... particularly those of the Scythians, from whom he claimed descent ... Civic indifference had been much more marked in the early days of European history than he found [amongst] these backward people.[25]

In Murray's administration, foot patrols by Australian colonial officials were to be conducted according to a clearly articulated plan of colonial rule: each village and district was to be visited at least once in three months. Rules governing hygiene and orderly behavior were to be drawn up for each district and explained to inhabitants. In addition, patrols were to collect items that would provide material evidence of the life of the people. This "evidence" duly became "The Official Papua Collection," now held as a discrete collection in the National Museum of Australia.[26]

In 1906, when the planned ethnographic collection had not yet been gazetted by the Australian state, but was still just a personal interest of Murray's, in a letter to his brother, a professor of Greek at Oxford University, Murray noted that a Motuan man named Dona had followed his instructions to bring back items procured in the course of a native trading expedition. These would provide evidence on which the life of the Elema people might be inferred. At that time, the Elema peoples, having their own established and functioning governance arrangements and trade networks well in hand, were hostile toward the intrusions of the Australian administration. Indeed, it is likely that Dona was traveling west as part of the well-established and highly valued *Hiri* trading cycle.[27] No record remains of what exactly Dona persuaded his Elema trading partners to hand over for him to transport back to colonial administrative headquarters in Port Moresby. However, in 2022 using the National Museum of Australia Collection website,[28] it is possible for me to infer what it was that Dona carried back to Hubert Murray to initiate the collection process. The item(s) Dona transported in the trading *lakatoi* canoe were likely the first items to enter what is now the Australian Official Papuan Collection. Among those was the zoomorphic wickerwork figure that lies at the core of this essay.

In 1907, Murray formally announced his plan to use the outcomes of anthropological research as a tool for planning his colonial administration, and in a letter to the Australian Minister for Home and Territories sought permission to turn the collection into an "official" one, and to establish an ethnological museum in Port Moresby. Murray gained permission from the Australian state for both requests. Plans and budgets were drawn up for building a museum and caring for the emerging collection.[29] The state saw its gathering collection of everyday items as investment in assembling a knowledge base to plan a new Papuan society suitable for a world of industry, trade, and capital accumulation through extractivism.

Glimpsing Historical and Political Moments through a Zoomorphic Wickerwork Figure

My essay has elaborated two historical tableaux situating an object that is now part of a colonial era ethnographic collection, in two distinct planning moments involving Papua New Guinea's Elema peoples. The first plan concerned governance of what we might think of as an Elema people-place. In this involvement the zoomorphic wickerwork figure took up an active role mediating passage between an other-worldly locale of place-beings, and a locale of human village men, women, and children along with domesticated village pigs and garden plants. The active work of the figure's mediation was the happening of what we might name as a *Hevehe* parlement—a meeting and mixing up across a set period of time of the other-than-human place-beings and the human inhabitants of a particular place, lubricated with dance, song, feasting, and storytelling.

The second plan concerned governance of Elema peoples and their places by officers of a far-flung imperium, the British Empire. In this plan the newly hatched Australian state, acting as intermediate, set about generating knowledge of the Elema society and its natural setting, to better furnish the archive of the British Empire. The plan imagined this work as providing a knowledge base to better design the means of forcing the institution of a new modern governance order upon the Elema peoples and their places.

My stories have shown how and where divergently disparate historical and political moments were entangled in and as the being of the object at the core of the essay.

Notes

1. https://collectionsearch.nma.gov.au/icons/images/kaui2/index.html#/home?usr=CE

2. Francis Edgar Williams, *The Drama of Orokolo: The Social and Ceremonial Life of the Elema* (1940; repr., Papua New Guinea: University of Papua New Guinea Press, 2015), 136.

3. Bruno Latour, "Postmodern? No Simply Amodern! Steps Towards an Anthropology of Science. An Essay Review," *Studies in the History and Philosophy of Science* 21, no. 1 (1990): 145–71.

4. C. R. Hallpike, *The Foundations of Primitive Thought* (Oxford: Clarendon Press, 1979), 36

5. Williams, *The Drama of Orokolo*, 136.

6. Marisol de la Cadena, *Earth Beings: Ecologies of Practice across Andean Worlds* (Durham: Duke University Press, 2015), 25.

7. Marisol de la Cadena and Mario Blaser, eds., *A World of Many Worlds* (Durham: Duke University Press, 2018), 5.

8. Bruno Latour, "From Realpolitik to Dingpolitik: Or How to Make Things Public," in *Making Things Public: Atmospheres of Democracy*, ed. Bruno Latour and Peter Weibel (Cambridge, MA: MIT Press, 2005), 14–43, on 14.

9. Helen Verran and Michael Christie, "Objects of Governance as Simultaneously Governed and Governing," in "Objects of Governance," special issue, *Learning Communities Journal* 15 (March 2015): 60–65, https://www.cdu.edu.au/sites/default/files/the-northern-institute/cdu_ni_learning_communities_journal_2015.pdf

10. Latour, "From Realpolitik to Dingpolitik," 39.

11. John Hubert Plunkett Murray, *The Scientific Aspect of the Pacification of Papua: Presidential Address at the Meeting of the Australian and New Zealand Association for the Advancement of Science* (Port Moresby, Territory of Papua: Edward George Baker, Government Printer, 1932), 6

12. Ben Dibley, "Assembling an Anthropological Actor: Anthropological Assemblage and Colonial Government in Papua," *History and Anthropology* 25, no. 2 (2014): 263–79, on 268.

13. For commentary on imperial extractivism in trade in ethnographic objects, see Dibley, "Assembling an Anthropological Actor," 267; see also Rainer Buschmann, "Exploring Tensions in Material Culture. Commercialising Ethnography in German New Guinea, 1870–1904," in *Hunting the Gatherers: Ethnographic Collectors, Agents and Agency in Melanesia, 1870s–1930s*, ed. Michael O'Hanlon and Robert Louis Welsch (Oxford: Berghan Books, 2000), 55–80.

14. Williams, *The Drama of Orokolo*, 269

15. Chris Urwin, "Excavating and Interpreting Ancestral Action: Stories from the Subsurface of Orokolo Bay, Papua New Guinea," *Journal of Social Archaeology* 19, no. 3 (2019): 279–306, on 283.

16. Herbert Brown, *Three Elema Myths* (Canberra: Research School Pacific Studies, Australian National University, 1988).

17. Morea Pekoro, *Orokolo Genesis* (Port Moresby, Territory of Papua: Niugini Press, 1973).

18. Williams, *The Drama of Orokolo*, 5

19. Williams, *The Drama of Orokolo*., 141

20. Williams, *The Drama of Orokolo*, 265–87

21. Quoted in Lewis Lett, *Sir Hubert Murray of Papua* (Sydney: Collins, 1949), 95.

22. Lett, *Sir Hubert Murray*, 66.
23. Lett, *Sir Hubert Murray*.
24. Lett, *Sir Hubert Murray*, 309.
25. Lett, *Sir Hubert Murray*, 84.
26. Sylvia Schaffarczyk, "Australia's Official Papuan Collection: Sir Hubert Murray and the How and Why of a Colonial Collection," *reCollections Journal of the National Museum of Australia* 1, no. 1 (2006): 41–58.
27. Tom Dutton, ed., *The Hiri in History: Further Aspects of Long Distance Motu Trade in Central Papua* (Canberra: The Australian National University, 1982).
28. https://collectionsearch.nma.gov.au/
29. Schaffarczyk, "Australia's Official Papuan Collection."

The Planning Moment: Avenues for Analysis

Sarah Blacker, Emily Brownell, Anindita Nag, Martina Schlünder, Sarah Van Beurden, and Helen Verran

The introduction to this book asks what might constitute a history of colonial and postcolonial planning. We then described three warp threads (coloniality, knowledge production, and materiality,) that we asked our contributors to highlight in their essays, as a way to give shape to this history. Warp threads are the backbone of any textile, holding the tension as the weft threads are woven through. In considering the prospect of identifying a few centuries of history and a broad swath of the globe as the "planning moment," we must also ask, what is accomplished by doing so, when planning lives on in the specific conditions of its everyday enactment? Indeed, the life of plans does not persist on paper, but in the lived experience of them. In invoking the "planning moment" we are shifting between the invocation of a moment as an historical era and a moment as an instant, unique and recalcitrant to analysis. There is a remainder left behind in reading these entries: they do not always or neatly scale up.

The moment—as in the actual time in which something happens—has always been a struggle for historians to access. But there is a good reason to try. As Scott Hancock writes, "Moments resonate. Moments help us care about specific people and events from our pasts. Moments make the past matter for the present."[1] The 27 essays at the heart of this volume move us closer to a history of planning that is constituted through acts like buying an alarm clock, ordering a bag of seeds, or stealing beehives. Moments can of course accrue into a historical event, but they also work against a focus on events as the preoccupation of history. Each and every planning moment is different from each and every other planning moment, yet these disparate happenings can be connected through stories. Placing materiality in the center also confounds our expectations of time: the qualities and constraints of objects can fail or succeed in an instant, but that can also work to undo or rework plans long after the fact (and we are left to volley between an instant and a historical epoch). The knowledge produced through such moments of planning also stretches long past their deployment. But of course: planning is itself an ardent faith in the future. So here in a brief coda to this collection of essays, we want to offer a few examples of how to "make sense" of these entries. And in this instance, we want to think of the historical aftermaths of the planning moment and note four ongoing processes that emerge from colonial and postcolonial planning and have long historical afterlives. These are processes of infrastructuralizing, enculturating, spatializing, and temporalizing.

Infrastructuralizing

The construction of physical infrastructures served in the colonies as proof of development and remains one of the clearest physical manifestations of imperial plans. Infrastructures determined the distribution of ostensibly "public" goods while also justifying colonial rule, reinforcing difference and distinction, and remaking space. As infrastructural studies have grown in the past few decades, more scholars are also thinking beyond "fixed facilities." Some of this work quite explicitly engages with the colonial and postcolonial world in particular to consider new ways that less sublime or even immaterial infrastructures constituted cities, publics, and labor and how central they were to establishing both power and belonging.[2] Among other things, these infrastructures remade relationships with natural resources, often entrenching and defining the very terms by which the Global South could engage with the rest of the world. Not all of these infrastructures emerged through explicit state planning efforts. Many pre-date colonial imposition and endure in sly ways; some were tacitly encouraged as alternative channels of distribution by the austere economics of colonial rule while other infrastructures successfully thwarted colonial attempts to control the distribution of goods.[3] Colonial plans frequently built on the past in ways they alternately fictionalized, valorized, and obscured. Colonial economies could be adept at capturing the infrastructures and entrepreneurship of "native" producers, even while considering these producers "backward."[4] Yet other examples abound of finding ways to exploit the exploiter, syphoning off or redistributing resources to which they were otherwise denied access.

Many of these essays capture both kinds of infrastructure—those constituted most obviously through their physical presence and those that functioned as knowledge infrastructures. Of course, many infrastructures do both—rearrange space and resources and produce networks of expertise.[5] What remains of these infrastructures and how do they live on? One way to consider these remnants is through the persistence of path dependencies. It is not always intentional that such networks persist, but rather the inertia of well-worn grooves, such as the Euclidian layout of colonial towns ("Grid") or a programming language ("COBOL") that outlasts the vision of its planners. At the heart of uneven development so often lies the crisis of path dependency.[6] How to remake the past into the future was a key question of decolonization and whether these path dependencies were enduring obstacles or opportunities. We can also see this in the autarkic visions of Francoist Spain, seeking to remake agriculture and industry as a form of political legitimacy ("Dodecahedral Silo").

Other articulations of path dependency are more unforeseen and can be considered opportunistic while still emerging from official infrastructural plans: this might be the establishment and endurance of charcoal as the predominant fuel in urban Africa alongside the skeletal public energy utilities distributed to European neighborhoods ("Charcoal"). Or, the ongoing reliance on Japanese expertise in post-colonial Korea ("Dam"), despite the bitterness of war and occupation.

Perhaps most interestingly of all, are the moments when path dependencies leave these infrastructures open to co-optation by human and non-human forces. We can see this in Liberia where it can be hard to stop a parasite after you have cleared forests, built roads, and created a labor infrastructure as vast as Firestone ("Parasite"). Or, in late colonial Belgian Congo where workshops designed to be spaces for honing a colonial vision of "traditional craft" instead functioned as spaces of innovation and experimentation, ironically more in the spirit of long standing local dynamics ("Kishikishi"). Of course, it was the hubris of colonial planners that their efforts would outlast them. In moving between these snapshots of planning in action and a larger accounting for the planning moment, we aim to parse those aspirations from what it means to actually live in their remains.

Enculturating and Differentiating

As this volume demonstrates, planning practices are central to the processes of the invention of the other and the self. The differentiation of cultures and groups -in other words, the hierarchies and ranking that emerge from processes of enculturation- can be the explicit part of a plan in attempts to define and identify. Classification exercises are often undertaken with the explicit aim of intervention and effecting change. The context of colonialism is particularly salient to these processes, both because of its penchant for planning and because of its unequal power relations. And despite the decolonizing rhetoric of many postcolonial leaders, their technologies of rule, including processes of identification and classification, often relied on or built from colonial structures. In other words, the ethos of postcolonial plans is difficult to disentangle from their colonial epistemics even though they might be deployed in opposition.

Crucial to the work of differentiating have been the logics and policies that were put into place to ensure the success and longevity of colonial rule: the strategies of "divide and conquer" and "divide and rule" through which the colonized were divided into separate groups physically, symbolically, and epistemically.[7] By far the most important category of difference that structured colonial planning and bears the impact of differentiation through science was race. "Surnames," for example, reveals a neocolonial set of relations between US and Latin-America undergirded by a racial ideology but also the role of realities on the ground in confounding these plans.

Rarely fully hegemonic, plans tend to have layered, multiple, limited, and at times unintentional effects on subjectivities, from their intended audience to the intermediaries executing or applying plans. For example, the fertility surveyors in Puerto Rico, the education of whom was one of the more important—but not planned- outcomes of the plan ("Fertility Survey Workforce"). Or the teachers administering the census in the New Hebrides, whose societal position was shaped by this work ("Census").[8] This "slippage" can be considered as a condition of the manifestation or enactment of power through plans.[9] "National Budget" reveals a top-down example of postcolonial economic planning in Sudan against colonial categories which nevertheless inadvertently reinforced them. Plans and planning practices in postcolonial contexts equally further worldviews that effect certain identities and subjectivities in ways that reproduce the structures of coloniality. "Orangutan" displays a similar entanglement between the colonial and postcolonial in Borneo. Animals held in custody of the state came to stand in for the subjects of a colonial experiment, forging an ethos of freedom and decolonization from within colonial confines. And as this entry demonstrates, processes of enculturation are not limited to human subjects.

Hierarchically ordered categories of difference are both epistemic and material, and this work is often lent authority through scientific discourses and practices.[10] Discussions of "ontological conflicts" as not merely involving different sets of beliefs or knowledges that remain contained, but instead as "conflicting stories about 'what is there' and how they constitute realities in power-charged fields" reveal the stakes of the work of enculturating and differentiating.[11] The differentiating enacted by settler colonial planning is often carried out through the development of technologies and standards that reflect and express colonial settler interests.[12] "Water Samples," which considers an Indigenous-led environmental monitoring program in Canada, reveals that these technologies and standards often turn out to be flexible and co-optable.

Spatializing

Over the course of the past three decades, space and spatiality as an analytic concept has emerged as the core of critical analysis on planning. Many planners see their work as "managing space," expressing an understanding of space as an a priori knowledge, external to human thinking. Such a static understanding of space has its roots in Euclidean geometry and seventeenth century Newtonian physics. The legacy of Euclid has remained so influential that it has led John Friedmann to suggest that, "the conventional concept of planning is so deeply linked to the Euclidian mode that it is tempting to argue that if the traditional model has to go, then the very idea of planning must be abandoned."[13]

The idea of planning, however, was not abandoned with the rise of a non-Euclidean relational view of space. Most famously attributed to Einstein's theory of relativity and Leibniz's philosophy of space, the relational view of space undercut the belief in the universality of the traditional absolutist view of space. An inevitable outcome of this was a renewed appreciation of space and an increased pressure on planning systems to become more spatial, and on planners to act more spatially.[14] The spatial turn has subsequently produced profound effects on colonial and postcolonial planning histories. In her influential study on space, geographer Doreen Massey argues that space—its control and administration—functions not as an auxiliary to colonial conquest, but as a central component enacted through such conquest.[15] Thus, the question of imperialism's geographies, both "imagined," in Edward Said's terms, and material, have long remained a central concern of postcolonial studies and of colonial discourse analysis in general.[16]

The essays in the book do not dwell on conquest, but rather explore how spatiality is instantiated in practices of planning. To paraphrase Massey: if one is taking a train journey across a landscape, one is cutting across a myriad of ongoing stories. Spatiality is like a pincushion of a million stories, and if one stops at any point and walks about, there will be a house, a settlement with many stories being told, and many more stories that could be told. If you "walk about" in the essays, they offer glimpses of how new forms of spatiality emerged in colonial planning, often clashing, but also meshing to bring postcolonial worlds into being. In this regard, the Cold War construction of the Hindustan Steel Plant at Rourkela ("Steel Plant"), illustrates how the twin tendencies of colonial spatial segregation and technological modernity were the primary forms of legitimacy for a newly formed postcolonial state. Such industrial spaces are revealed as "planned material embodiments of the future, . . . written into the spatial present in order for the state to claim success in its project of postcolonial transformation." How postcolonial states built on colonial practices of erasure is also evident in "Riverbed." The essay highlights how such practices of remaking nature "makes room for [contemporary Korea's] economic development" in a highly contested enactment of modernity's fantasy of the "dead flat surface."

In the context of contemporary Canada, this sort of reordering of space is now being made again, through the reemergence of alternative relational spatialities co-constituting landscapes. "Water Samples," which tells the story of Indigenous versus state use of water sampling techniques, reveals the often overlooked, diverse processes of place-making and spatial tactics, and demonstrates how the dialectical tension between "objective" and "situated" knowledge remains unresolved and continues to manifest itself in contemporary planning practices.

Temporalizing

The twentieth-century rise of economic planning as a vehicle for national economic development posed a challenge to the presumption that planning was a way of arranging space.[17] What characterizes these planning regimes is a way of conceptualizing time, where planning is future-oriented insofar as it seeks to normalize the future, reducing risk and uncertainty. In recent years, a body of anthropological work has called for a rigorous rethinking of time and temporality, pointing to the specificities of time-space relationship in planning.[18]

In contrast, for planners, time and temporalities seem to have remained blank spaces, even blind spots.[19] Much of the work within planning theory has placed the future as the sole object of planning while planning practice remains technocratically wedded to the logics of probabilistic forecasting. This future, it seems can be sufficiently described through the content and intention of a plan; it is a time that is modernizing, linear, and irreversible. This is surprising, particularly since a close attention to planning practices suggests that time can appear as a fragile and precarious category, something that is established and produced through collective, social processes.[20] Time has been worked upon, tweaked, and planned for a long time: it has been standardized, synchronized, measured, broken up into small portions, and reassembled.[21]

A familiar narrative of time and its regulation is found in the colonial context in which time functioned as a powerful legitimizing discourse for colonial and missionary projects. Recent work in colonial history has emphasized the dark role of narratives of progress in underwriting global and racial hierarchies. "Progress," so often seen as the outcome of planning, reframed the grubby real-time politics of colonial domination and exploitation as an orderly natural process of evolution toward modernity. Time figures in many narratives of colonial planning as a culturally constructed instrument of power.[22] Yet historians of colonialism, like planners, have been slow to pick up on the analytic utility of multiple temporalities.

In the texts collected here we see planning as constitutive of entangled temporalities rather than as a category with specific established and fixed models. The consequential nature of temporality becomes clear in an essay like "Constitution." Bhimrao Ramji Ambedkar, the chairman of the drafting committee of the Indian constitution, saw citizenship as emergent in an Indian future that was continually in the process of becoming the present. This contrasted with more conventional understandings of citizenship, originating in the past which the present would inevitably become. This originating past was seen as determined by religion, culture, and territory.

Often unwittingly, colonial planning and its aftermaths are non-linear. Several contributions in this book show that nonhuman animals and plants like Tsetse flies, orangutans, gorillas, bees, and weeds are particularly good at cutting across plans once they unfold their own "undisciplined" and elliptical trajectories and temporalities. For example, the unpredictable cyclic temporality of "Seeds" and the promise of plants they contain cut across the linear time of nineteenth-century economic planning that ends abruptly in war. The seed here is both a relic and a promise of the future.

In postcolonial societies, past and future remain bound together and the present is filled with figures of loops and returns. In "EMES Sonochron," a ghostly tension between latency and urgency is produced by German society's refusal to face its Nazi past. In other stories, different sorts of temporal couplings are articulated. In "Hackathon," the event described expresses not planning but rather speculation. Here, the ends of economic planning are projected into an imagined future in some form of product design, usually, but not always, in the form of an algorithm. This also reveals how much temporalities are entangled in affect. Time anxieties are omnipresent: one has no time to speed up, or else, a little time, but still not enough to jump to another scale of modernity.

Notes

1. Scott Hancock, "What Good Is a Moment?" *J19: The Journal of Nineteenth-Century Americanists* 9, no. 1 (2021): 61–68, 62.

2. To explore the increasingly robust field of Global South infrastructures, a good starting point is this conversation with Gabrielle Hecht about designing a syllabus for a course on the topic as well as the syllabus itself. Isabel M. Salovaara, "Teaching Infrastructures: A Conversation with Gabrielle Hecht," *Society for Cultural Anthropology*, accessed August 5, 2022, https://culanth.org/fieldsights/teaching-infrastructures-a-conversation-with-gabrielle-hecht; Gabrielle Hecht, "Infrastructure and Power in the Global South Syllabus," published online, December 2019, https://gabriellehecht.files.wordpress.com/2019/10/303-infrastructure-power-global-south-v4-1.pdf.

3. For more on colonial austerity, see Emma Park's forthcoming book, *Infrastructural Attachments* (Duke University Press, 2023). Also relevant, Emily Brownell, *Gone to Ground: A History of Environment and Infrastructure in Dar es Salaam*, Pittsburgh University Press, 2020; Sheetal Chhabria, *Making the Modern Slum: The Power of Capital in Colonial Bombay*, Seattle: University of Washington Press, 2019.

4. One now canonical example of this is cocoa farmers in West Africa; see Sara S. Berry, "The Concept of Innovation and the History of Cocoa Farming in Western Nigeria," *Journal of African History* 15, no. 1 (1974): 83–95.

5. Pointing out the duality or multiplicity of form and function in infrastructure here is not novel. Several scholars have noted the different ways infrastructure is experienced or made manifest. This is in part, as scholars have noted, because infrastructure is fundamentally relational. See Penny Harvey and Hannah Knox, "The Enchantments of Infrastructure," *Mobilities* 7, no. 4 (2012): 521–36; Brian Larkin, "The Politics and Poetics of Infrastructure," *Annual Review of Anthropology* 42, no. 1 (2013): 327–43; Susan Leigh Star, "The Ethnography of Infrastructure," *American Behavioral Scientist* 43, no. 3 (1999): 377–91.

6. Samir Amin, *Unequal Development: An Essay on the Social Formations of Peripheral Capitalism* (New York: Monthly Review Press, 1976); Walter Rodney, *How Europe Underdeveloped Africa*, rev. ed., (Washington, DC: Howard University Press, 1981); Albert Adu Boahen, *African Perspectives on Colonialism* (Baltimore, MD: Johns Hopkins University Press, 1989).

7. The processes of enculturation that accompanies colonialism and its planning was central in the "cultural turn" in scholarship and has for several decades been a popular subject in scholarly literature. While initially the focus was on the ways in which colonialism generated and imposed cultural change in the colonies, often via

processes of colonial invention and reimagination, scholars also turned toward the history of colonial cultures in the metropole, and lately the focus has been more on movement, circulation, borderlands, and transnationalism.

8. In fact, contrary to the hegemonic nature of the plans in James Scott's description of high modernist contexts, there is often a decided difference between the prescriptive ethos of plans and their material and subjective effects. James C. Scott, *Seeing Like a State: How Certain Schemes to Improve the Human Condition Have Failed*, New Haven, CT: Yale University Press, 1999. The literature on colonial intermediaries is large. See, for example, Roger Sanjek, "Anthropology's Hidden Colonialism: Assistants and Their Ethnographers," *Anthropology Today* 9, no. 2 (1993): 13–18; Benjamin N. Lawrence, Emily Lynn Osborn, and Richard L. Roberts, eds., *Intermediaries, Interpreters, and Clerks: African Employees in the Making of Colonial Africa* (Madison: University of Wisconsin Press, 2015); Lynn Schumaker, *Africanizing Anthropology: Fieldwork, Networks, and the Making of Cultural Knowledge in Central Africa* (Durham: Duke University Press, 2001); Hervé Jezequal, "Voices of Their Own? African Participation in the Production of Knowledge in French West Africa, 1910–1950," in *Ordering Africa: Anthropology, European Imperialism and the Politics of Knowledge*, eds. Helen Tilley and Robert Gordon (Manchester: University of Manchester Press, 2007), 145–73.

9. On slippage and the manifestation of power, see Homi Bhabha, *The Location of Culture* (London: Routledge, 1994), 85–86.

10. Kapil Raj, "Beyond Postcolonialism . . . and Postpositivism: Circulation and the Global History of Science," *Isis* 104, no. 2 (2013): 337–47; Suman Seth, "Colonial History and Postcolonial Science Studies," *Radical History Review* 127 (2017): 63–85.

11. Mario Blaser, "Ontological Conflicts and the Stories of Peoples in Spite of Europe: Toward a Conversation on Political Ontology," *Current Anthropology* 54, no. 5 (2013): 547–568, 548.

12. See, for example, Audra Simpson, *Mohawk Interruptus: Political Life across the Borders of Settler States* (Durham: Duke University Press, 2014).

13. John Friedmann, "Toward a Non-Euclidean Mode of Planning," *Journal of the American Planning Association* 59, no. 4 (2007): 482–86.

14. For some, a focus on integrating space with the socio-political landscape of society served as a stepping stone to understand contemporary spatial transformations such as globalization and related theorizations of power, space, and place. Initiated by the pioneering work of Henri Lefebvre and Michel Foucault, efforts to interlace planning with its production process was developed further in the works of David Harvey, Edward Soja, Manuel Castells, and Peter Marcuse, who challenged the characterization of planning as a benevolent act. Henri Lefebvre, *The Production of Space* (Oxford: Blackwell Publishers, 1991); David Harvey, *The Condition of Postmodernity: An Enquiry into the Origins of Cultural Change* (Malden, MA: Blackwell Publishers, 1990); David Harvey, *Spaces of Hope* (Berkeley, CA: University of California Press, 2000); Edward W. Soja, "Foreword," in *Postcolonial Spaces: The*

Politics of Place in Contemporary Culture, eds. Andrew Teverson and Sara Upstone (Basingstoke: Palgrave Macmillan, 2011), ix–xiii.

15. Doreen Massey, *The Doreen Massey Reader*, eds. Brett Christophers, Rebecca Lave, Jamie Peck, and Marion Werner (Newcastle upon Tyne: Agenda Publishing, 2017); Marion Werner, Jamie Peck, Rebecca Lave, and Brett Christophers, eds., *Doreen Massey: Critical Dialogues* (Newcastle upon Tyne: Agenda Publishing, 2018); Nigel Thrift, "Space: The Fundamental Stuff of Geography," in *Key Concepts in Geography*, eds. Nicholas J. Clifford, Sarah L. Holloway, Stephen P. Rice, and Gill Valentine, 2nd ed. (London: SAGE Publications, 2009), 85–96; "Doreen Massey on Space," Social Science Bites (podcast), February 1, 2013, accessed August 9, 2020, https://www.socialsciencespace.com/2013/02/podcastdoreen-massey-on-space/.

16. Edward W. Said, *Orientalism* (New York: Pantheon Books, 1978), 49–73.

17. For example, Cambridge University Press established the series Cambridge Economic Handbooks soon after World War II. In the 1980s the series continued as Cambridge Economics. See Michael Ellman, *Socialist Planning*, 2nd ed. (Cambridge: Cambridge University Press, 1989). This book is now in a 3rd edition.

18. Laura Bear, "Doubt, Conflict and Mediation: The Anthropology of Modern Time," in "Doubt, Conflict, Mediation: The Anthropology of Modern Time," ed. Laura Bear, special issue, *Journal of the Royal Anthropological Institute* 20, S1 (2014): 3–30; Jane Guyer, "Prophecy and the Near Future: Thoughts on Macroeconomic, Evangelical, and Punctuated Time," *American Ethnologist* 34, no. 3 (2007): 409–21; Simone Abram and Gisa Weszkalnys, "Introduction: Anthropologies of Planning; Temporality, Imagination, and Ethnography," *Focaal: Journal of Global and Historical Anthropology* 61 (2011): 3–18; Simone Abram, "The Future of Planning?" online article, metropolitics.org, November, 13 2013, accessed August 10, 2020, http://www.metropolitiques.eu/The-future-of-planning.html.

19. The exception here is a recent book inspired by Bruno Latour that focuses on planning through materialities, including temporalities' materialities, Robert A. Beauregard, *Planning Matter: Acting with Things* (Chicago: University of Chicago Press, 2015), 151–71.

20. Henri Lefebvre, *Rhythmanalysis*, trans. Stuart Elden and Gerald Moore (New York: Continuum, 2004).

21. John Postill, "Clock and Calendar Time: A Missing Anthropological Problem," *Time & Society* 11, nos. 2/3 (2002): 251–70; Postill 2000, now as John Postill, *Media and Nation Building: How the Iban Became Malaysian* (Oxford and New York: Berghahn Books, 2006); Barbara Adam, *Time and Social Theory* (Cambridge: Polity Press, 1990); and, analyzing a counter example, Zara Mirmalek, "Working Time on Mars," *KronoScope* 8, no. 2 (2009): 158–78.

22. Dipesh Chakrabarty, "Postcoloniality and the Artifice of History: Who Speaks for the 'Indian' Pasts?" *Representations* 37 (1992): 1–26.

Acknowledgments

The plan for this book was born at the Max Planck Institute for the History of Science (MPIWG) in Berlin, where we were all fellows. It would be impossible to imagine this project coming together intellectually and materially without this institution, and specifically its Department III: "Artifacts, Action, Knowledge," which has brought a focus on the non-Western world to the center of the history of science and technology. Being part of this department in its early days was an exciting opportunity for us to shape a workshop and a book project explicitly aimed to address the colonial context and its lingering presence—both of which are too often missing from the history of science.

We are deeply grateful to the department's director, Dagmar Schäfer, for giving us an intellectual home for this project and supporting it from inception to publication. MPIWG fellow Kavita Philip was also an early group member and was foundational to conceptualizing this project—both the idea that we must look at planning through historically situated empirical examples and as a particularly colonial and postcolonial technology.

We would also like to thank the logistical staff at the MPIWG, as well as Department III's office staff, all of whom made the immense task of organizing international workshops possible. Spencer Forbes, Melanie Glienke, Gina Grzimek, Lennart Holst, Sarah Kuehne, and Alison Kraft offered crucial editing support.

Our project also benefited immensely from those we shared it with as we developed it further. We are thankful for the invitation to present the project at the Science, Technology, and Race Speakers' Series at the Newkirk Center for Science and Society at the University of California, Irvine, and for the feedback offered there. Orit Halpern and the anonymous reviewers of the manuscript gave valuable insights, which made this a much better book. Nick DiLiberto offered editorial support and a close reading of the introduction and conclusion. The Ohio State University provided financial support for the editing process.

We thank Thomas Lay for his brilliant editorial suggestions, for his enthusiastic support of this book, and for shepherding it to publication. The editorial team at Fordham University Press—particularly Kem Crimmins, Mark Lerner, and Courtney Adams—offered invaluable support.

Finally, we are grateful for the sincere, challenging, and enlivening engagement of our workshop participants and authors, who generously engaged with our questions about planning, the colonial, the postcolonial, the settler colonial, and knowledge making. As with all good intellectual endeavors, this book emerged from the process, not according to the plan.

—The Editors

Archival Sources

African Studies Collections. Special and Area Studies Collections. George A. Smathers Libraries. University of Florida, Gainesville.
Bureau of Applied Social Research Papers. Columbia University Archives.
Center for History of Medicine. Francis A. Countway Library of Medicine. Harvard University, Boston.
Central Zionist Archives, Jerusalem, Israel.
Cisler, Walker L. Papers. Bentley Historical Library. University of Michigan, Ann Arbor.
Colonial Archive. Ministry of Foreign Affairs (Ministerie van Buitenlandse Zaken), The Hague, Netherlands.
Cultural Anthropology and History. Archives. Royal Museum for Central Africa, Tervuren, Belgium.
Duraffourd, Camille. Fonds. Centre des Archive diplomatiques de Nantes, France.
Family Study Center Collection. University of Minnesota Archives, Minneapolis.
Fondo Informes Anuales. University of Puerto Rico Archive, San Juan.
Foundation Projects. Rockefeller Archive Center, Sleepy Hollow, New York.
Fundación Luis Muñoz-Marín, Archivo Histórico, San Juan, Puerto Rico.
Hill, Reuben. Papers. University of Minnesota Archives, Minneapolis.
Holberton, Frances E. Papers. Charles Babbage Institute Archives. University of Minnesota, Minneapolis.
Marín, Luis Muñoz. Papers. The New York Public Library, New York.
Records of the Department of State. U.S. National Archives.
Royal Palace Archives, Brussels, Belgium.
Sarawak Museum, Kuching, Sarawak.
Stycos, J. Mayone. Papers. Division of Rare and Manuscript Collections. Cornell University Library, Ithaca.
State Archives, Calcutta, West Bengal.

Sudan Archive. Archives and Special Collections. Durham University.
UNOG Registry. Records and Archives. League of Nations Secretariat, Geneva, Switzerland.
Western Pacific Archives. University of Auckland.
West Bengal State Archives, Calcutta.
Whitman, Loring. Collection. Indiana University Libraries, Bloomington.

Bibliography

Abraham, Itty. "The Andamans as a 'Sea of Islands': Reconnecting Old Geographies through Poaching." *Inter-Asia Cultural Studies* 19, no. 1 (2018): 1–20.
——. *How India Became Territorial: Foreign Policy, Diaspora, Geopolitics.* Stanford, CA: Stanford University Press, 2014.
Abram, Simone, and Gisa Weszkalnys. "Introduction: Anthropologies of Planning; Temporality, Imagination, and Ethnography." *Focaal: Journal of Global and Historical Anthropology* 61 (2011): 3–18.
——. "The Future of Planning?" Online article (metropolitics.org), November, 13 2013. Accessed August 10, 2020. http://www.metropolitiques.eu/The-future-of-planning.html.
Acosta, Ivonne. *La Mordaza: Puerto Rico, 1948–1957.* Río Piedras: Edil, 1989.
Adam, Barbara. *Time and Social Theory.* Cambridge: Polity Press, 1990.
Advisory Committee on National Agenda Planning. "The Five-Year Plan of the Moon Jae-in Government."
Agnew, John A., and Stuart Corbridge. *Mastering Space: Hegemony, Territory and International Political Economy.* London: Routledge, 1995.
Agrawi, Arun. *Environmentality: Technologies of Government and the Making of Subjects.* Durham: Duke University Press, 2005.
Akeley, Carl. *Ethan In Brightest Africa.* Garden City, NY: Garden City, 1923.
Alavi, Hamza. "The State in Post-Colonial Societies: Pakistan and Bangladesh." *New Left Review* 74 (July 1972): 59–81.
Alexander, Ernest R. "There is No Planning—Only Planning Practices." *Planning Theory* 15, no. 1 (February 2016): 91–103.
Allais, Lucia. "Disaster as Experiment: Superstudio's Radical Preservation." *Log*, no. 22, (Spring/Summer 2011): 125–29.
Allen F. Roberts, "Peripheral Visions." In *Memory: Luba Art and the Making of History.* Mary Nooter Roberts and Allen F. Roberts, ed. Munich: Prestel, 1996.

Allen, Michael. "The Establishment of Christianity and Cash-Cropping in a New Hebridean Community." *Journal of Pacific History* 3, no.1 (1968): 25–46.

Allman, Jean. "Phantoms of the Archive: Kwame Nkrumah, a Nazi Pilot Named Hanna, and the Contingencies of Postcolonial History-Writing." *The American Historical Review* 118, no. 1 (2013): 104–29.

Allmendinger, Philip, and Mark Tewdwr-Jones. "The Communicative Turn in Urban Planning: Unravelling Paradigmatic, Imperialistic and Moralistic Dimensions." *Space and Polity* 6, no. 1 (April 2002): 5–24.

———. eds. *Planning Futures: New Directions for Planning Theory.* London: Routledge, 2002.

Alvey, Mark. "The Cinema as Taxidermy: Carl Akeley and the Preservative Obsession." *Framework: The Journal of Cinema and Media* 48, no. 1 (2007): 23–45.

Ambedkar, Bhimrao R. [Babasaheb]. "Annihilation of Caste: With a Reply to Mahatma Gandhi" [1936]. In Ambedkar, *Writings and Speeches*, vol. 1, 23–25.

———. *Dr. Babasaheb Ambedkar: Writings and Speeches.* Vol. 1. Edited by Vasant Moon. 1979. Reprint. New Delhi: Dr. Ambedkar Foundation, 2014.

———. "Draft Constitution—Discussion: Motion re Draft Constitution" [1948]. In *Dr. Babasaheb Ambedkar. Writings and Speeches*. Vol. 13. Edited by Vasant Moon, 49–70. 1994. Reprint. New Delhi: Dr. Ambedkar Foundation, 2014.

———. "Evidence Before the Southborough Committee" [1919]. In Ambedkar, *Writings and Speeches*, vol. 1, 243–78.

———. "The Hindu Social Order: Its Essential Principles." In *Dr. Babasaheb Ambedkar. Writings and Speeches*. Vol. 3. Edited by Hari Narake, 95–129. Reprint. 1987. New Delhi: Dr. Ambedkar Foundation, 2014.

———. "Ranade, Gandhi and Jinnah" [1943]. In Ambedkar, *Writings and Speeches*, vol. 1, 205–240.

———. "States and Minorities" [1947]. In Ambedkar, *Writings and Speeches*, vol. 1, 381–449.

———. *The Untouchables: Who Were They and Why Did They Become Untouchables?* New Delhi: Amrit Book Co., 1948.

Amin, Samir. *Unequal Development: An Essay on the Social Formations of Peripheral Capitalism.* New York: Monthly Review Press, 1976.

Anderson, Warwick. *Colonial Pathologies: American Tropical Medicine, Race, and Hygiene in the Philippines.* Durham: Duke University Press, 2006.

Appadurai, Arjun. "Number in the Colonial Imagination." In *Orientalism and the Postcolonial Predicament: Perspectives on South Asia*, edited by Carol A. Breckenridge and Peter van der Veer, 314–36. Philadelphia: University of Pennsylvania Press, 1993.

Appel, Hannah. "Toward an Ethnography of the National Economy." *Cultural Anthropology* 32, no. 2 (2017): 294–322.

Arnold-Foster, H. O. ed. *The Queen's Empire: A Pictorial and Descriptive Record.* London: Cassell, 1897.

Arnold, David. *Colonizing the Body: State Medicine and Epidemic Disease in Nineteenth-Century India*. Berkeley: University of California Press, 1993.
Asian-African Conference. *Asia-Africa Speaks from Bandung*. Djakarta: Ministry of Foreign Affairs, Republic of Indonesia, 1955.
Atul Kohli, *Poverty Amid Plenty in the New India*. Cambridge: Cambridge University Press, 2012.
Ayala, César and Rafael Bernabe. *Puerto Rico in the American Century: A History since 1898*. Chapel Hill: University of North Carolina, 2007.
Aytekin, Attila E. "Agrarian Relations, Property and Law: An Analysis of the Land Code of 1858 in the Ottoman Empire." *Middle Eastern Studies* 45, no.6 (2009): 935–51. https://doi.org/10.1080/00263200903268694.
Azevêdo, Eliana. *Características antropogenéticas da população da Bahia, Brazil*. Research report presented to Organization of American States, 1979.
Bahl, Ekta. "An Overview of CSR Rules under Companies Act, 2013." *Business Standard India*, March 10, 2014. http://www.business-standard.com/article/companies/an-overview-of-csr-rules-under-companies-act-2013-114031000385_1.html.
Baker, John R. *Man and Animals in the New Hebrides*. London: Routledge & Sons, 1928.
——. "On Sex-Integrate Pigs." *British Journal of Experimental Biology* 2, no. 2 (1925): 247–63.
Balakrishna, Sitamraju. *Family Planning, Knowledge, Attitude and Practice, A Sample Survey in Andhra Pradesh*. Hyderabad: National Institute of Community Development, 1971.
Barandiaran, Javiera. "Chile's Environmental Assessments: Contested Knowledge in an Emerging Democracy." *Science as Culture* 24, no. 3 (2015): 251–75.
Barbara Harrisson, *Orang-utan*. London: Collins, 1987.
Barber, Marshall A., Justus B. Rice, and James Y. Brown. "Malaria Studies on the Firestone Rubber Plantation in Liberia, West Africa *The American Journal of Hygiene* 15, no. 3 (May 1932): 601–33.
Bardhan, Pranab K. *The Political Economy of Development in India*. Delhi: Oxford University Press, 1985.
Barker, Adam J., Toby Rollo, and Emma Battell Lowman. "Settler Colonialism and the Consolidation of Canada in the Twentieth Century." In *The Routledge Handbook of the History of Settler Colonialism*, edited by Edward Cavanagh and Lorenzo Veracini, 153–68. London: Routledge, 2016.
Barney, Keith. "China and the Production of Forestlands in Lao PDR: A Political Ecology of Transnational Enclosure." In *Taking Southeast Asia to Market: Commodities, Nature, and People in the Neo-Liberal Age*, edited by Joseph Nevins and Nancy Lee Peluso, 91–107. Ithaca: Cornell University Press, 2008.
Baxi, Upendra. "Outline of a 'Theory of Practice' of Indian Constitutionalism." In *Politics and Ethics of the Indian Constitution*, edited by Rajeev Bhargava, 92–118. New Delhi: Oxford University Press, 2008.

Bear, Laura. "Doubt, Conflict and Mediation: The Anthropology of Modern Time." In "Doubt, Conflict, Mediation: The Anthropology of Modern Time," edited by Laura Bear. Special issue, *Journal of the Royal Anthropological Institute* 20, S1 (2014): 3–30.

Beauregard, Robert A. *Planning Matter: Acting with Things.* Chicago: University of Chicago Press, 2015.

Benítez, Jaime. *Junto a la Torre: Jornadas de un Programa Universitario.* Río Piedras: UPR, 1962.

Berman, Nina, Klaus Mühlhahn, and Patrice Nganang, eds. *German Colonialism Revisited: African, Asian, and Oceanic Experiences.* Ann Arbor: University of Michigan Press, 2014.

Bernstein, J. M., Adi Ophir, and Ann Laura Stoler, eds. *Political Concepts: A Critical Lexicon.* New York: Fordham University Press, 2018.

Berry, Sara S. "The Concept of Innovation and the History of Cocoa Farming in Western Nigeria." *Journal of African History* 15, no. 1 (1974): 83–95.

Beyer, Kurt. *Grace Hopper and the Invention of the Information Age.* Cambridge, MA: MIT Press, 2009.

Bhabha, Homi. *The Location of Culture,* London: Routledge, 1994.

Bhan, Gautam. "Notes on a Southern Urban Practice." *Environment and Urbanization* 31, no. 2 (2019): 639–54.

Bilger, Burkhard. "Hearth Surgery." *New Yorker,* December 13, 2009. https://www.newyorker.com/magazine/2009/12/21/hearth-surgery.

Biswas, K. "Some Foreign Weeds and their Distribution in India and Burma." *Indian Forester* 60, no. 12 (1934): 861–65

Blacker, Sarah. "Strategic Translation: Pollution, Data, and Indigenous Traditional Knowledge." *Journal of the Royal Anthropological Institute* 27, no. S1 (2021): 142–58.

Blacker, Sarah. "Technologies of Quiescence: Measuring Biodiversity, 'Intactness,' and Extractive Industry in Canada." *Catalyst: Feminism, Theory, Technoscience* 8 (2) (2022): 1–26.

Blackstock, Cindy. "67-Million Nights in Foster Care." *Maclean's,* October 16, 2019. https://www.macleans.ca/opinion/67-million-nights-in-foster-care/.

Blake, Stanley S., and Stanley E. Blake. "The Medicalization of Nordestinos: Public Health and Regional Identity in Northeastern Brazil, 1889–1930." *The Americas* 60, no. 2 (2003): 217–48.

Blaser, Mario. "Ontological Conflicts and the Stories of Peoples in Spite of Europe: Toward a Conversation on Political Ontology." *Current Anthropology* 54, no. 5 (2013): 547–568.

Blomley, Nicholas. "Land Use, Planning, and the 'Difficult Character of Property.'" *Planning Theory & Practice* 18, no. 3 (2017): 351–64. https://doi.org/10.1080/14649357.2016.1179336.

Blondeel, William. "Settlement-Policy of the Missionaries of Africa (White Fathers) in Kivu, Belgian Congo, Phase 1910–1914." *Belgisch Tijdschrift Voor Nieuwste Geschiedenis* 3/4 (1975): 329–62.

Boahen, Albert Adu. *African Perspectives on Colonialism*. Baltimore, MD: Johns Hopkins University Press, 1989.

Bodry-Sanders, Penelope. *African Obsession: The Life and Legacy of Carl Akeley*. Rev. 2nd ed. Jacksonville, FL: Batax Museum, 1998.

Boelens, Luuk. "Theorizing Practice and Practising Theory: Outlines for an Actor-Relational-Approach in Planning." *Planning Theory* 9, no. 1 (2010): 28–62.

Bonilla, Eduardo Seda. *Interacción Social y Personalidad en una Comunidad de Puerto Rico*. San Juan: Juan Ponce de Leon, 1964.

Bonilla, Eduardo Seda. *Requiem por una Cultura*. Río Piedras: Edil, 1970.

Bowker, Geoffrey C. *Memory Practices in the Sciences*. Cambridge, MA: MIT Press, 2008.

———. "Sustainable Knowledge Infrastructures." In *The Promise of Infrastructure*, edited by Nikhil Anand, Akhil Gupta, and Hannah Appel, 203–22. Durham, NC: Duke University Press 2018.

Boyd, Danah, and Kate Crawford. "Critical Questions for Big Data." *Information, Communication, and Society* 15, no. 5 (2012): 662–79.

Bratton, Benjamin. *The Stack: On Software and Sovereignty*. Cambridge: MIT Press, 2015.

Braun, Carl P. J. G. "Die Agaven, ihre Kultur und Verwendung: Mit besonderer Berücksichtigung von Agave Rigida Var. Sisalana Engelm." *Der Pflanzer* 14, no. 2 (1906): 209–24.

Briggs, Laura. *Reproducing Empire: Race, Sex, Science and US Imperialism in Puerto Rico*. Berkeley: University of California Press, 2002.

Brown, Gary De Ward. "COBOL: The Failure That Wasn't." *The Cobol Report* Object-Z Systems Inc., May 15, 2001. www.cobolreport.com/columnists/gary/05152000.htm.

Brown, George W. *The Economic History of Liberia*. Washington, DC: Associated Publishers, 1941), 231.

Brown, Herbert. *Three Elema Myths*. Canberra: Research School Pacific Studies, Australian National University, 1988.

Brown, Peter. "Microparasites and Macroparasites." *Cultural Anthropology* 2, no. 1 (1987): 155–71.

Brownell, Emily. *Gone to Ground: A History of Environment and Infrastructure in Dar es Salaam*. Pittsburgh University Press, 2020.

Buschmann, Rainer. "Exploring Tensions in Material Culture. Commercialising Ethnography in German New Guinea, 1870–1904." In *Hunting the Gatherers: Ethnographic Collectors, Agents and Agency in Melanesia, 1870s–1930s*, edited by Michael O'Hanlon and Robert Louis Welsch, 55–80. Oxford: Berghan Books, 2000.

Butler, Octavia E. *Parable of the Sower*. New York: Warner Books, 1995.

Buxton, Patrick. "The Depopulation of the New Hebrides and Other Parts of Melanesia." *Transactions of the Royal Society of Tropical Medicine and Hygiene* 19, no. 8 (1926): 420–58.

Caldwell, John. "Fertility Attitudes in Three Economically Contrasting Rural Regions of Ghana." *Economic Development and Cultural Change* 15, no. 2 (1967): 217–38.
Camprubí, Lino. *Engineers and the Making of the Francoist Regime*. Cambridge, MA: MIT Press, 2014.
———. "Resource Geopolitics: Cold War Technologies, Global Fertilizer, and the Fate of Western Sahara." *Technology and Culture* 57, no. 3 (2015): 676–703.
———. "Whose Self-Sufficiency? Energy Dependency in Spain from 1939." *Energy Policy* 125, no. 2 (2019): 227–34.
Carman, Jillian. *Uplifting the Colonial Philistine: Florence Phillips and the Making of the Johannesburg Art Gallery*. Johannesburg: Witwatersrand University Press, 2006.
Carr Saunders, Alexander. "Review of Essays on the Depopulation of Melanesia," *The Eugenics Review* 14, no. 4 (1923): 282–83.
Chakrabarty, Dipesh. "Postcoloniality and the Artifice of History: Who Speaks for the 'Indian' Pasts?" *Representations* 37 (1992): 1–2.
Chalk, Frank. "The Anatomy of an Investment: Firestone's 1927 Loan to Liberia." *Canadian Journal of African Studies* 1, no. 1 (1967): 12–32.
Chandavarkar, Rajnarayan. *Imperial Power and Popular Politics: Class, Resistance and the State in India, 1850–1950*. Cambridge: Cambridge University Press, 1998.
Chang, Ha-Joon. *Kicking Away the Ladder: Development Strategy in Historical Perspective/* London: Anthem Press, 2002.
Chatterjee, Partha. *The Nation and Its Fragments: Colonial and Postcolonial Histories*. Princeton: Princeton University Press, 1993.
Chhabria, Sheetal. *Making the Modern Slum: The Power of Capital in Colonial Bombay*. Seattle: University of Washington Press, 2019.
Chibber, Vivek. *Locked in Place: State-Building and Late Industrialization in India*. Princeton University Press, 2006.
Chin, James, and Jayl Langub. "The Sarawak Administrative Service: Its Origins." In *Reminiscences: Recollections of Sarawak Administrative Service Officers*, edited by James U. H. Chin and Jayl Langub, 1–20. Subung Jaya: Pelanduk.
Cho, Myung-Rae. "The Politics of Urban Nature Restoration: The Case of Cheonggyecheon Restoration in Seoul, Korea." *International Development Planning Review* 32, no. 2 (2010): 145–65.
Chung, Daejeon. "Foreign Things no Longer Foreign: How South Koreans Ate US Food." PhD diss., Columbia University, 2015.
Cleland, John. "A Critique of KAP Studies and Some Suggestions for Their Improvement." *Studies in Family Planning* 4, no. 2 (1973): 42–47.
Cohen-Cole, Jamie. *The Open Mind: Cold War Politics and the Sciences of Human Nature*. Chicago: University of Chicago Press, 2015.
Cohn, Bernard. *Colonialism and Its Forms of Knowledge: The British in India*. Princeton: Princeton University Press, 1996.

Collard J., and C. Lamote. "Les ateliers d'art du Kasaï/Ba ateliers Ya Art na Kasai." *Nos Images* 10, no. 169 (1957).

Collins, Robert O. "The Sudan Political Service: A Portrait of the 'Imperialists.'" *African Affairs* 71, no. 284 (1972): 293–303.

Conrad, Sebastian. *German Colonialism: A Short History*. Translated by Sorcha O'Hagan. Cambridge: Cambridge University Press, 2012.

Conzen, M. R. G. "Towards a Systematic Approach in Planning Science: Geoproscopy." *Town Planning Review* 18, no. 1 (July 1938): 1–26.

Cook, Harold J. *Matters of Exchange: Commerce, Medicine, and Science in the Dutch Golden Age*. New Haven, CT: Yale University Press, 2008.

Cooper, Frederick. *Citizenship between Empire and Nation: Remaking France and French Africa, 1945–1960*. Princeton, NJ: Princeton University Press, 2014.

Cooper, Frederick, and Randall M. Packard. "Introduction." In *International Development and the Social Sciences: Essays on the History and Politics of Knowledge*, edited by Frederick Cooper and Randall M. Packard, 1–142. Berkeley: University of California Press, 1997.

Coopman, Colin. *How We Became Our Data: A Genealogy of the Informational Person* Chicago: University of Chicago Press, 2019.

Corbridge, Stuart, and John Harriss. *Reinventing India: Liberalization, Hindu Nationalism, and Popular Democracy*. Cambridge: Blackwell Publishers, 2000.

Cullather, Nick. *The Hungry World: America's Cold War Battle Against Poverty in Asia*. Cambridge: Harvard University Press, 2013.

Cupers, Kenny. "Editorial: Coloniality of Infrastructure." *E-flux Architecture*, September 2021. https://www.e-flux.com/architecture/coloniality-infrastructure/412386/editorial/.

Cwiertka, Katarzyna. *Cuisine, Colonialism, and Cold War: Food in Twentieth Century Korea*. London: Reaktion Books, 2013.

Dalloul, Motasem A. "Gaza Honey Production Stung by Israeli Policies." *Middle East Eye*, May 21, 2015. https://www.middleeasteye.net/features/gaza-honey-production-stung-israeli-policies.

Daly, M. W. *Darfur's Sorrow: A History of Destruction and Genocide*. Cambridge: Cambridge Univesity Press, 2007.

———. "The Development of the Governor-Generalship of the Sudan, 1899–1934." *The Journal of African History* 24, no. 1 (1983): 77–96.

———. *Imperial Sudan: The Anglo-Egyptian Condominium 1934–1956*. Cambridge: Cambridge University Press, 1991.

Daston, Lorraine. "Objectivity and the Escape from Perspective." *Social Studies of Science* 22, no. 4 (November 1992).

Daston, Lorraine and Peter Galison. "The Image of Objectivity." *Representations* 40 (October 1992): 81–128.

Davidoff Paul, and Thomas A. Reiner. "A Choice Theory of Planning." *Journal of the American Institute of Planning* 28, no. 2 (1962): 103–15.

Davis, Kingsley. "Latin America's Multiplying Peoples." *Foreign Affairs*, July 1, 1947. https://www.foreignaffairs.com/articles/central-america-caribbean/1947-07-01/latin-americas-multiplying-peoples.

Davoudi, Simin. "Planning as Practice of Knowing." *Planning Theory* 14, no. 3 (August 2015): 316–331.

de Arellano, Annette Ramírez, and Conrad Seipp. *Colonialism, Catholicism, and Contraception: A History of Birth Control in Puerto Rico*. Chapel Hill: University of North Carolina Press, 1983.

de Casseres, J. M. "Principles of Planology: A Contribution to the Scientific Foundation of Town and Country Planning." *Town Planning Review* 17, no. 2 (February 1937): 103.

de Certeau Michel. *The Practice of Everyday Life*. Berkeley: University of California Press, 2013. First published 1984 by University of California Press.

de la Bellacasa, Maria Puig. "Matters of Care in Technoscience: Assembling Neglected Things." *Social Studies of Science* 41, no. 1 (2011): 85–106.

de la Cadena, Marisol. *Earth Beings: Ecologies of Practice across Andean Worlds*. Durham: Duke University Press, 2015.

de la Cadena, Marisol, and Mario Blaser, eds. *A World of Many Worlds*. Durham: Duke University Press, 2018.

de Saille, Stevienna. *Knowledge as Resistance: The Feminist International Network of Resistance to Reproductive and Genetic Engineering*. London: Palgrave Macmillan, 2017.

de Silva, G. S. "The East Coast Experiment," Conference on Conservation of Nature and Natural Resources in Tropical Southeast Asia, Bangkok, Thailand, 1965.

de Soto, Hernando. *The Mystery of Capital: Why Capitalism Triumphs in the West and Fails Everywhere Else*. New York: Basic Books, 2000.

de Sousa Santos, Boaventura. "The Heterogeneous State and Legal Pluralism in Mozambique." *Law and Society Review* 40, no. 1 (2006):42–44.

———. *Toward a New Common Sense: Law, Science and Politics in the Paradigmatic Transition*. New York: Routledge, 1995.

de Sousberghe, Léon. *L'art pende*. Brussels: Pailais des Académies, 1959.

Dean, Mitchell. *The Constitution of Poverty: Toward a Genealogy of Liberal Governance*. New York: Routledge, 1991.

Deloria, Vine, Jr. *God is Red: A Native View of Religion*. 30th anniversary ed. Golden, CO: Fulcrum Publishing, 2003.

Denis, Manuel Maldonado. *Puerto Rico: Una Interpretación Histórico-Social*. Mexico: Siglo XXI, 1969.

Denning, Peter, and Craig Martell. *Great Principles of Computing*. Cambridge, MA: MIT Press, 2015.

Depelchin, Jacques. *From the Congo Free State to Zaire (1885–1974): Towards a Demystification of Economic and Political History*. Dakar: Codeseria, 1992.

Deshpande, Satish. *Contemporary India: A Sociological View*. Delhi: Penguin Books, 2004.

Despret, Vinciane. *What Would Animals Say If We Asked the Right Questions?* Translated by Brett Buchanan. Minneapolis: University of Minnesota Press, 2016.

DeWinter, Amos. "Falestinaim yilmedu legadel dvorim bekavarot sheganvu meitanu" פלסטינאים ילמדו לגדל דבורים בכוורות שגנבו מאיתנו [Palestinians will learn to keep bees in hives they stole from us]. *Mashov Hakla'I* 55 (2000).

Dibley, Ben. "Assembling an Anthropological Actor: Anthropological Assemblage and Colonial Government in Papua." *History and Anthropology* 25, no. 2 (2014): 263–79.

Dickson, Courtney, and Bridgette Watson. *Remains of 215 Children Found Buried at Former B.C. Residential School, First Nation Says.* CBC News, May 27, 2021.

Donaldson, S. G., J. Van Oostdam, C. Tikhonov, M. Feeley, B. Armstrong, P. Ayotte, O. Boucher, W. Bowers, L. Chan. F. Dallaire, E. Dewailly, J. Edwards, G. M. Egeland, J. Fontaine, C Furgal, T. Leech, E. Loring, G. Muckle. T. Pereg, P. Plusquellec, M. Potyrala, O. Roceveur, and R. G. Shearer. "Environmental Contaminants and Human Health in the Canadian Arctic." *Science of The Total Environment* 408, no. 22 (2010): 5165–234.

"Doreen Massey on Space." Social Science Bites (podcast). February 1, 2013. Accessed August 9, 2020. https://www.socialsciencespace.com/2013/02/podcastdoreen-massey-on-space/.

Dove, Michael R. "The Practical Reasons of Weeds in Indonesia: Peasant vs. State Views of *Imperata* and *Chromolaena*." *Human Ecology* 14, no. 2 (1986): 163–90

Dowdeswell, Liz, Peter Dillon, Subhasis Ghoshal, Andrew Miall, Joseph Rasmussen, and John P. Smol. *A Foundation for the Future: Building an Environmental Monitoring System for the Oil Sands.* Ottawa: Environment Canada, 2010.

Downey Gary, and Teun Zuiderent-Jerak. "Making and Doing: Engagement and Reflexive Learning in STS." In *The Handbook of Science and Technology Studies*, edited by Ulrike Felt, Rayvon Fouché, Clark A. Miller, and Laurel Smith-Doerr. 4th ed. Cambridge, MA: MIT Press, 2017.

Dutton, Tom, ed. *The Hiri in History: Further Aspects of Long Distance Motu Trade in Central Papua.* Canberra: The Australian National University, 1982.

Dupuy, Pierre. *The Mechanization of the Mind: On the Origins of Cognitive Science.* Cambridge, MA: MIT Press, 2009.

Edwards, Paul. *The Closed World: Computers and the Politics of Discourse in Cold War America.* Cambridge, MA: MIT Press, 1997.

Elden, Stuart. *The Birth of Territory.* Chicago: University of Chicago Press, 2013.

Ellman, Michael. *Socialist Planning.* 2nd ed. Cambridge: Cambridge University Press, 1989.

Elyachar, Julia. "Next Practices: Knowledge, Infrastructure, and Public Goods at the Bottom of the Pyramid." *Public Culture* 24, no. 1 (66) (2012): 109–29.

Engler, Adolf. "Das Biologisch-landwirtschaftliche Institut zu Amani in Ost-Usambara." *Notizblatt des königlichen Botanischen Gartens und Museums zu Berlin* 4, no. 31 (1903): 63–66.

Ensmenger, Nathan. *The Computer Boys Take Over: Computers and the Politics of Discourse in Cold War America* Cambridge, MA: MIT Press, 2010.

"Erstmal wegschließen." *Der Spiegel*, May 20, 1990), 68–73.
Escobar, Arturo. *Encountering Development*. Princeton, NJ: Princeton University Press, 1994.
Fabian, Johannes. *Time and the Other: How Anthropology Makes it Subject*. New York: Columbia University Press, 2002.
"Falsch bombadiert" [sic]. *Der Spiegel*, February 12, 1989, 64–65.
Fanon, Frantz. *Black Skin, White Masks*. Reprint. New York Grove Press, 2008.
———. *A Dying Colonialism*. New York: Grove Press, 1967.
Fawaz, Mona. "Hezbollah as Urban Planner? Questions to and from Planning Theory." *Planning Theory* 8, no. 4 (2009): 323–34. https://doi.org/10.1177/1473095209341327.
———. "The Politics of Property in Planning." *IJURR* 38, no. 3 (2014): 922–34.
Feichtinger, Mortiz, and Stephan Malinowski. "Transformative Invasions: Western Post-9/11 Counterinsurgency and the Lessons of Colonialism." *Humanity* 3, no. 1 (2014): 35–63.
Feldman, Ronnie. "Legal Actions and Protection of Farmers from Thefts Originating in the (Palestinian) Autonomy, and the Government's Demand of the Palestinian Authority to Cooperate for the Purposes of Preventing Theft and for Law Enforcement." Protocol of the State Controller Committee Meeting, Israeli Parliament, March 7, 2000. https://www.nevo.co.il/law_html/law103/15_ptv_497726.htm. Accessed August 19, 2022.
Finnie, Richard. *Korea's New Energy: The Construction of Three Steam-Electric Plants*. San Francisco, CA: Bechtel, 1956.
Fischler, Raphaël. "Fifty Theses on Urban Planning and Urban Planners." *Journal of Planning Education and Research* 32, no. 1 (September 2011): 107–14, https://doi.org/10.1177/0739456X114204.
Flanders, Judith. *A Place for Everything: The Curious History of Alphabetical Order*. New York: Basic Books, 2020.
Fleck, Ludwik. *Genesis and Development of a Scientific Fact*, edited by Thaddeus J. Trenn and Robert K. Merton. Translated by Fred Bradley and Thaddeus J. Trenn. Chicago: University of Chicago Press, 1979.
Forsyth, Tim, and Andrew Walker, *Forest Guardians, Forest Destroyers: The Politics of Environmental Knowledge in Northern Thailand*. Seattle: University of Washington Press, 2008).
Foucault, Michel. "Governmentality." In *The Foucault Effect: Studies in Governmentality: With Two Lectures by and an Interview with Michel Foucault*, edited by Michel Foucault, Graham Burchell, Colin Gordon, and Peter Miller, 67–104. Chicago: University of Chicago Press, 1991.
Frank, Rüdiger. *Die DDR und Nordkorea: Der Wiederaufbau der Stadt Hamhùng in Nordkorea von 1954–1962*. Shaker Verlag: Aachen, 1996.
Freedman, Ronald, and John Y. Takeshita. *Family Planning in Taiwan*. Princeton: Princeton University Press, 1969.
Friedmann, John. "Toward a Non-Euclidean Mode of Planning." *Journal of the American Planning Association* 59, no. 4 (2007): 482–86.

———. *Planning in the Public Domain*. Princeton, NJ: Princeton University Press, 1987.
Fromont, Cecile. *The Art of Conversion: Christian Visual Culture in the Kingdom of Kongo*. Raleigh: University of North Carolina Press, 2014.
Früchte des Zorns: Texte und Materialien zur Geschichte der Revolutionären Zellen und der Roten Zora. Amsterdam: Edition ID, 1993.
Galdikas, Birute. "Orangutan Reproduction in the Wild." In *Reproductive Biology of the Great Apes: Comparative and Biomedical Perspectives*, edited by C.E. Graham, 281–300. New York: Academic Press, 1981
Galinato, Marita Ignacio, Keith Moody, and Colin M. Piggin. *Upland Rice Weeds of South and Southeast Asia*. Los Banos, Philippines: International Rice Research Institute, 1999.
Gardener, Martin. *The Mind's New Science: A History of the Cognitive Revolution*. New York: Basic Books, 1985.
Gargan, Edward A. "International Report; Tanzania's 'Green Gold' Woes." *New York Times*, June 23, 1986.
Gerber, Jan. "'Schalom und Napalm': Die Stadtguerilla als Avantgarde des Antizionismus." In *Rote Armee Fiktion*, edited by Joachim Bruhn and Jan Gerber, 39–84. Freiburg: ça ira, 2007.
Getachew, Adom. *Worldmaking After Empire: The Rise and Fall of Self-Determination*. Princeton, NJ: Princeton University Press, 2019.
Ghosh, Amitav. *The Great Derangement: Climate Change and the Unthinkable, The Randy L. and Melvin R. Berlin Family Lectures*. Chicago: University of Chicago Press, 2016.
Gitelman, Lisa. *Paper Knowledge: Toward a Media History of Documents*. Durham: Duke University Press, 2014.
Gollom, Mark. "How Radar Technology Is Used to Discover Unmarked Graves at Former Residential Schools." *CBC News*. June 14, 2021.
Gongyŏng, Ilbon. *Taehanmin'guk suryŏk chosa bogosŏ*. Seoul (?): Ilbon Gongyŏng Chusik Hoesa, 1962.
"Gottverdammter Zufall." *Der Spiegel*, February 21, 1988.
Government-General of Chosen, ed. *Thriving Chosen: A Survey of Twenty-Five Years'Administration*. Seoul: Government-General of Chosen, 1935.
Graham, Loren, and Michel Kantor. *Naming Infinity: A True Story of Religious Mysticism and Mathematical Creativity*. Cambridge, MA: Harvard University Press, 2009.
Grin, Mônica. "Mito de excepcionalidade? O caso da nação miscigenada brasileira." In *O Brasil em dois tempos: História, pensamento social e tempo presente*, edited by Eliana de Freitas Dutra, 321–40. Belo Horizonte: Autêntica, 2013.
Grove, Richard H. *Green Imperialism: Colonial Expansion, Tropical Island Edens and the Origins of Environmentalism, 1600–1860*. New York: Cambridge University Press, 1996.
Guiart, Jean. "Les mouvement coopératif aux Nouvelles-Hébrides." *Journal de la Société des Océanistes* 12 (1956): 242–46.

Guru, Gopal. "Liberal Democracy in India and the Dalit Critique." *Social Research* 78, no. 1 (2011): 99–122.

Gururani, Shubhra. "Flexible Planning: The Making of India's 'Millennium City,' Gurgaon." In *Ecologies of Urbanism in India: Metropolitan Civility and Sustainability*, edited by Anne M. Rademacher and K. Sivaramakrishnan. Hong Kong: Hong Kong University Press, 2013.

Guyer, Jane. "Prophecy and the Near Future: Thoughts on Macroeconomic, Evangelical, and Punctuated Time." *American Ethnologist* 34, no. 3 (2007): 409–21.

Haigh, Thomas, and Mark Priestly. "Innovators Assemble: Ada Lovelace, Walter Isaacson, and the Superheroines of Computing." *Communications of the ACM* 58, no. 9 (2015): 20–27.

"Hakavranim dorshim ezrat hamemshala" הכוורנים דורשים עזרת הממשלה [The beekeepers demand the help of the government]. *Al-Hamishmar* May 18, 1954.

Halevi, Joseph. "The Accumulation Process in Japan and East Asia as Compared with the Role of Germany in European Post-War Growth." In *Post-Keynseian Essays from Down Under: Volume II*, edited by Joseph Halevi, G. C. Harcourt, Peter Kriesler, and J. W. Nevile, 355–67. London: Palgrave McMillan, 2016.

Hallpike, C. R. *The Foundations of Primitive Thought*. Oxford: Clarendon Press, 1979.

Han, Heejin. "Authoritarian Environmentalism under Democracy: Korea's River Restoration Project," *Environmental Politics* 24, no. 5 (2015): 810–29.

Hancock, Scott. "What Good Is a Moment?" *J19: The Journal of Nineteenth-Century Americanists* 9, no. 1 (2021): 61–68.

Hansen, Viveka, and Lars Hansen. *The Linnaes Apostles: Global Science and Adventure*. Whitby, UK: IK Foundation, 2007.

Haraway, Donna. "A Manifesto for Cyborgs: Science, Technology, and Socialist Feminism in the 1980s." *Socialist Review* 80 (1985): 65–107.

Haraway, Donna Jeanne. *Modest-Witness@Second-Millennium. FemaleMan-Meets-OncoMouse: Feminism and Technoscience*. New York: Routledge, 1997.

———. "Teddy Bear Patriarchy: Taxidermy in the Garden of Eden, New York City, 1908–1936." *Social Text*, no. 11 (December 1984): 20–64. https://doi:10.2307/466593.

"Harvard Expedition Off to Africa with Cures for Tropical Diseases." *Boston Traveler*, May 15, 1926.

Harvey, David. *The Condition of Postmodernity: An Enquiry into the Origins of Cultural Change*, Malden, MA: Blackwell Publishers, 1990.

———. *The New Imperialism*. Oxford: Oxford University Press, 2003.

———. *Spaces of Hope*. Berkeley, CA: University of California Press, 2000.

Harvey, Penny, and Hannah Knox. "The Enchantments of Infrastructure." *Mobilities* 7, no. 4 (2012): 521–36.

Hau'ofa, Epeli. "Our Sea of Islands." *The Contemporary Pacific* 6, no. 1 (1994): 148–61.

Hawthorne, Julian. "India Starving," *Cosmopolitan*, August, 1897.

Hecht, Gabrielle. "Infrastructure and Power in the Global South Syllabus." Published online, December 2019. https://gabriellehecht.files.wordpress.com/2019/10/303-infrastructure-power-global-south-v4-1.pdf.

Herzig, Rebecca M. *Suffering for Science: Reason and Sacrifice in Modern America*. Piscataway: Rutgers University Press, 2005.
Hicks, Mar. *Programmed Inequalities: How Britain Discarded Women Technologists and Lost Its Edge*. Cambridge, MA: MIT Press, 2016.
Hill, Reuben, Joseph Stycos, and Kurt Back. *The Family and Population Control: A Puerto Rican Experiment in Social Change*. Chapel Hill: University of North Carolina Press, 1959.
Hilton, Boyd. *Age of Atonement: The Influence of Evangelicalism on Social and Economic Thought, 1785–1865*. Oxford: Oxford University Press, 1991.
Himmelfarb, Gertrude. *The Idea of Poverty: England in the Early Industrial Age*. London: Faber, 1984.
Hochschild, Adam. *King Leopold's Ghost: A Story of Greed, Terror, and Heroism in Colonial Africa*. Boston: Houghton Mifflin, 1998.
Hong, Young-Sun. *Cold War Germany, the Third World, and the Global Humanitarian Regime*. Cambridge: Cambridge University Press, 2015.
Hopper, Grace Murray. "Keynote Address." In *History of Programming Languages*, edited by Richard L. Wexelblat. New York: Academic Press, 1981.
Hose, Charles, and W. McDougall. *The Relations Between Men and Animals in Sarawak*. London: Anthropological Institute of Great Britain and Ireland, 1901.
Hosier, Richard H. "The Economics of Deforestation in Eastern Africa." *Economic Geography* 64, no. 2 (1988):121–36.
Howard, Michael. *Fiji: Race and Politics in an Island State*. Vancouver: University of British Columbia Press, 2011.
Hughes, Christina, and Celia Lury. "Re-Turning Feminist Methodologies: From a Social to an Ecological Epistemology." *Gender and Education* 25, no. 6 (2013): 786–99.
Hughes, Theodore. *Literature and Film in Cold War South Korea: Freedom's Frontier*. New York: Columbia University Press, 2012.
Hughes, Thomas P. "The Evolution of Large Technological Systems." In *The Social Construction of Technological Systems: New Directions in the Sociology and History of Technology*, edited by Wiebe E. Bijker, Thomas Parke Hughes, and Trevor Pinch, 51–82. Cambridge, MA: MIT Press, 1987.
Hurst, Andrew. "State Forestry and Spatial Scale in the Development Discourses of Post-Colonial Tanzania: 1961–1971." *The Geographical Journal* 169, no. 4 (2003): 358–69.
Hussain, Nasser. *The Jurisprudence of Emergency: Colonialism and the Rule of Law*. Ann Arbor: University of Michigan Press, 2003.
Hwang, Jin-Tae. "A Study of State–Nature Relations in a Developmental State: The Water Resource Policy of the Park Jung-Hee Regime, 1961–79." *Environment and Planning A* 47, no. 9 (2015): 1926–43.
"India@Davos: Batting for Inclusive Growth." *NDTV*, January 29, 2011. Video. https://www.ndtv.com/video/news/the-big-fight/india-davos-batting-for-inclusive-growth-189493.

Irani, Lilly. *Chasing Innovation: Making Entrepreneurial Citizens in Modern India.* Princeton: Princeton University Press, 2019.
Isaacson, Steven. *The Innovators: How a Group of Hackers, Geniuses, and Geeks Created the Digital Revolution.* New York: Simon & Schuster, 2015.
İslamoğlu, Huri. "Property as a Contested Domain: A Reevaluation of the Ottoman Land Code of 1858." In *New Perspectives on Property and Land in the Middle East,* edited by Roger Owen. Cambridge, MA: Harvard Center for Middle Eastern Studies, 2000.
Janer, José, Guillermo Arbona, and J. S. McKenzie-Pollock. "The Place of Demography in Health and Welfare Planning in Latin America." *Milbank Memorial Fund Quarterly* 42, no. 2 (1964): 328–45.
Janet Abbate, *Inventing the Internet.* Cambridge, MA: MIT Press, 2000.
Janik, Allan, and Stephen Toulmin. *Wittgenstein's Vienna.* 2nd ed. Chicago: Elephant Paperbacks, 1996.
Jáuregui, Carlos A. "Oswaldo Costa, Antropofagia, and the Cannibal Critique of Colonial Modernity." *Culture & History Digital Journal* 4, no. 2 (2015): e017.
Jencks, Charles. *Modern Movements in Architecture.* New York: Penguin Books, 1985.
Jeon, Chihyung, and Yeonsil Kang. "Restoring and Re-Restoring the Cheonggyecheon: Nature, Technology, and History in Seoul, South Korea." *Environmental History* 24, no. 4 (2019): 736–65.
Jeon, Chihyung. "A Road to Modernization and Unification: The Construction of the Gyeongbu Highway in South Korea." *Technology and Culture* 51, no. 1 (2010): 55–79.
Jewsiewicki, Bogumil. "Rural Society and the Belgian Colonial Economy." In *History of Central Africa,* edited by David Birmingham and Phyllis M. Martin, 123–24. Vol. 2. London: Longman, 1983.
Jezequal, Hervé. "Voices of Their Own? African Participation in the Production of Knowledge in French West Africa, 1910–1950." In *Ordering Africa: Anthropology, European Imperialism and the Politics of Knowledge,* edited by Helen Tilley and Robert Gordon, 145–73. Manchester: University of Manchester Press, 2007.
Jolly, Margaret. "The Anatomy of Pig Love: Substance, Spirit and Gender in South Pentecost, Vanuatu." *Canberra Anthropology* 7, no. 1/2 (1984): 78–108.
Jones, Gareth Stedman. *An End to Poverty? A Historical Debate.* London: Colombia University Press, 2004.
Joseba de la Torre, Joseba, and María del Mar Rubio-Varas. "Nuclear Power for a Dictatorship: State and Business Involvement in the Spanish Atomic Program, 1950–85." *Journal of Contemporary History* 51, no. 2 (2015): 385–411.
Kab-je, Cho. "Ch'ongdokpu gogwan dŭl ŭi kŭdwi." *Wŏlgan Chosŏn* (August 1984): 294.
Kaiser, Katja. "Exploration and Exploitation: German Colonial Botany at the Botanic Garden and Botanical Museum Berlin." In *Sites of Imperial Memory: Commemorating Colonial Rule in the Nineteenth and Twentieth Centuries,* edited by Dominik Geppert and Frank Lorenz Müller, 225–42. Manchester: Manchester University Press, 2015.

———. *Wirtschaft, Wissenschaft und Weltgeltung: Die Botanische Zentralstelle für die deutschen Kolonien am Botanischen Garten und Museum Berlin (1891–1920)*. Vienna: Peter Lang, 2021.

Karcher, Katharina. *Sisters in Arms. Militant Feminisms in the Federal Republic of Germany since 1968*. Oxford: Berghahn, 2017.

Kaviraj, Sudipta. "Global Intellectual History: Meanings and Methods." In *Global Intellectual History*, edited by Samuel Moyn and Andrew Sartori. New York: Columbia University Press, 2015.

Kaya, Alp Yücel. "Les Villes Ottomanes Sous Tension Fiscale: Les Enjeux de l'Évaluation Cadastrale Au XIXe Siècle." In *La Mesure Cadastrale: Estimer La Valeur Du Foncier En Europe Aux XIXe et XXe Siècles*, edited by Florence Bourillon and Nadine Vivier. Rennes: Presses universitaires de Rennes, 2012.

Kazuo, Kawai. "Dai niji suiryoku chōsa to Chōsen sōtokufu kanryō no suiryoku ninshiki." In *Nihon no Chōsen Taiwan shihai to shokuminchi kanryō*, edited by Matsuda Toshihiko and Yamada Atsushi, 304. Tokyo: Shibunkaku, 2009.

Kent, Michael, and Peter Wade. "Genetics against Race: Science, Politics and Affirmative Action in Brazil." *Social Studies of Science* 45, no. 6 (2015): 816–38.

Kertzer, David I., and Dominique Arel. *Census and Identity: The Politics of Race, Ethnicity, and Language in National Census*. Cambridge: Cambridge University Press, 2002.

Kidambi, Prashant. "'The Ultimate Masters of the City': Police, Public order and the Poor in Colonial Bombay, c. 1893–1914." *Crime, Histoire & Sociétés / Crime, History & Societies* 8, no. 1 (2004): 27–47.

Kilroy, David. "Extending the American Sphere to West Africa: Dollar Diplomacy in Liberia, 1908–1926." PhD diss., University of Iowa, 1995.

Kim, Cheehyung Harrison. "The Furnace is Breathing Work as Life in Postwar North Korea." PhD diss., Columbia University, 2010.

———. *Heroes and Toilers: Work as Life in Postwar North Korea, 1953–1961*. New York: Columbia University Press, 2018.

Kim, Chŏng-uk. *Na nŭn pandae handa: 4-tae kang t'ogŏn kongsa e taehan chinsil pogosŏ*. Seoul: Nŭrin Kŏrŭm, 2010.

Kim, Janice. *To Live to Work: Factory Women in Colonial Korea, 1910–1945*. Stanford, CA:

Kim, Kyung-Soo. "Mother, Mother, Our Mother. . . ." *Naeil Shinmun*, April 27, 2010, http://www.naeil.com/upload/cartoon/n100427.jpg.

Kim, Tae-ho. "Miracle Rice for Korea: Tong-il and South Korea's Green Revolution." In *Engineering Asia: Technology, Colonial Development, and the Cold War Order*, edited by Hiromi Mizuno, Aaron S. Moore, and John DiMoia, 189–208. London: Bloomsbury, 2018.

———. "김태호, 리승기의 북한에서의 '비날론' 연구와 공업화: 식민지시기와의 연속과 단절을 중심으로." MA Thesis, Seoul National University 2001.

———. "통일벼"와 1970년대 쌀 증산체제의 형성 (New rice "Tongil" and the technology system of rice production in South Korea in the 1970s)". PhD diss., Seoul National University, 2009.

Kim, Yong-Min. "6월 30일 [June 30]." *Kyunghyang Shinmun*, June 30, 2009. http://news.khan.co.kr/kh_cartoon/khan_index.html?artid=200906292054192&code=361101.

King, Hayden. 2016. "New Treaties, Same Old Dispossession: A Critical Assessment of Land and Resource Management Regimes in the North" in *Canada: The State of the Federation 2013: Aboriginal Multilevel Governance* (pp. 83-98), edited by M. Papillion and A. Juneau. Montreal: McGill-Queen's University Press.

Kingori, Patricia, and René Gerrets. "The Masking and Making of Fieldworkers and Data in Postcolonial Global Health Research Contexts." *Critical Public Health* 29, no. 4 (2019): 494–507.

Kinsey, Alfred, Wardell Pomeroy, and Clyde Martin. *Sexual Behavior in the Human Male*. Philadelphia: W.B. Saunders, 1948.

Kirk, Jay. *Kingdom under Glass: A Tale of Obsession, Adventure, and One Man's Quest to Preserve the World's Great Animals*. New York: Henry Holt, 2010.

Kloppenburg, Jack Ralph. *Seeds and Sovereignty: The Use and Control of Plant Genetic Resources*. Durham, NC: Duke University Press, 1988.

Kochanek, Stanley A. "The Transformation of Interest Politics in India." *Pacific Affairs* 68, no. 4 (1995): 529–50.

Koerner, Lisbet. *Linnaeus: Nature and Nation*. Cambridge, MA: Harvard University Press, 1999.

Koolhaas, Rem. *Delirious New York: A Retroactive Manifesto for Manhattan*. New York: The Monacelli Press, 1994.

Korea Dredging Corporation. *A Seven-Year History of Dredging*. Seoul: Korea Dredging Corporation, 1975.

Krauss, Rosalind. "Grids." *October* 9 (Summer 1979): 51–64.

Kuklick, Henrika. "The Colour Blue: From Research in the Torres Strait to an Ecology of Human Behavior." In *Darwin's Laboratory: Evolutionary Theory and Natural History in the Pacific*, edited by Roy M. Macleod and Philip F. Rehbock, 339–70. Honolulu: University of Hawai'i Press, 1994.

———. "Islands in the Pacific: Darwinian Biogeography and British Anthropology." *American Ethnologist* 23, no. 3 (1996): 611–38

Kyu-Jung Kim, "[I Am a Spade-Fish]." http://www.ohmynews.com/NWS_Web/View/at_pg.aspx?CNTN_CD=A0001371874

Lah, T. J., Yeoul Park, and Yoon Jik Cho. "The Four Major Rivers Restoration Project of South Korea: An Assessment of Its Process, Program, and Political Dimensions," *The Journal of Environment & Development* 24, no. 4 (2015): 375–94.

Lang, Peter. "Suicidal Desires." In *Superstudio: Life without Objects*, edited by Peter Lang and William Menking, 31–51. Milan: Skira, 2003.

Lapp, Michael. "The Rise and Fall of Puerto Rico as a Social Laboratory, 1945–1965." *Social Science History* 19, no. 2 (1995): 169–99.

Larkin, Brian. "The Politics and Poetics of Infrastructure." *Annual Review of Anthropology* 42, no. 1 (2013): 327–43.

LASU-LAWS. "Indira Gandhi's Speech at the Stockholm Conference in 1972." LASU-LAWS Environmental Blog. Accessed May 15, 2017. http://lasulawsenvironmental.blogspot.de/2012/07/indira-gandhis-speech-at-stockholm.html.

Latham, Gwynneth, and Michael C Latham. *Kilimanjaro Tales: The Saga of a Medical Family in Africa*. London: The Radcliffe Press, 1995.

Latour, Bruno. *Conversations on Science, Culture, and Time*. Translated by Roxanne Lapidus. Ann Arbor: University of Michigan Press, 1992.

———. "From Realpolitik to Dingpolitik: Or How to Make Things Public." In *Making Things Public: Atmospheres of Democracy*, edited by Bruno Latour and Peter Weibel, 14–43. Cambridge, MA: MIT Press, 2005.

———. "Postmodern? No Simply Amodern! Steps Towards an Anthropology of Science. An Essay Review." *Studies in the History and Philosophy of Science* 21, no. 1 (1990): 145–71.

———. "Why Has Critique Run out of Steam? From Matters of Fact to Matters of Concern." *Critical Inquiry* 30, no. 2 (2004): 225–48.

Lavrakas, Paul J. *Encyclopedia of Survey Research Methods*. Thousand Oaks, CA: Sage Publications, 2008.

Law, John, and Annemarie Mol. "Notes on Materiality and Sociality." *The Sociological Review* 43, no. 2 (1995): 274–94.

Annemarie Mol, *The Body Multiple: Ontology in Medical Practices*. Durham, NC: Duke University Press, 2002.

Lawrence, Benjamin N., Emily Lynn Osborn, and Richard L. Roberts, eds. *Intermediaries, Interpreters, and Clerks: African Employees in the Making of Colonial Africa*. Madison: University of Wisconsin Press, 2015.

Leach Gerald, and Robin Mearns. *Beyond the Woodfuel Crisis: People, Land and Trees in Africa*. London: Earthscan, 2016.

Lee, Juyoung. "Chemical Fertilizer and the Making of 'Scientific' Farmers in 1960s South Korea." Paper presented at Association for Asian Studies 2022, Graduate Student Paper Prizes, Northeast Asia Council, Hawai'i Convention Center, Honolulu, HI, March 26, 2022.

———. "한국의 제 1차 국토종합개발계획 수립을 통해서 본 발전국가론 '계획 합리성' 비판 [Making of the Developmental State's 'Plan Rational': The Case of South Korea's First Comprehensive National Physical Development Plan, 1963–1971]." 공간과 사회 [*Space and Society*] 25, no. 3 (2015).

Lefebvre, Henri. *Rhythmanalysis*. Translated by Stuart Elden and Gerald Moore. New York: Continuum, 2004.

———. *The Production of Space*. Oxford: Blackwell, 1991.

Leigh, Michael, B. *The Rising Moon: Political Change in Sarawak*. Sydney: Sydney University Press, 1974.

Lett, Lewis. *Sir Hubert Murray of Papua*. Sydney: Collins, 1949.

Li, Fabiana. *Unearthing Conflict: Corporate Mining, Activism, and Expertise in Peru*. Durham: Duke University Press, 2015.

Lopez, Iris. *Matters of Choice: Puerto Rican Women's Struggle for Reproductive Freedom*. Piscataway: Rutgers University Press, 2008.

López, Raúl Necochea. "Fertility Surveyors and Population-Making Technologies in Latin America." *Perspectives on Science* 25, no. 5 (2017): 631–54.

———. *A History of Family Planning in Twentieth Century Peru*. Chapel Hill: University of North Carolina Press, 2014.

Low, D. A., and J. M. Lonsdale. "Introduction: Towards the New Order 1945–1963." In *History of East Africa*, edited by D. A. Low and Alison Smith, 1–64. Oxford: Clarendon Press, 1976.

Lowenhaupt, Tsing Anna. *Friction: An Ethnography of Global Connection*. Princeton: Princeton University Press, 2005.

Ludden, David. "Development Regimes in South Asia: History and the Governance Conundrum." *Economic and Political Weekly* 40, no. 37 (2005): 4048.

Lunnebach, Edith. "Der Weckerkauf und seine Folgen—'Beschäftigung mit anschlagsrelevanten Themen oder geistige Nähe zum Terrorismus.'" In *Staatssicherheit. Die Bekämpfung des politischen Feindes im Innern*, edited by Helmut Janssen and Michael Schubert, 140–50. Bielefeld: AJZ, 1990.

Macekura, Stephen. *Of Limits and Growth*. New York: Cambridge University Press, 2015.

Makdisi, Saree. "Laying Claim to Beirut: Urban Narrative and Spatial Identity in the Age of Solidere." *Critical Inquiry* 23, no. 3 (1997): 661–705. https://doi.org/10.1086/448848.

Maloneym, Michael. "The History of Computing in the History of Technology." *Annals of the History of Computing* 10, no. 2 (1988): 113–25.

Malthus, T. R. *An Essay on the Principle of Population, or a view of Its Past and Present Effects on Human Happiness: With an Inquiry into Our Prospects Respecting the Future Removal or Mitigation of the Evils which it Occasions*, 7th ed. London: Reeves and Turner, 1872.

Mamdani, Mahmood. "Beyond Settler and Native as Political Identities: Overcoming the Political Legacy of Colonialism." *Comparative Studies in Society and History* 43, no. 4 (2001): 651–64.

Mascarenhas, Adolfo. "Resistance and Change in the Sisal Plantation System of Tanzania." PhD diss., University of California, 1970.

Massey, Doreen. *The Doreen Massey Reader*. Edited by Brett Christophers, Rebecca Lave, Jamie Peck, and Marion Werner. Newcastle upon Tyne: Agenda Publishing, 2017.

McAllister, Karen Elisabeth. "Rubber Rights and Resistance: The Evolution of Local Struggles against a Chinese Rubber Concession in Northern Laos." *Journal of Peasant Studies* 42, no. 3/4 (2015):817–37.

McAllister, Karen Elisabeth. "Shifting Rights, Resources and Representations: Agrarian Transformation of Highland Swidden Communities in Northern Laos." PhD diss., McGill University, 2016.

McArthur, Norma, and John F. Yaxley. *Condominium of the New Hebrides: A Report on the First Census of the Population 1967*. Sydney: New South Wales Government Printer, 1968.

McDuie-Ra, Duncan. "Fifty-Year Disturbance: The Armed Forces Special Powers Act and Exceptionalism in a South Asian Periphery." *Contemporary South Asia* 17, no. 3 (2009): 255–70.

McLachlan, Stéphane M. *Deaf in One Ear and Blind in the Other: Science, Aboriginal Traditional Knowledge, and Implications of the Keeyask Hydro Dam for the Socio-Environment. A Report for the Manitoba Clean Environment Commission on Behalf of the Concerned Fox Lake Grassroots Citizens*. Winnipeg: Environmental Conservation Laboratory, University of Manitoba, 2013.

McLachlan, Stéphane M. *Water Is a Living Thing: Environmental and Human Health Implications of the Athabasca Oil Sands for the Mikisew Cree First Nation and Athabasca Chipewyan First Nation in Northern Alberta*. Winnipeg: Environmental Conservation Laboratory, University of Manitoba, 2014.

Mead, Margaret. "The Cybernetics of Cybernetics." In *Purposive Systems*, edited by H. von Forester, J. D. White, L. J. Peterson, and J. Russell, 1–11. New York: Spartan Books, 1968.

Mead, Margaret, and Paul Byers. *Small Conferences: An Innovation in Communication*. The Hague: Mouton, 1968.

Mehta, Uday Singh. *Liberalism and Empire: A Study in Nineteenth-Century British Liberal Thought*. Chicago: University of Chicago Press, 1999.

Mertia, Sandeep. "FCJ-217 Socio-Technical Imaginaries of a Data-Driven City: Ethnographic Vignettes from Delhi." *The Fibreculture Journal*, no. 29 (2017). https://doi.org/10.15307/fcj.29.212.2017.

Meyerhoff, Miriam. "A Vanishing Act: Tonkinese Migrant Labour in Vanuatu in the Early 20th Century." *Journal of Pacific History* 37, no. 1 (2002): 45–56

Mies, Maria, and Vandana Shiva. *Ecofeminism*. 2nd ed. London: Zed Books, 2014.

Ministry of Land and Ocean. *The Masterplan for Saving the Four Rivers*, July 2009.

Ministry of Public Information, ed. *Korea and Vietnam*. Seoul: Ministry of Public Information (Republic of Korea), 1967.

Ministry of the Interior and Safety, "To Save the Four Rivers Is to Save Life," June 18, 2010. Video. Accessed December 1, 2018, https://goo.gl/c7XmrH.

Mir, Farina. *The Social Space of Language*. Berkeley: University of California Press, 2010.

Mirmalek, Zara. "Working Time on Mars." *KronoScope* 8, no. 2 (2009): 158–78.

"'Mistanenim' yehudim chashudim bigneiva" "מסתננים" יהודים חשודים בגניבה [Jewish "infiltrators" are accused of theft]. *Ma'ariv*, April 12, 1950.

Mitchell, Laura J. "Close Encounters of the Methodological Kind: Contending with Enlightenment Legacies in World History." In *Encounters Old and New in World History*, edited by Alan Karras and Laura J. Mitchell, 165–80. Honolulu: University of Hawaii Press, 2017.

———. "The Natural World." In *A Cultural History of Western Empires: Volume 4; The Age of Enlightenment (1650–1800)*, edited by Ian Coller, 69–91. London: Bloomsbury Press, 2018).
Mitchell, Margaret Jean. "'Comrades,' 'Trouble-Makers' and French '*Ressortissants*': The Repatriation of the Tonkinese from New Hebrides to North Vietnam." Paper presented at Association for Asian Studies Annual Conference, Toronto, ON, March 16, 2017.
Mitchell, Timothy. "The Crisis That Never Happened." In *Carbon Democracy*. New York: Verso, 2011.
———. *Rule of Experts: Egypt, Techno-Politics, Modernity*. Berkeley: University of California Press, 2002.
Mitman, Gregg. *Empire of Rubber: Firestone's Scramble for Land and Power in Liberia*. New York: The New Press, 2021.
Monier, Monier-Williams, ed. *Original Papers Illustrating the History of the Application of the Roman Alphabet to the Languages of India*. London: Longman, Brown, Green, Longmans, and Roberts, 1859.
Moore, Aaron S. *Constructing East Asia: Technology, Ideology, and Empire in Era, 1931–1945*. Stanford: Stanford University Press, 2013.
———. "'The Yalu River Era of Developing Asia': Japanese Expertise, Colonial Power and the Construction of Sup'ung Dam, 1937–1945." *Journal of Asian Studies* 72, no. 1 (2013): 115–139.
Morgan, Mary S. "'On a Mission' with Mutable Mobiles." *The Nature of Evidence: How Well do 'Facts' Travel?* Working paper no. 34/08. August, 2008, 6–7. http://dx.doi.org/10.2139/ssrn.1497107
Morton, Newton E. "Genetic Studies of Northeastern Brazil." *Cold Spring Harbor Symposia on Quantitative Biology* 29 (1964): 69–80.
Mosby, Ian. "Administering Colonial Science: Nutrition Research and Human Biomedical Experimentation in Aboriginal Communities and Residential Schools, 1942–1952." *Histoire Sociale / Social History* 46, no. 1 (2013): 145–72.
Moumtaz, Nada. "Gucci and the Waqf: Inalienability in Beirut's Postwar Reconstruction." *American Anthropologist*, forthcoming.
Muehlenbach, Andrea. "Building an Archive of Vulnerability: #GuerrillaArchiving at #UofT." EDGI, 2017. Accessed January 2, 2018. http://flolab.org/wp19/building-an-archive-of-vulnerability-guerrillaarchiving-at-uoft/.
Mullaney, Thomas S., Benjamin Peters, Mar Hicks, and Kavita Philip, eds. *Your Computer Is on Fire*. Cambridge, MA: MIT Press, 2021.
Müller-Wille, Staffan. *Botanik und weltweiter Handel: Zur Begründung eines natürlichen Systems der Pflanzen durch Carl von Linné (1707–78)*. Berlin: Verlag für Wissenschaft und Bildung, 1999.
Murphy, Michelle. *The Economization of Life*. Durham, NC: Duke University Press, 2017.
Murray, John Hubert Plunkett. *The Scientific Aspect of the Pacification of Papua: Presidential Address at the Meeting of the Australian and New Zealand Associa-*

tion for the Advancement of Science. Port Moresby, Territory of Papua: Edward George Baker, Government Printer, 1932.

Nadim, Tahani. "Haunting Seedy Connections." In *Routledge Handbook of Interdisciplinary Research Methods* edited by Celia Lury, Rachel Fensham, Alexandra Heller-Nicholas, Sybille Lammes, Angela Last, Mike Michael, and Emma Uprichard. New York: Routledge, 2018.

Nair, Janaki. *Mysore Modern: Rethinking the Region Under Princely Rule.* Hyderabad: Orient Black Swan, 2012.

Natalini, Adolfo. "Inventory, Catalogue, Systems of Flux . . . a Statement." In *Superstudio: Life without Objects*, edited by Peter Lang and William Menking, 164–67. Milan: Skira, 2003.

Netto, José Tavares, and Eliana Azevêdo. "Family Names and ABO Blood Group Frequencies in a Mixed Population Of Bahia, Brazil," *Human Biology* 50, no. 3 (September 1978): 361–67.

———. "Racial Origins and Historical Aspects of Family Names in Bahia, Brazil." *Human Biology* 49, no. 3 (1977): 287–99.

Neumann, Roderick P. "Forest Rights, Privileges and Prohibitions: Contextualising State Forestry Policy in Colonial Tanganyika." *Environment and History* 3, no.1 (1997): 45–68.

Nguyen, Vinh-Kim. "Government-by-Exception: Enrolment and Experimentality in HIV Treatment Programmes in Africa." *Social Theory & Health* 7, no. 3 (2009): 196–217.

"Nignav eder mekibbutz gaviv banegev" נגנב עדר מקיבוץ גביב בנגב [A herd was stolen from Kibbutz Gaviv in the Negev]. *HaTsofeh*, September 2, 1949.

Normile, Dennis. "Restoration or Devastation?" *Science* 327, no. 5973 (2010): 1568–70.

North, Douglass C. *Institutions, Institutional Change, and Economic Performance.* Cambridge: Cambridge University Press, 1990.

Notestein, Frank. "The Population Council and the Demographic Crisis of the Less Developed World." *Demography* 5, no. 2 (1968): 553–60.

Nowell, William. "Agave Amaniensis: A New Form of Fibre-Producing Agave from Amani." *Bulletin of Miscellaneous Information, Kew*, no. 10 (1933): 465–67.

———. "Supplement: The Agricultural Research Station at Amani." *Journal of the Royal African Society* 33, no. 131 (1934): 1–20.

Nyamunda, Tinashe. "Money, Banking and Rhodesia's Unilateral Declaration of Independence." In *The Journal of Commonwealth and Imperial History* 45, no. 5 (2017): 746–76.

Nyerere, Julius K. "A New Order." Speech, Arusha, February 12, 1979. http://www.juliusnyerere.org/uploads/unity_for_a_new_order_1979.pdf.

Oldenziel, Ruth. "Islands: The United States as Networked Empire." In *Entangled Geographies: Empire and Technopolitics in the Global Cold War*, edited by Gabrielle Hecht, 13–42. Cambridge: MIT Press, 2011.

"Once called 'golden eggs,' dredged sands from four rivers became a white elephant." *JTBC News*, February 14, 2017. http://news.jtbc.joins.com/article/article.aspx?news_id=NB11422526.

Open Development Laos, ed. *Forest Law*, No. 06/NA, 2007. Accessed September 2022. https://data.laos.opendevelopmentmekong.net/laws_record/forestry-law-2007-lao-pdr.

Orlikowski, Wanda. "Sociomaterial Practices: Exploring Technology at Work." *Organization Studies* 28, no. 9 (2007): 1435–44.

Paiva, Odair da Cruz, and Soraya Moura. *Hospedaria de Imigrantes de São Paulo*. São Paulo: Paz e Terra, 2008.

Pantojas-García, Emilio. *Development Strategies as Ideology: Puerto Rico's Export-Led Industrialization Experience*. Boulder: Lynne Rienner, 1990.

Park, Emma. *Infrastructural Attachments*. Duke University Press, forthcoming in 2023.

Park, Melany Sun-Min. "The Truss and the Cave: Architecture, Industrial Expertise, and Scientific Knowledge in Postwar Korea, 1953–1974." PhD diss., Harvard University, 2020.

Park, Tae-gyun. "W. W. Rostow and Economic Discourse in South Korea in the 1960s." *Journal of International and Area Studies* 8, no. 2 (2001): 55–66.

Parreñas, Juno S. *Decolonizing Extinction: The Work of Care in Orangutan Rehabilitaiton* (Durham: Duke University Press, 2018).

Pasternak, Shiri, Hayden King, and The Yellowhead Institute. 2019. "Land Back: A Yellowhead Institute Red Paper." https://redpaper.yellowheadinstitute.org/wp-content/uploads/2019/10/red-paper-report-final.pdf.

Patton, Arthur. "Liberia and Containment Policy Against Colonial Take-Over: Public Health and Sanitation Reform." *Liberian Studies Journal* 30, no. 2 (2005): 40–65.

Pekoro, Morea. *Orokolo Genesis*. Port Moresby, Territory of Papua: Niugini Press, 1973.

Peluso, Nancy Lee, and Peter Vandergeest, "Genealogies of the Political Forest and Customary Rights in Indonesia, Malaysia and Thailand." *The Journal of Asian Studies* 60, no. 3 (2001): 761–812.

Pemberton, Robert W., and Hong Liu. "Marketing Time Predicts Naturalization of Horticultural Plants." *Ecology* 90, no. 1 (2009): 69–80.

Peters, John Durham & Peter Simonson. *Mass Communication & American Social Thought*. Chicago: University of Chicago, 2004.

Philip, Kavita. *Civilizing Natures: Race, Resources, and Modernity in Colonial South India*. New Brunswick, NJ: Rutgers University Press, 2004.

Pias, Claus, ed. *Cybernetics: The Macy Conferences, 1948–1953: The Complete Transactions*. Züruch: Diaphanes, 2016.

Pickering, Andrew. *The Cybernetics Brain: Sketches of Another Future*. Chicago: University of Chicago Press, 2014.

Pine, Kathleen H. and Max Liboiron. "The Politics of Measurement and Action." In *CHI 2015—Proceedings of the 33rd Annual CHI Conference on Human Factors in*

Computing Systems, 3147–57. New York: Association for Computing Machinery, 2015.
Pinson, Daniel. "Urban Planning: An 'Undisciplined' Discipline?" In "Transdisciplinarity," edited by R. Lawrence and C. Despres. Special issue, Futures 36, no. 4 (May 2004): 503–13.
Pitt-Rivers, George, H. L. F. "The Effect on Native Races of Contact with European Civilization." Man 27, no. 2 (1927): 2–10.
Planning Commission (Government of India), Eleventh Five Year Plan. Vol 1. New Delhi: Oxford University Press, 2007.
Polanyi, Karl. The Great Transformation : The Political and Economic Origins of Our Time. Boston, MA: Beacon Press, 2001.
Porter, Libby. Unlearning the Colonial Cultures of Planning. Farnham: Ashgate, 2010.
Porter, Theodore M. Trust in Numbers: The Pursuit of Objectivity in Science and Public Life. Princeton, NJ: Princeton University Press, 1995.
Postill, John. "Clock and Calendar Time: A Missing Anthropological Problem." Time & Society 11, no. 2/3 (2002): 251–70.
———. Media and Nation Building: How the Iban Became Malaysian. Oxford and New York: Berghahn Books, 2006.
Power, Margaret, and Andor Skotnes, eds. "Puerto Rico: A US Colony in a Postcolonial World?" Issue, Radical History Review, no. 128 (2017).
Prahalad, C. K. The Fortune at the Bottom of the Pyramid: Eradicating Poverty through Profits. Upper Saddle River, NJ: Wharton School, 2005.
Pretel, David, and Lino Camprubí. Technology and Globalisation: Networks of Experts in World History. London: Palgrave MacMillan, 2018.
Pribilsky, Jason. "Developing Selves: Photography, Cold War Science and 'Backwards' People in the Peruvian Andes, 1951–1966." Visual Studies 30, no. 2 (2015): 131–50.
Price, Jennifer. Flight Maps: Adventures with Nature in Modern America. New York: Basic Books, 1999.
Priel, Aharon. "Cavarot gnuvot yimtaku" כוורות גנובות ימתקו [Stolen hives are sweeter]. Ma'ariv, January 5, 1979.
Procida, Mary A. Married to the Empire: Gender, Politics and Imperialism in India, 1883–1947. Manchester: Manchester University Press, 2002.
Quinn, Stephen Christopher. Windows on Nature: The Great Habitat Dioramas of the American Museum of Natural History. New York: American Museum of Natural History, 2006.
Raj, Kapil. "Beyond Postcolonialism . . . and Postpositivism: Circulation and the Global History of Science." Isis 104, no. 2 (2013): 337–47.
Rana, Aziz. The Two Faces of American Freedom. Cambridge, MA: Harvard University Press, 2011.
Rao, Parimala V. Foundations of Tilak's Nationalism: Discrimination, Education and Hindutva. Hyderabad: Orient Blackswan, 2011.

RCA. "Now RCA Removes More of the Mystery from Data Processing." Advertisement. *Datamation*, July/August 1960.
Rees, J. D. *Famine Facts and Fallacies*. London: Harrison and Sons, 1901.
Ressler, Oliver. "Die Rote Zora." 2000. Video. http://www.ressler.at/de/die_rote_zora/.
Rethmann, Petra. "On Militancy, Sort of." *Cultural Critique* 62 (Winter 2006): 67–91.
Review of Contracts Dated May 13, 1955, and March 27, 1959, with McGraw-Hydrocarbon, a Joint Venture for the Construction and Operation of a Fertilizer Plant in Korea, International Cooperation Administration, Department of State; report to the Congress of the United States by the Comptroller General of the United States. GAO: Washington, DC, 1960.
"Reviews and Summaries: Morogoro Fuelwood Stove Project." *Boiling Point*, no. 10 (August 1986). http://www.nzdl.org/cgi-bin/library?e=q-00000-00---off-ohdl--00-0----0-10-0---0---odirect-10---4-----ste--0-1l--11-en-50---20-about-materials+for+mud+stoves--00-0-1-00-0-0-11-1--0-0-&a=d&c=hdl&srp=0&srn=0&cl=search&d=HASH110a57bad34a66995a8aa1.15.2.
Rivers, William H. R. "The Psychological Actor." In *Essays on the Depopulation of Melanesia*, edited by William H. R. River, 84–113. Cambridge: Cambridge University Press, 1922.
Robert R. Nathan Associates, UNKRA (United Nations Korea Reconstruction Agency), *Preliminary Report on the Economic Reconstruction of Korea*. Washington, DC: Robert R. Nathan Associates, 1952.
Roberts, Mary Nooter, and Allen F. Roberts. "Audacities of Memory." In *Memory: Luba Art and the Making of History*, 17–48. Munich: Prestel, 1996.
Roberts, Stephen. *Population Problems of the Pacific*. London: Routledge, 1927.
Roder, W, S. Phengchanh, B. Keoboulapha, and S. Maniphone "*Chromolaena odorata* in Slash-and-Burn Rice Systems of Northern Laos." *Agroforestry Systems* 31 (1995): 79–92.
Rodney, Walter. *How Europe Underdeveloped Africa*. Rev. ed. Washington, DC: Howard University Press, 1981.
———. "Migrant Labour and the Colonial Economy." In *Migrant Labour in Tanzania during the Colonial Period: Case Studies of Recruitment and Conditions of Labour in the Sisal Industry*, edited by Walter Rodney, Kapepwa Tambila, and Laurent Sago, 7. Hamburg: Institut für Afrika-Kunde, 1983).
Rolandsen, Oystein, and Cherry Leonardi. "Discourses of Violence in the Transition from Colonialism to Independence in Southern Sudan, 1955–1960." *Journal of Eastern African Studies* 8, no. 4 (2014): 609–25.
Romo, Anadelia. *Brazil's Living Museum: Race, Reform, and Tradition in Bahia*. Chapel Hill: University of North Carolina Press, 2010.
Rosenberg, Emily S. *Spreading the American Dream: American Economic and Cultural Expansion, 1890–1945*. New York: Hill and Wang, 1982.
Rosin, Orit. "Infiltration and the Making of Israel's Emotional Regime in the State's Early Years." *Middle Eastern Studies* 52, no. 3 (2016): 448–72.

Ross, Matthew S., Alberto dos Santos Pereira, Jon Fennell, Martin Davies, James Johnson, Lucie Silva, and Jonathan W. Martin. "Quantitative and Qualitative Analysis of Naphthenic Acids in Natural Waters Surrounding the Canadian Oil Sands Industry." *Environmental Science and Technology* 46, no. 23 (2012): 12796–805.

Rote Zora. *Mili's Tanz auf dem Eis: Von Pirouetten, Schleifen, Einbrüchen, doppelten Saltos und dem Versuch, Boden unter die Füße zu kriegen*, 1993, 22, http://www.freilassung.de/div/texte/rz/milis/milis1.htm

Roy, Srirupa. *Beyond Belief: India and the Politics of Postcolonial Nationalism*. Durham, NC: Duke University Press, 2007.

Royce, Josiah. *The Problem of Christianity*. 1913; repr., Washington, DC: Catholic University of America Press, 2001.

Rydin, Yvonne. "Re-Examining the Role of Knowledge Within Planning Theory." *Planning Theory* 6, no. 1 (2007): 52–68.

Said, Edward W. *Orientalism*. New York: Pantheon Books, 1978.

Sakai, Naoki. "Theory and Asian Humanity: On the Question of *Humanitas* and *Anthropos*." *Postcolonial Studies* 13, no. 4 (2010): 441–64. https://doi.org/10.1080/13688790.2010.526539.

Salovaara, Isabel M. "Teaching Infrastructures: A Conversation with Gabrielle Hecht." *Society for Cultural Anthropology*. Accessed August 5, 2022. https://culanth.org/fieldsights/teaching-infrastructures-a-conversation-with-gabrielle-hecht.

Saltzman, Cynthia. *Old Masters, New World: America's Raid on Europe's Great Pictures*. New York: Viking Adult, 2008.

Salzano, Francisco, and Newton Freire-Maia. *Populações brasileiras: Aspectos demográficos, genéticos e antropológicos*. São Paulo: Companhia Editora Nacional, 1967.

Sammet, Jean. "Early History of COBOL." In *History of Programming Languages*, edited by Richard L. Wexelblat, 199–201. New York: Academic Press, 1981.

Sandercock, Leonie, ed. *Making the Invisible Visible: A Multicultural Planning History*. Berkeley: University of California Press, 1998.

Sanjek, Roger. "Anthropology's Hidden Colonialism: Assistants and Their Ethnographers." *Anthropology Today* 9, no. 2 (1993): 13–18.

Santos, Ricardo Ventura, Susan Lindee, and Vanderlei S. de Souza. "Varieties of the Primitive: Human Biological Diversity Studies in Cold War Brazil (1962–1970)." *American Anthropologist* 116, no. 4 (2014): 723–35.

Sanyal, Bishwapriya, Lawrence J. Vale, and Christina D. Rosan, eds. *Planning Ideas That Matter: Livability, Territoriality, Governance, and Reflective Practice*. Cambridge, MA: MIT Press, 2012.

Sanyal, Kalyan K. *Rethinking Capitalist Development: Primitive Accumulation, Governmentality and Post-Colonial Capitalism*. London: Routledge, 2007.

Saraiva, Tiago. *Fascist Pigs: Technoscientific Organisms and the History of Fascism*. Cambridge, MA: MIT Press, 2016.

Sarini, and Adivasi-Koordination, eds. "Adivasis of Rourkela: Looking Back on 50 Years of Indo-German Cooperation: Documents, Interpretations, International Law." Bhubaneshwar: CEDEC, 2006.

Saunders, Alexander Carr. "Review of Essays on the Depopulation of Melanesia." *The Eugenics Review* 14, no. 4 (1923): 282–83.

Schaffarczyk, Sylvia. "Australia's Official Papuan Collection: Sir Hubert Murray and the How and Why of a Colonial Collection." *reCollections Journal of the National Museum of Australia* 1, no. 1 (2006): 41–58.

Schaffer, Simon, Lissa Roberts, Kapil Raj, and James Delbourgo. *The Brokered World: Go-Betweens and Global Intelligence, 1770–1820*. Sagamore Beach: Science History Publications, 2009.

Schiebinger, Londa L. *Plants and Empire: Colonial Bioprospecting in the Atlantic World*. Cambridge, MA: Harvard University Press, 2004.

Schiebinger, Londa L., and Claudia Swan, eds. *Colonial Botany: Science, Commerce, and Politics in the Early Modern World*. Philadelphia: University of Pennsylvania Press, 2005.

Schildkrout, Enid, and Curtis A. Keim. *The Scramble for Art in Central Africa*. Cambridge: Cambridge University Press, 1998.

Schudson, Michael. "Cultural Studies and the Social Construction of 'Social Construction:' Notes on 'Teddy Bear Patriarchy.'" In *From Sociology to Cultural Studies: New Perspectives*, edited by Elizabeth Long, 379–98. Malden, MA: Blackwell, 1997.

Schumaker, Lynn. *Africanizing Anthropology: Fieldwork, Networks, and the Making of Cultural Knowledge in Central Africa*. Durham: Duke University Press, 2001.

Scott, James C. *Seeing Like a State: How Certain Schemes to Improve the Human Condition Have Failed*. New Haven: Yale University Press, 1998.

Se Gun, Kim. "Soyanggangdaemgwa chiyŏkchumindŭrŭi ilssang - hŭrŭgirŭl mŏmch'un kang kŭrigo chŏngch'ŏ ŏpsi hŭrŭnŭn sam." *Sahoegwahagyŏn'gu* 54, no. 2 (2015): 26, 29.

Sefu, Anne "Morogoro Fuelwood Stove Project Tanzania." *Boiling Point*, no. 13 (August 1987). http://www.nzdl.org/cgi-bin/library?e=d-00000-00---off-ohdl--00 -0---0-10-0---0---odirect-10---4------0-1l--11-en-50---20-about---00-0-1-00-0---4---0-0-11-10 -outfZz-8-00&cl=CL2.20.8&d=HASH3ec3d846525090878a0abe.11&gt=1.

Sennett, Richard. "The Neutral City." Chap. 2 in *The Conscience of the Eye: The Design and Social Life of Cities*. New York: Alfred A. Knopf, 1990.

Seong-Hyo Yoon, "Where Are the Abandoned Dredging Vessels?" *Ohmynews*, October 18, 2012. http://www.ohmynews.com/NWS_Web/View/at_pg.aspx?CNTN _CD=A00017.

Seth, Suman. "Colonial History and Postcolonial Science Studies." *Radical History Review* 127 (2017): 63–85.

Shapin, Steven, Simon Schaffer, and Thomas Hobbes. *Leviathan and the Air-Pump: Hobbes, Boyle, and the Experimental life: Including a Translation of Thomas*

Hobbes, Dialogus physicus de natura aeris by Simon Schaffer. Princeton, NJ: Princeton University Press, 1985.

Shechambo, Fanuel C. "Urban Demand for Charcoal in Tanzania: Some Evidence from Dar es Salaam and Mwanza." *Research Report No 67*. [Dar es Salaam?]: Institute of Resource Assessment University of Dar es Salaam, 1986.

Shepherd, George W., Jr. "National Integration and the Southern Sudan." *The Journal of Modern African Studies* 4, no.2 (1966): 193–212.

Shin, Gi-Wook and Michel Robinson, *Colonial Modernity in Korea*. Cambridge, MA: Harvard East Asia, 2001.

Siegert, Bernhard. *Cultural Techniques: Grids, Filters, Doors and Other Articulations of the Real*. New York: Fordham University Press, 2015.

Simone, AbdouMaliq. "People as Infrastructure: Intersecting Fragments in Johannesburg." *Public Culture* 16, no. 3 (2004): 407–429.

Simpson, Audra. *Mohawk Interruptus: Political Life across the Borders of Settler States*. Durham: Duke University Press, 2014.

Simpson, Audra. "The State Is a Man: Theresa Spence, Loretta Saunders and the Gender of Settler Sovereignty." *Theory & Event* 19 (4) 2016.

Simpson, C. L. *The Memoirs of C. L. Simpson: The Symbol of Liberia* (London: Diplomatic Press & Publishing, 1961.

Sivasundaram, Sujit. "Science." In *Pacific Histories: Ocean, Land, People*, edited by David Armitage and Alison Bashford, 237–62. London: Palgrave Macmillan, 2013.

Slobodian, Quinn. *Foreign Front: Third World Politics in Sixties West Germany*. Durham: Duke University Press, 2012.

Slobodian, Quinn. "How to See the World Economy: Statistics, Maps, and Schumpeter's Camera in the First Age of Globalization." *Journal of Global History* 10, no. 2 (2015): 307–32.

Smith, Wally. "Theatre of Use: A Frame Analysis of Information Technology Demonstrations." *Social Studies of Science* 39, no. 3 (2009): 449–80. https://doi.org/10.1177/0306312708101978.

Sohn, Dae-Sung. "33 Dredging Vessels Abandoned on the Nakdong River." *Yonhap News*, June 17, 2015. http://www.yonhapnews.co.kr/bulletin/2015/06/17/0200000000AKR20150617045900053.HTML.

Soja, Edward W. "Foreword." In *Postcolonial Spaces: The Politics of Place in Contemporary Culture*, edited by Andrew Teverson and Sara Upstone, ix–xiii. Basingstoke: Palgrave Macmillan, 2011.

Soluri, John. *Banana Cultures: Agriculture, Consumption, and Environmental Change in Honduras and the United States*. Austin: University of Texas Press, 2006.

Sōtokufu, Chōsen. *Chōsen Kasen chōsasho*. Keijō: Chōsen sōtokufu, 1929.

Speiser, Felix. *Ethnology of Vanuatu: An Early Twentieth Century Study*. Honolulu: University of Hawai'i Press, 1996.

Spencer, Michaela, Endre Dányi, and Yasunori Hayashi. "Asymmetries and Climate Futures: Working with Waters in an Indigenous Australian Settlement." *Science,*

Technology, and Human Values 44, no. 5 (2019): 786–813. https://doi.org/10.1177/0162243919852667.

Sperling, Jan Bodo. *The Human Dimension of Technical Assistance: The German Experience at Rourkela, India*. Translated by Gerald Onn. Ithaca: Cornell University Press, 1969.

Squier, Susan. *Poultry Science, Chicken Culture: A Partial Alphabet*. New Brunswick, NJ: Rutgers University Press, 2011.

Star, Susan Leigh. "The Ethnography of Infrastructure." *American Behavioral Scientist* 43, no. 3 (1999): 377–91.

Starosielski, Nicole. *The Undersea Network*. Durham: Duke University Press, 2015.

"Statement of Harvey S. Firestone." In *Crude Rubber, Coffee, etc., Hearings before the Committee on Interstate and Foreign Commerce House of Representatives, 69th Congress, First Session on H. Res. 59*, 247–272. Washington: Government Publishing Office, 1926.

Stoler, Ann Laura. *Along the Archival Grain: Epistemic Anxieties and Colonial Common Sense*. Princeton, NJ: Princeton University Press, 2009.

———. *Duress: Imperial Durabilities in Our Times*. Durham, NC: Duke University Press, 2016.

Strobl, Ingrid. *Vermessene Zeit: Der Wecker, der Knast und ich*. Hamburg: Edition Nautilus, 2020.

Strong, Richard P. "The Relationship of Certain 'Free-Living' and Saprophytic Microorganisms to Disease." *Science* 61, no. 1570 (1925): 97–107.

Strother, Zoë S. "A Terrifying Mimesis: Problems of Portraiture and Representation in African Sculpture (Congo-Kinshasa)." *Res: Journal of Anthropology and Aesthetics* 65–66 (2014/2015):141–42.

———. *Inventing Masks: Agency and History in the Art of the Central Pende*. Chicago: University of Chicago Press, 1998.

———. *Pende: Visions of Africa*. Milan: 5 Continents Editions, 2008.

Sturm, Sean. "Terra (In)cognita: Mapping Academic Writing." *TEXT: Journal of Writing and Writing Courses* 16, no. 2 (October 2012). http://www.textjournal.com.au/oct12/sturm.htm.

Stycos, Joseph. "Further Observations on the Recruitment and Training of Interviewers in Other Cultures." *Public Opinion Quarterly* 19, no. 1 (1955): 68–78.

———. "Interviewer Training in Another Culture." *Public Opinion Quarterly* 16, no. 2 (1952): 236–46.

Stycos, Joseph. *Human Fertility in Latin America*. Ithaca: Cornell University Press, 1968.

Suchman, Lucy. *Human-Machine Reconfiguration: Plans and Situated Actions*. Cambridge: Cambridge University Press, 2007.

Sudan Government. *Five Year Plan for Post-War Development*. Khartoum: Department of Finance, 1946.

Sundiata, Ibrahim. *Brothers and Strangers: Black Zion, Black Slavery, 1914–1940*. Durham: Duke University Press, 2003.

Superstudio. "The Continuous Monument: An Architectural Model for Total Urbanization." *Superstudio: Life without Objects*, edited by Peter Lang and William Menking, 122–47. Milan: Skira, 2003.

———. "Description of the Microevent/Micorenvironment." In *Italy: The New Domestic Landscape*, edited by Emilio Ambasz, 242–51. New York: Museum of Modern Art, 1972.

———. "Evasion Design and Invention Design." In *Superstudio: Life without Objects*, edited by Peter Lang and William Menking, 116–17. Milan: Skira, 2003.

———. "Histograms." *Superstudio: Life without Objects*, edited by Peter Lang and William Menking, 114–15. Milan: Skira, 2003.

Supersurface: An Alternative Model for Life on Earth, 1972.

Szeman, Imre, Patricia Yaeger, and Jennifer Wenzel, eds. *Fueling Culture: 101 Words for Energy and Environment*.New York: Fordham University Press, 2017.

Szreter, Simon. "The Idea of Demographic Transition and the Study of Fertility Change: A Critical Intellectual History." *Population and Development Review* 19, no. 4 (1993): 659–701.

"Technology is not neutral." *Daily News*, May 11, 1977.

Teishinkyoku, Chōsen Sōtokufu .*Chōsen suiryoku chōsasho dai ikkan (sōron)*. Keijō: Chōsen sōtokufu teishinkyoku, 1930.

Tesselaar, Suzanne, and Annet Scheringa. *Storytelling handboek: Organisatieverhalen voor managers, trainers en onderzoekers*. Quoted in Iris De Boer, "Storytelling and Its Potential for Planning Practice." Bachelor's thesis, Wageningen University, 2012.

Thienpont, Joshua R., Cyndy M. Desjardins, Linda E. Kimpe, Jennifer B. Korosi, Steven V. Kokelj, Michael J. Palmer, Derek C. G. Muir, Jane L. Kirk, John P. Smol, and Jules M. Blais. "Comparative Histories of Polycyclic Aromatic Compound Accumulation in Lake Sediments near Petroleum Operations in Western Canada." *Environmental Pollution* 231 (December 2017): 13–21.

Thompson, A. K., Kelly Fritsch, and Clare O'Connor, eds. *Keywords for Radicals: The Contested Vocabulary of Late-Capitalist Struggle*.Chico, CA: AK Press, 2016.

Thoms, Eva-Maria. "Der Freiheit eine Falle," *Die Zeit*, February 3, 1989.

Thrift, Nigel. "Space: The Fundamental Stuff of Geography." In *Key Concepts in Geography*, edited by Nicholas J. Clifford, Sarah L. Holloway, Stephen P. Rice and Gill Valentine, 2nd ed., 85–96. London: SAGE Publications, 2009.

Throgmorton, James A. "Planning as Persuasive Storytelling About the Future: Negotiating an Electric Power Rate Settlement in Illinois." *Journal of Planning Education and Research* 11, no. 3 (October 1992): 17–31.

Tilak, Bal Gangadhar. *Bal Gangadhar Tilak: His Speeches and Writings*. Madras: Ganesh & Co. 1922.

Tilak, Bal Gangadhar. *Sri Bhagavatgita-Rahasya, or, Karma-Yoga Sastra*. Vol. 1. First Edition. Pune and Bombay: Tilak Bros., 1965.

Tilley, Helen. *Africa as a Living Laboratory: Empire, Development and the Problem of Scientific Knowledge, 1870–1950*. Chicago: University of Chicago Press, 2011.

Tokihiko, Satō. *Doboku jinsei gojūnen*. Tokyo: Chūō kōron jigyō shupan, 1969.
Toledo, Franciza, and Clifford Price. "A Note on Tropical, Hot, and Humid Museums." *JCMS Journal of Conservation and Museum Studies* 4 (1998): 11.
Tolmein, Oliver. "Da haben alle mitgezogen." *konkret*, April 12, 1989
Tomihisa, Shimizu. "Kubota Yutaka-Nippon Kōei to Chōsen-Betonamu." *Shisō* 14 (November 1973): 35.
Torroja, Eduardo. *Philosophy of Structures*. Translated by J. J. Polivka and Milos Polivka. Berkeley: University of California Press, 1967.
Trevor-Roper, Hugh. "The Past and Present: History and Sociology," *Past and Present* 42, no. 1 (February 1969): 3–17.
——. *The Rise of Christian Europe*. London: Thames and Hudson, 1965.
Truth and Reconciliation Commission of Canada, *Honouring the Truth, Reconciling for the Future: Summary of the Final Report of the Truth and Reconciliation Commission of Canada*. Ottawa: Truth and Reconcilliation Commission of Canada, 2015.
Tuan, Yi-fu. *Dominance and Affection: The Making of Pets*. New Haven: Yale University Press, 1984.
Unger, Corinne M. "Rourkela, ein 'Stahlwerk im Dschungel: Industrialisierung, Modernisierung und Entwicklungshilfe im Kontext von Dekolonisation und Kaltem Krieg (1950–1970)." *Archiv für Sozialgeschichte* 48 (2008): 367–88.
Urwin, Chris. "Excavating and Interpreting Ancestral Action: Stories from the Subsurface of Orokolo Bay, Papua New Guinea." *Journal of Social Archaeology* 19, no. 3 (2019): 279–306.
Van Beurden, Sarah. *Authentically African: Arts and the Transnational Politics of Congolese Culture*. Athens, OH: Ohio University Press, 2015.
van Hulst, Merlijn. "Storytelling, a Model *of* and a Model *for* Planning." *Planning Theory* 11, no. 3 (August 2012): 299–318.
Van Reybrouck, David. *Congo: The Epic History of a People*. Translated by Sam Garrett. New York: Ecco, 2014.
Varadarajan, Latha. *The Domestic Abroad: Diasporas in International Relations*. New York: Oxford University Press, 2010.
Verly, Robert. "L'art africain et son devenir." *Problèmes d'Afrique Centrale*, 13, no. 44 (1959): 145–51.
Verran, Helen. *Science and an African Logic*. Chicago: University of Chicago Press, 2001.
——. "A Postcolonial Moment in Science Studies: Alternative Firing Regimes of Environmental Scientists and Aboriginal Landowners." *Social Studies of Science* 32, nos. 5/6 (2002): 729–62.
——. "Transferring Strategies of Land Management: Indigenous Land Owners and Environmental Scientists." In *Research in Science and Technology Studies, Knowledge and Society*, edited by Marianne de Laet, 155–81. Oxford: Elsevier, 2002.

Verran, Helen, and Michael Christie. "Objects of Governance as Simultaneously Governed and Governing." In "Objects of Governance." Special issue, *Learning Communities Journal* 15 (March 2015): 52–59.
Vidler, Anthony. "Diagrams of Diagrams: Architectural Abstraction and Modern Representation." *Representations*, no. 72 (Autumn 2000): 1–20.
von Oertzen, Christine, Maria Rentetzi, and Elizabeth S. Watkins. "Finding Science in Surprising Places: Gender and the Geography of Scientific Knowledge. Introduction to 'Beyond the Academy: Histories of Gender and Knowledge.'" *Centaurus* 55, no.2 (2013): 73–80. https://doi.org/10.1111/1600-0498.12018.
von Werlhof, Claudia. "Leserbrief an die taz (13.2.1981)." In *Die Neue Frauenbewegung in Deutschland: Abschied vom kleinen Unterschied; Eine Quellensammlung*, edited by Ilse Lenz, 277–79. Wiesbaden: VS, 2008.
Vukandinović, Saša. "Spätreflex: Eine Fallstudie zu den Revolutionären Zellen, der Roten Zora und zur verlängerten Feminismus-Obsession bundesdeutscher Terrorismusfahnder." In *Der Linksterrorismus der 1970er-Jahre und die Ordnung der Geschlechter*, edited by Irene Bandhauer-Schöffmann and Dirk van Laak, 139–61. Trier: Wissenschaftlicher Verlag, 2013.
Wade, Peter. *Degrees of Mixture, Degrees of Freedom: Genomics, Multiculturalism, and Race in Latin America*. Durham: Duke University Press, 2017.
Wallace, Alfred Russel. *The Malay Archipelago: The Land of the Orang-utan and the Bird of Paradise*. London: Macmillan, 1890.
Waltz, Kenneth. *Man, the State, and War: A Theoretical Analysis*. New York: Columbia University Press, 1959.
——. *Theory of International Politics*. Reading, MA: Addison-Wesley, 1979.
Walter, Maggie, and Russo-Carroll, Stephanie. "Indigenous Data Sovereignty, Governance and the Link to Indigenous Policy," in *Indigenous Data Sovereignty and Policy*, eds. Maggie Walter et al. (London: Routledge, 2021), 1–20.
Weber, Bob. "Showdown Looming for Alberta's Oil Sands over Cleanup of Tailings Ponds: Report." *The Globe and Mail*, March 30, 2017.
Werner, Marion, Jamie Peck, Rebecca Lave, and Brett Christophers, eds. *Doreen Massey: Critical Dialogues*. Newcastle upon Tyne: Agenda Publishing, 2018.
White, Richard. *The Organic Machine*. New York: Hill and Wang, 1995.
Wich, Serge A., S. Suci Utami Atmoko, Tatang Mitra Setia, and Carel P. van Schaik, eds. *Orangutans: Geographic Variation in Behavioral Ecology and Conservation*. Oxford: Oxford University Press, 2009.
Widmer, Alexandra. "Genealogies of Biomedicine: Formations of Modernity and Social Change in Vanuatu." PhD diss., York University, 2007.
——. "Making People Countable: Analyzing Paper Trails and the Imperial Census." *Sources and Methods in Histories of Colonialism: Approaching the Imperial Archive*, edited by Kirsty Reid and Fiona Paisley, 96–112. New York: Routledge, 2017.
——. *Moral Figures: Making Reproduction Public in Vanuatu*. Toronto: University of Toronto Press, 2023.

Wildavsky, Aaron. "If Planning is Everything, Maybe It's Nothing." *Policy Sciences* 4, no. 2 (June 1973): 127–53.
Williams, Francis Edgar. *The Drama of Orokolo: The Social and Ceremonial Life of the Elema.* Reprint. Papua New Guinea: University of Papua New Guinea Press, 2015.
Williams, Raymond. *Keywords: A Vocabulary of Culture and Society.* Fontana: Croom Helm, 1976.
Winichakul, Thongchai. *Siam Mapped: A History of the Geo-Body of a Nation.* Honolulu: University of Hawai'i Press, 1994.
Wohlberg, Meagan. "Court Case Claims Suncor Tailings Pond Leaking into Athabasca." *Northern Journal,* July 8, 2013.
Wolfe, Audra. *Competing with the Soviets: Science Technology, and the State in Cold War America.* Baltimore: John Hopkins University Press, 2013.
Wolfe, Tom. *From Bauhaus to Our House.* New York: Picador, 1981.
Woo-Cumings, Merideth. *Race to the Swift: State and Finance in Korean Industrialization.* New York: Columbia University Press, 1991.
Woodward, Keith. "Historical Note." In *Tufala Gavman: Reminiscences From the Anglo-French Condominium of the New Hebrides,* edited by Brian Bresnihan and Keith Woodward, 16–72. Suva: Institute of Pacific Studies, 2002.
World Bank, *World Bank Report No. 4969-TA Tanzania; Issues and Options in the Energy Sector.* November, 1984.
Young, Alden. "African Bureaucrats and the Exhaustion of the Developmental State: Lessons from the Pages of the Sudan Economist." *Humanity* 8, no. 1 (2017): 49–75.
———. *Transforming Sudan: Decolonization, Economic Development, and State Formation.* Cambridge: Cambridge Univearsity Press, 2018.
Yu, Xiangqin, Tianhua He, Jianli Zhao, and Qiaoming Li. "Invasion Genetics of Chromolaena odorata (Asteraceae): Extremely Low Diversity across Asia." *Biological Invasions* 16, no. 11 (2014): 2351–66.
Zapata, Carlos. *De Independentista a Autonomista: La Transformación del Pensamiento Político de Luis Muñoz Marín, 1931–1949.* San Juan: Fundación Luis Muñoz Marín, 2003.
Zhang, Yifeng, William Shotyk, Claudio Zaccone, Tommy Noernbern, Rick Pelletier, Beatriz Bicalho, Duane G. Froese, Lauren Davies, and Jonathan W. Martin. "Airborne Petcoke Dust Is a Major Source of Polycyclic Aromatic Hydrocarbons in the Athabasca Oil Sands Region." *Environmental Science & Technology* 50, no. 4 (2016): 1711–20.
忠肥 / 忠州肥料工場運營株式會社. Seoul: 忠州肥料工場運營株式會社, 1963.

Contributors

Itty Abraham is Professor in the School for the Future of Innovation in Society at Arizona State University.

Benjamin Allen is a Lecturer in the School of Information at the University of California, Berkeley.

Sarah Blacker is a Sessional Assistant Professor in the Department of Social Science at York University, Toronto.

Emily Brownell is a Senior Lecturer in Environmental History at the University of Edinburgh.

Lino Camprubí is Professor of History and Philosophy of Science at the Universidad de Sevilla and PI of ERC-CoG DEEPMED.

John DiMoia is Professor of Modern Korean History in the College of Humanities at Seoul National University.

Mona Fawaz is Professor in Urban Studies and Planning at the American University of Beirut.

Lilly Irani is Associate Professor of Communication, Science Studies, Computer Science, and Critical Gender Studies at the University of California, San Diego.

Chihyung Jeon is Associate Professor at the Graduate School of Science and Technology Policy of the Korea Advanced Institute of Science and Technology (KAIST) in South Korea.

Robert J. Kett is Associate Professor of Design Anthropology at ArtCenter College of Design and Adjunct Curator at the Palm Springs Art Museum.

Monika Kirloskar-Steinbach is Professor at the Department of Philosophy at the Vrije Universiteit, Amsterdam.

Karen McAllister is Assistant Professor in the Department of Global Development Studies at Saint Mary's University in Halifax, Canada.

Laura J. Mitchell is Associate Professor of History at the University of California, Irvine.

Gregg Mitman is the Vilas Research and William Coleman Professor of History, Medical History, and Environmental Studies at the University of Wisconsin-Madison.

Aaron S. Moore(†) was Associate Professor in the History of Technology in Modern Japan at Arizona State University.

Nada Moumtaz is Associate Professor in the Department for the Study of Religion at the University of Toronto.

Tahani Nadim is Junior Professor of Socio-Cultural Anthropology at the Department for European Ethnology in a joint appointment between the Humboldt-Universität zu Berlin and the Museum für Naturkunde (Museum for Natural History), Berlin.

Anindita Nag is Associate Professor and the Associate Dean of International Affairs at the Jindal School of Art and Architecture, New Delhi.

Raúl Necochea López is Associate Professor of Social Medicine and Adjunct Associate Professor of History at the University of North Carolina.

Tamar Novick is a Senior Research Scholar at the Max Planck Institute for the History of Science in Berlin.

Juno Salazar Parreñas is Associate Professor in the Department of Science and Technology Studies at Cornell University.

Benjamin Peters is the Hazel Rogers Associate Professor of Media Studies at the University of Tulsa.

Dagmar Schäfer is Director of Department III, "Artifacts, Action, Knowledge," at the Max Planck Institute for the History of Science in Berlin.

Martina Schlünder is a Research Scholar at the Max Planck Institute for the History of Science and a visiting associate professor of Science and Technology Studies at the Center for Technology, Innovation, and Culture at the University of Oslo.

Sarah Van Beurden is Associate Professor History and African American and African Studies at the Ohio State University.

Helen Verran taught history and philosophy of science at University of Melbourne Australia, for nearly twenty-five years. Since 2012 she has been Research Professor at Charles Darwin University. Verran's book *Science and an African Logic* (University of Chicago Press, 2001) was awarded the Society for the Social Studies of Science's Ludwik Fleck Prize in 2003.

Ana Carolina Vimieiro is Professor for the History of Science in the Department of History at the Federal University of Minas Gerais, Brazil.

Alexandra Widmer is Associate Professor in the Department of Anthropology at York University, Toronto.

Alden Young is Assistant Professor in the Department of African American Studies at the University of California, Los Angeles.

Index

abecedaries, 3, 17n13
ACM History of Programming Languages conference (1978), 37, 39, 42, 43
Action Committee, 230–32
Adivasi people, 138, 204–5, 210n4
afforestation, 33, 34
Afghanistan, 97, 102n3
Africa Museum, 150
Agave amaniensis, 199, 200
Agave sisalana (sisal agave), 195–99, 201
Age of Atonement, 103n7
agriculture, 10, 12, 22, 32, 197–200, 216, 267; beehive thefts, 230–33, 234n8, 265; grain, 100–101, 102n3, 114, 117–19, 162, 255; rice, 101, 113, 117–19, 245, 247, 250–51, 252n9; South Korea and, 64, 66, 113, 116, 118–19; swidden cultivation and, 246–47, 249, 251, 252nn8–9, 253n24. *See also* fertilizer
AIMACO, 38
Akeley, Carl, 220–28, 221
Akeley, Delia, 223, 225
Alexander, Ernest, 2
ALGOL, 39
Alliès, Paul, 209
alphabets, 4–6
Amani Institute (East African Agricultural Research Station), 198–99, 200
Ambedkar, Bhimrao Ramji (1891–1956), 56–61, 62n3, 62n9, 63nn13–14, 271
American Museum of Natural History (AMNH), 224–27
"amodern" analytic framing, 255

Anglo-Egyptian Condominium (1899), 160, 164
animals, xi, 114, 162, 171, 236, 247; bees, ix, 230–33, 234n8; gorillas, 15, 220–21, 221, 224–28, 229n13, 271; orangutans, 168–74, 175n21, 268; pigs, 22, 25–26, 78, 108, 257, 259, 262
anschlagsrelevante topics, 92
architecture, 124–28. *See also* grids
artifacts, xii, 2, 4, 6, 13, 201, 222–23, 227
"Artifacts, Action, Knowledge," Department III, ix
artisanal workshops, colonialism, 144–50, 267
ASAI (Ateliers Sociaux d'Art Indigène), 147–49
Asian-African Conference, 172
Asian financial crisis (1997–1998), 114
Association of al-Azhar Graduates, 155
Ateliers Sociaux d'Art Indigène (ASAI), 147–49
Athabasca Chipewyan First Nation, 235–37, 240–41
Athabasca oil industry, 235, 236, 240
Australia, 205, 256–57, 260–62. *See also* zoomorphic wickerwork figure, Australian Administered British New Guinea
autarky, 76–79, 77, 81–82, 267
Azevêdo, Eliane, 212, 213, 214, 216

Babbage, Charles, 80
Babel, 43–45

Bahian population, 212–13, 216
Bako National Park, 168, 173
Barber, Marshall, 179, 180
BASR (Bureau of Applied Social Research), Columbia University, 104, 107, 109n1
Baxi, Upendra, 62n2
Beauregard, Robert, 14–15
Beccari, Oduardo, 170
beehives, thefts, 230–33, 234n8, 265
bees, humans and, ix, 230–33, 234n8
Beirut, Lebanon, 153, 153–56
Belgian Congo (Zaire), 10, 144, 147, 149, 224–28. *See also* kishikishi, Belgian Congo
Belgium, 146, 150
Bennett, Carolyn, 243n15
Bentham, Jeremy, 99–100
Berelson, Bernard, 104
Bhan, Gautham, 14
biological miscegenation, 212–14, 217, 218
biomass, 31–32, 34
biomedical research, 178–80, 215, 217
biomedicine, 184n11
Bislama, 25
bitumen, 235–36
BKA (Federal Criminal Police Office), 85, 89–92
Bodry-Sanders, Penelope, 223
Boiling Point (magazine), 33
bombings, 85, 87–92, 95n21
Borneo. *See* orangutans, Bornean
Botanical Museum, Berlin, 199, 200
Botanic Central Office for the German Colonies (*Botanische Zentralstelle für die Deutschen Kolonien*), 198, 202n16
botany, 178, 196–97, 200–201
"bottom of the pyramid," 135, 136, 141
Bouvard et Pécuchet (Flaubert), 12
Brachystegia (*miombo*) genus of trees, 30
Braun, Carl Philipp Johann Georg (1870–1935), 199
Brazil, 135, 177, 202n4, 215. *See also* surnames, Brazil
British India Company, 245
British Museum, 224
British (Papua) New Guinea. *See* zoomorphic wickerwork figure, Australian Administered British New Guinea
bronze plate, xi
Brooke, Charles, 170
Brooke, James, 170
Brown, George, 181–82

Brown, James, 179, 180
Brown, Peter, 176
Bulletin of Miscellaneous Information (Royal Botanic Gardens), 197
Bureau of Applied Social Research (BASR), Columbia University, 104, 107, 109n1
Burma, 67, 248
business computing, 37, 41
business programming languages. *See* COBOL
Butler, Octavia E., 203n22

cadastral maps, 152, 153, 156–58
cadastral records, 251
Cambridge Economic Handbooks, 274n17
Canada, 236–40. *See also* water samples, in Treaty 8 Territory
cancer rates, 237
Capital (Marx), ix
capitalism, xii, 2, 24, 134, 143n18, 182, 184n11
Carmichael, John, 162, 163, 164, 166n14
"Cartesian perspectivalism," 127
Castells, Manuel, 273n14
categorization, 22, 25–26, 222
Catholicism, 79, 149, 226
cement, 76, 80, 120–21
census, 97, 180
census, New Hebrides: economic categories, 25; labor and, 21–22, 23; teachers as administrators of, 20, 268; Vietnamese plantation laborers, 23–24
Center for Social Research (Centro de Investigaciones Sociales, CIS), University of Puerto Rico, 106–9
ceremonies, Elema peoples, 255, 257, 258, 259
chaebol, in South Korea, 122n13
chance, x, 90, 91
Chang, Ha-Joon, 78
charcoal, 29–35, 35n5, 267
Cheonggyecheon "restoration" project (2003–2005), 188
Chick, Arthur L., 163, 164, 166n14
children, 94n5, 96, 107–8, 141, 149, 179, 184n10; labor and, 100, 180; at residential schools, 237–38, 243n15
China, x–xii, 66, 98, 114, 121n4, 135, 195
Chokwe people, 148
Chosŏn Dynasty (1392–1897), 71
Christianity, 5–6, 21, 76, 79–80, 89, 149, 204–5

INDEX   317

The Christian Concept of Autarky (Pérez del Pulgar), 79
Chromolaena odorata (Siam weed, *Nya Kiloh*), 245–49, 251–52, 252nn5–6
Chung, Dajeon, 118
Ch'ungju Fertilizer Factory (충주 비료 공장), 113–14, 117, 119, 121n1
Churchill, Winston, 79
CIS (Centro de Investigaciones Sociales, Center for Social Research), University of Puerto Rico, 106–9
civic nation, India as, 56–61
civil unrest, in Sudan, 162–63
classification, 7, 12, 13, 268
Clifford, George, 196–97
climate change, 74n21, 77
The Closed Commercial State (Fichte), 78
coal, 81, 82, 207. *See also* dodecahedral silo, in Spain
COBOL (Common Business-Oriented Language), 15; Babel and, 43–45; CODASYL and, 37–39, 41, 43; committees, 38–41, 43, 44; FLOW-MATIC and, 38, 43; keywords and, 41, 43; programmers and, 38–45; universal plain English and, 40–42
coconuts, 22, 25
CODASYL (Committee on Data Systems Languages), 37–39, 41, 43
coding, 40, 47–48, 142n3, 248
collective action, x, 6, 7, 12
colonialism, 4, 21, 50, 172, 271; artisanal workshops, 144–50, 267; differentiating and, 267–68; enculturation and, 267–68, 272n7; experiments, 169, 170, 171, 268; planning and, 1–2, 7, 8–11, 14; postcolonial spaces and, 206, 270; seeds and, 195–201, 202n4, 202n16; settler-, 10, 52, 128, 233, 236–40, 269; taxonomy and, 221–22, 227–28; violence with, 169, 195, 205, 237, 238; weeds and, 245, 246; worldwide, 5–6, 16n15, 20, 29, 56, 60, 64–66, 79, 85, 110n7, 114–18, 149, 157–58, 160, 200, 215, 226, 232, 255–62. *See also* postcolonialism
Colonial Office, Great Britain, 102n2, 260
colonial planning, 1–2, 25, 72–73, 144, 196, 198, 222; with alphabet project of Monier-Williams, 5–6; differentiation and, 268–69; postcolonial and, 4, 265–66, 269; spatiality and, 270; time and, 271
Columbia University, 56, 104, 107, 109n1

Commission for the Protection of Indigenous Arts and Crafts (COPAMI, La Commission pour la Protections des Arts et Métiers Indigènes), 146–48
Committee on Data Systems Languages (CODASYL), 37–39, 41, 43
Common Business-Oriented Language. *See* COBOL
Companies Act (2013), India, 135
comprehensive national land planning policy, South Korea, 64, 66, 69–70, 72–73
computation, 47–48, 50–54
computer science, 37, 47
computing, 15, 37, 41, 134; as keyword, 46–49, 53; small groups, 47–48, 50–54
COMTRAN, 38
conflicts, ontological, 268
Congress, US, 110n7
Consejo Superior de Investigaciones Científicas (CSIC), 79–80
Constituent Assembly, India, 56, 62n9
constitution, of India, 11, 61n1; Ambedkar and, 56–61, 62n9, 271; Drafting Commission, 56, 60; Indian Committee on Franchise and, 59–60, 63n14; with moral and social equality, 58–59
constitution, of South Korea, 123n23
The Continuous Monument (Superstudio), 127, 128
contraceptives, 105, 107
Conversations on Science, Culture, and Time (Serres and Latour), 94n4
Conzen, M. R. G., 3
Coopman, Colin, 49
COPAMI (Commission for the Protection of Indigenous Arts and Crafts, La Commission pour la Protections des Arts et Métiers Indigènes), 146–48
Cornish, Robert, 101
corporate social responsibility (CSR), 135, 138
Cosmopolitan (magazine), 96
Costillares Coal silo, 76–77, 77, 80–81
Costillares laboratory, 76, 80, 81
cotton complex, 162–63
cotton yarn, sugar, wheat flour ("three whites"), 113, 116, 117, 122n13
counter-plans, 6, 7, 11, 13–14, 94, 218, 240
COVID pandemic, 77
creepiness (*geli*), 171
CSIC (Consejo Superior de Investigaciones Científicas), 79–80

CSR (corporate social responsibility), 135, 138
culture, 10, 94n4, 119, 195, 212–14, 217, 237
Cuo (Chinese King), xi
cybernetics, 50–52

"Dalits" ("downtrodden"), 56, 62n3
Daly, M. W., 166n14
dams, 65, 81, 147, 207, 236
dams, in South Korea: with colonial and post-colonial power relations, 64–65; developmentalist planning, 72–73; intellectual foundation for water resources planning, 67–68; Japanese ex-colonial engineers and, 65–73, 74n5, 74n8; river and hydropower investigation reports, 66–71; Soyanggang River, 67–68, 72, 74n21
Dar es Salaam, Tanzania: charcoal in, 29–35, 35n5; livelihoods, 33
Dartmouth Conferences on Artificial Intelligence, 52
Dartmouth Summer Research Project on Artificial Intelligence, 50
Darwin, Charles, 170, 220
Datamation (journal), 41–42
David Taylor Model Basin, 39
Davis, Kingsley, 109n1
Davos (World Economic Forum) (2011), 136
Davoudi, Simin, 11
Dayak people, 171
De Casseres, J. M., 3
"Decolonial Planning," ix
decolonization, 2, 4, 11, 21, 56, 61, 110n7; orangutans and, 168–70, 172–74; political, 8, 60; Sudan and, 160–61
deforestation, 32, 34, 168, 182
Delirious New York (Koolhaas), 128
demos, in information technology, 133–34, 138–39, 141n1
Denning, Peter, 47
Department III, "Artifacts, Action, Knowledge," ix
Department of Defense, US, 38
Descartes, Rene, 126
De Silva, Gananath Stanley, 170, 173
De Soto, Hernando, 159n12
De Sousa Santos, Boaventura, 167n20
Deutsch Ostafrikanische Gesellschaft (DOAG, German East Africa Company), 197
DevDesign, 136–37

devotional names, 214
Dickens, Charles, 97
differentiating, enculturating and, 267–69
differentiation, identification and, 224
Digby, William, 96
Dijkstra, Edsger, 43
Discourse on Method and Geometry (Descartes), 126
DOAG (*Deutsch Ostafrikanische Gesellschaft*, German East Africa Company), 197
dodecahedral silo, in Spain: autarky and, 76–79, 77, 81–82, 267; Costillares Coal, 76–77, 77, 80–81
Dove, Michael R., 252n5
"downtrodden" ("Dalits"), 56, 62n3
Doyle, Arthur Conan, 97
Drafting Commission, Indian constitution, 56, 60
dredging, in South Korea: to cure dying rivers, 189–90; for economic growth, 187–89; Four Rivers Restoration Project and, 186–93; to kill rivers, 190–91, 190–92
drought, 32, 100, 189, 223
DuPont, 38
Durand, Jean-Nicholas-Louis, 126

"Earthseed," 203n22
East Africa, 32. *See also* seeds, German East Africa
East African Agricultural Research Station (Amani Institute), 198–99, 200
economic nationalism, 76, 78, 82
economics, 10, 56, 78, 82, 98, 181, 266, 274n17
The Economic History of Liberia (Brown, G.), 181–82
Economist (magazine), 141
economy, 31, 34, 73n3, 102n3, 114, 122n8, 206–7; South Korea, 72, 121, 187–89; subsistence, 22–23, 25, 29, 199; Sudan and national, 161–65, 268; US aid and, 113, 117–18, 120. *See also* political economy
education, 10, 21, 62n8, 134, 237–38, 243n15
Edwards, Paul, 53
e-Government, 138
Egypt, 160, 162, 167n16
Egyptian Pound (LE), 163, 167n16
Eharo, 257
Einstein, Albert, 269
Eisenhower, Dwight, 39

INDEX 319

electricity, 29, 31, 34, 67, 69–70, 122n6, 207
Elema peoples, 255–62
Elliot, Daniel Giraud, 223
Ellis Island, New York, 219n15
Elyachar, Julia, 141
EMES Sonochron, FRG and, 84, 85, 271; BKA and, 85, 89–92; as legal evidence, 85, 86, 92–93; as material trap, 85, 86, 89–90, 92; public prosecutor and, 91–92; Rote Zora and, 85, 87–93, 94n5; terrorism and, 85, 89, 90–92; as time fuse, 85–89, 91, 93, 95n21
EMMA (magazine), 90
enculturation, 266, 267–69, 272n7
English, 40–44, 48, 176
Enlightenment, 1, 62n2
environments, 25, 30–32, 235–41
Equatorial Guinea, 161
erasure, of prior existence, 208, 209, 270
eravo (men's houses), 258, 259
Esso Standard Oil, 38
ethnic identities, 20, 212
Euclid, 130, 267, 269
eugenics, 88, 238

Facebook, 53, 134
FACT, 38
Family Life Study (FLS) (1948–1959), 104–9
family planning, 11, 106–7. *See also* fertility survey workforce, in Puerto Rico
famine, in India: camps, 98–99, 101, 102; Great Britain with, 96–102; relief, 97–99, 99, 100–102, 102n3
famine, in Tanzania, 32
Famine Commission Report (1878–1880), 102
farmers, 88, 136, 231–32, 272n4; South Korean, 113–14, 117, 119; weeds and, 245–52
farming, dredging and, 189, 192
fascism, 76, 78–79, 85, 128
Fascist Pigs (Saraiva), 78
Federal Criminal Police Office (BKA), 85, 89–92
Federal Republic of Germany (FRG, West Germany), 85, 88, 115, 121n4, 122n9, 205, 207. *See also* EMES Sonochron, FRG and
feminists, 22, 85, 87–88, 90–93
fertility research, in US, 106–7
fertility survey workforce, in Puerto Rico, 111n16; CIS, 106–9; FLS, 104–9; KAP studies, 105; respondents and, 107, 108

fertilizer, 31, 65–66, 79, 122n6
fertilizer, South Korea and: Ch'ungju Fertilizer Factory, 113–14, 117, 119, 121n1; with colonial echoes, 115–18; economic aid and, 113, 117–18, 120; Tongil rice and, 118–19; with transition to exports, 119–21
Fichte, Johan Gottlieb, 78
Field Museum, Chicago, 223
Film India (magazine), 204
Finance Corporation of America, 181
Firestone, Harvey, 177, 178
Firestone Company: as parasite, 176–77, 181–83; plantations, 177–81, 183; Tire & Rubber, 176, 177, 181–82, 267
fish-being of place (*Mara'ope*), 257, 258
fishing, 65, 257
Five-Year Plans: India, 134–35; Sudan, 163–64
Flaubert, Gustav, 12
Fleck, Ludwig, 52
FLOW-MATIC, 38, 43
FLS (Family Life Study) (1948–1959), 104–9
"Food for Peace" program (PL480), 117–18
"Foreign Things No Longer Foreign" (Chung), 118
forests, 33, 168, 172, 173, 183, 253n17; LFAP, 248–51; trees, 30, 32, 34, 177, 182, 198, 251, 253n24
Forth, Aidan, 98
The Fortune at the Bottom of the Pyramid (Prahalad), 135
Foucault, Michel, 273n14
Four Rivers Restoration Project (2008–2012), 186–93
The Four Seasons, 223
France, 20, 26n2, 79, 126, 157–58, 177–78, 181
Francis, William T., 176
Franco, Francisco, 78–79
Frassinelli, Gian Piero, 131n1
Frederick William University, 201n3
Free World Asia, 113
French, 3, 42–43
French Mandate (1920–1943), 155
Freyre, Gilberto, 215
FRG. *See* Federal Republic of Germany
Friedmann, John, 269
fuels, 31–34. *See also* charcoal; coal; oil; wood

Galbraith, John Kenneth, 106
Gandhi, Indira, 30–31

Gates, Bill and Melinda, 134
geli (creepiness), 171
gender, 23, 47, 57, 60, 109, 190–91, 205–6
genetics, 78, 88, 90, 200, 212–18
genocide, 89, 93, 237–38, 243n15
geopolitics, 24, 205, 207–8, 225
"Geoproscopy," 3
German, 42–43
German East Africa (1885–1919). *See* seeds, German East Africa
German East Africa Company (*Deutsch Ostafrikanische Gesellschaft*, DOAG), 197
Germany, 78–79, 87–89, 91, 93, 101, 178, 271
Ghana, 172
gift-giving relation, 256
global environment, 30, 32
globalization, xii, 77, 82, 273n14
Global North, 16n5, 94, 144, 146, 222, 224, 225
Global South infrastructure, 266, 272n2
glossaries, 2, 3, 4
golden shirt, xii
Goldman Sachs, 135
Google, 134, 140, 160
gorillas, 15, 220–21, 221, 224–28, 229n13, 271
governance, 135, 137, 169, 172, 225–26, 255–59, 261–62
grain, 100–101, 102n3, 114, 117–19, 162, 255. *See also* rice
Grand Korea Waterway, 188–89
Grant, Charles, 101
graves, unmarked mass, 237–38
Great Britain, 26n2, 177–78, 181, 184n10, 207, 248, 254; with census, 21, 24; colonialism, 20, 79; Colonial Office, 102n2, 260; with famine in India, 96–102
"Green Revolution," 122n15
grids, 11, 29, 31, 42, 114, 207; with architecture rejected, 125–28; as life between "A and B," 130–31; as "little squares," 129–30; Superstudio and, 124–31, 125; as uniform infrastructure, 127–29
Gropius, Walter, 126
groups: computing small, 47–48, 50–54; with India as civic nation, 56–61; problem solving with effort of, 49
Guidelines of Family rituals, China, xi
Gururani, Shubhra, 14

hacienda system, 197
hackathons, 15, 47, 271

hackathons, India, 133, 142n3, 271; from design studio, 136–41; five-year plans to managed speculation, 134–36
Hancock, Scott, 265
Haraway, Donna, 137, 223, 224–25
Haret Hreik, Lebanon, 153, 153, 156, 158
Hariri, Rafic, 153
Harrisson, Barbara, 168, 170–73
Harrisson, Tom, 168, 172, 173
Harvey, David, 273n14
Hawthorne, Julian, 96
Health Canada, 236
Hecht, Gabrielle, 272n2
Held, Kurt, 94n5
"The Heterogeneous State and Legal Pluralism in Mozambique" (de Sousa Santos), 167n20
Hezbollah, 153, 156
Hicks, Mar, 53
high modernism, 7, 9, 67–69, 72–73, 73n4
Hill, Reuben, 107, 108, 109
Hindorf, Richard (1863–1954), 197–98
Hinduism, 56, 57, 62n8, 101
Hindustan Steel Plant, India: colonial and postcolonial spaces, 206, 270; geopolitics, 205, 207–8; inter-racial relationships and, 204–5, 210; plan for future, 209–10; planning with past, present and future, 205–6; technological modernity, 206–7
hinterlands inhabitants (*sertanejo*), 217
Hitler, Adolf, 78, 79, 172
Ho Chi Minh, 24
Holberton, Betty, 39
honey, 230, 231, 233
Honeywell, 38
Hopper, Grace, 38, 39, 41, 43
Hornaday, William, 170
Hortus Cliffortianus (Linnaeus), 196
Hose, Charles, 171
Hospedaria de Imigrantes, 216, 219n15
hubris, 2, 8, 44, 54, 267
Hughes, Thomas, 140
Hugo, Victor, 126
humans, 3, 10, 23, 25, 78; bees and, ix, 230–33, 234n8; non-humans and, 14, 86
The Human Dimension of Technical Assistance (Sperling), 204
hunger, 88, 96–99, 171, 237. *See also* famine
hydro-electricity, 34, 207
hygiene, 10, 261
Hyndman, Henry, 97

Iban people, 171
IBM, 38, 134
identification, xiii, 7, 62n3, 85, 91, 93, 224, 268
identities, 20, 23, 49, 207, 212, 215–17
Ilo, Philip, 25
imagination, ix, xiii, 134, 205, 208, 272n7
Imperial Land Decree (1895), 198
indentured labor, 23, 24
India, 5–6, 31, 34, 62n2, 62n9, 97, 101; as civic nation, 56–61; Planning Commission, 134–36; postcolonialism and, 56–57, 205–10. See also constitution, of India; famine; hackathons, India; Hindustan Steel Plant, India
Indian Committee on Franchise, 59–60, 63n14
indigeneity, 23, 25
Indigenous people, 21, 22, 23, 25; children at residential schools, 237–38, 243n15; communities, 29–30, 98, 138, 144, 145, 146, 148–50, 151n13, 171, 176, 182, 198, 204–5, 210n4, 250–51, 255–62; with Traditional Knowledge, 236–41; in Treaty 8 Territory, 235–42
Indigenous values, ontologies and, 25, 241
Indonesia, 34, 67, 172, 252n5
industrialization, 65–67, 70, 74n21, 78, 80, 106, 117
Industrial Policy Resolution (1948), 206–7
inequality, 2, 135–36, 252
Infosys, 134
infrastructuralizing, 266–67, 272n5
infrastructure, 12, 29, 31, 34, 49, 87, 134; Global South, 266, 272n2; grids as uniform, 127–29; large-scale projects, 64–65; poverty and, 207; in South Korea, 122n6. See also dams
Institute for National Museums, 144, 150
intellectual decolonization, 56, 61
Intermediate Committee, COBOL and, 38, 39
International Association for Pre-stressed Concrete, 80
International Biological Program, WHO, 213
International Rice Research Institute (IRRI), 118
internet, 133, 139–40
inter-racial relationships, 204–5, 210
Irani, Lilly, 53

Ireland, 97–98
IRRI (International Rice Research Institute), 118
Islamic charitable endowments (waqfs), 154–56, 158
Islamic law, 154, 155
Israel, 153, 230–33
Italy, 78–79
Italy (MoMA), 124, 128

Jaden, Aggrey, 163
Jan Lokpal, 139
Jan Sabha, 139
Japan, 73n3, 78, 113, 119–20; colonialism and, 65, 114, 116–17; with ex-colonial engineers in post-colonial Korea, 65–73, 74n5, 74n8
Jay, Martin, 127
Jencks, Charles, 128
Johnson, Martin, 224
Johnson, Osa, 224

Kaiser, Katja, 200, 202n16
Kalsakau, John, 25
KAP (knowledge, attitudes, and practices) studies, 105
Kaviraj, Sudipta, 61
Kenya, 32, 98
keywords, 3, 4, 41, 43, 46–49, 53
Khartoum Conference (1965), 163
Kicking Away the Ladder (Chang), 78
Kikuyu people, 98
Kim, Kyung-Soo, 190
Kim, Tae-ho, 119
Kim, Yong-Min, 191
King, C. Dunbar, 179
kishikishi, Belgian Congo: ASAI and, 147–49; colonial artisanal workshops and, 144–50, 267; COPAMI and, 146–48; Makumbi and, 144, 145, 146, 148–50, 151n13; *politique esthétique* and, 146–50
Kissinger, Henry, 82
Kiswahili, 30
knowledge, ix, xiii, 1, 9, 220; categorization, 25–26; infrastructures, 4, 6, 11, 266; making, x, xii, 12, 61, 170; plant, 197, 202n16; "three-track" methodology, 238–39, 241–42; Traditional Knowledge, 236–41
knowledge, attitudes, and practices (KAP) studies, 105

knowledge production, xii, 1, 4, 8–9, 61, 62n2, 222, 226; with creativity and chance, x; ordering of power and, 221; planning moment and, 7, 11–14
Kocchar, Chanda, 136
Koerner, Lisbet, 78
Koolhaas, Rem, 128
Korea Dredging Corporation, 187–88
Korea Electric Power Company, 65, 66, 68
Korea Hydropower Investigation Report (1930), 69–70
Korean Central Intelligence Agency, 65
Korean War (1950–1953), 113, 114, 115
Korea River Investigation Report (1929), 67–70
Krauss, Rosalind, 128
Kubota Yutaka, 65

labor, 10, 22, 26, 49, 105, 142n3, 199; animal, 114; child, 100, 180; exploitation, 8, 87, 88, 89, 141; Firestone Company with, 177–83; poverty and, 97–98, 100; rationalization of, 76, 80; slavery and, 10, 178, 195, 198, 214–16; wage, 21, 23–25, 100, 198, 251; women and, 33, 87–89, 108, 204–5. See also farmers
Laboratory of Medical Genetics, 213
laissez-faire economics, 78, 98
land, 23, 25, 29, 189, 217; bees retrieving stolen, 233; dispossessions, 176, 177, 181, 182, 198, 205, 232, 247, 251; Indigenous people and, 239–40; national planning policy in South Korea, 64, 66, 69–70, 72–73; rights, 137, 248–51. See also agriculture; forests
Land and Forest Allocation Program (LFAP), 248–51
Land Code (1858), 157
land parcels, Laos, 246, 249, 250
land parcels, Lebanon: Beirut, 153, 153–56; cadastral maps, 152, 153, 156–58; Haret Hreik, 153, 153, 156, 158; ownership model, 152, 156, 158; Solidere, 153–57; Waad, 153–54, 157
Landy, David, 107
Lang, Peter, 127, 131
languages, 3–6, 20, 37–39, 135; English, 40–44, 48, 176; Indigenous people, 23, 25, 30, 237, 257; violence with, 237. See also COBOL
Lao people, 250

Laos, 10, 67, 246, 249, 250. See also weeds, Laos
Latour, Bruno, 94n4, 255, 274n19
Lawrence, Halcyon, 53
Lazarsfeld, Paul, 109n1
LE (Egyptian Pound), 163, 167n16
League of Nations, 178
Lebanon. See land parcels, Lebanon
Le Corbusier, 126, 207
Lee Myung-bak, 186, 188–89, 190, 192
Lefèbvre, Henri, 126, 209, 273n14
legal evidence, EMS Sonochron as, 85, 86, 92–93
legibility, 8, 46, 53, 250, 251
Leibniz, Gottfried Wilhelm, 269
Leopold II (King of Belgium), 226
Lerrer family, 231
LFAP (Land and Forest Allocation Program), 248–51
Liberia, 183n7. See also parasites, in Liberia
Liberian English, 176
Linnaean Gardens, 196
Linnaeus, Carl, 78, 196, 220, 222
List, Friedrich, 78
livestock, xi, 162, 232, 247
Long Range Committee, COBOL and, 38, 39
Luba people, 148
luck, 91, 225
Ludden, David, 142n10
Lue people, 250, 251
Lufthansa headquarters, 87, 89–92
Lulua people, 148
Lusitania graduate seminar, 51–52
Lusitania group, 50
Lytton (Lord), 100

Maathai, Wangari, 32
macroparasites, 176
Macy Conferences, 50, 52
Madras Famine (1877–1878), 98, 99, 100
"magico-religious impersonation," 255
Magris, Alessandro, 131n1
Magris, Roberto, 131n1
Mahar caste, 57
Mai Collection, MfN, 196, 201
Making Things Public (Latour), 255
Makumbi, Kaseya Tambwe, 144, 145, 146, 148–50, 151n13
malaria, 177–80, 183
Malaysia, 170, 172

INDEX

Malaysia Agreement (1963), 173
Malthus, Thomas, 99, 100, 102n6
Manchukuo, 65, 66, 117
manuals, 4, 180, 236, 239
maps, 152, 153, 156–58, 245, 246, 247–51
Mara'ope (fish-being of place), 257, 258
Marcuse, Peter, 273n14
Martell, Craig, 47
Marx, Karl, ix
Massey, Doreen, 269
materiality, 2, 14, 15, 86, 88, 90, 265–66
material trap, EMS Sonochron as, 85, 86, 89–90, 92
Max Planck Institute for Breeding Research, Cologne, 88
Max Planck Institute for the History of Science (MPIWG), Berlin, ix, 2
McArthur, Norma, 20–24
McKinsey management, 140
McLachlan, Stéphane, 237, 238–39
Mead, Margaret, 50
measurement, Traditional Knowledge and, 239–40
"mechanical objectivity," 8
medicine, xiii, 21, 179, 180, 184n11, 238, 245
Medicine in the Tropics (film), 182–83
membership identities, 49
men's houses (*eravo*), 258, 259
mental openness, 52–53
Merewether, Francis, 96
Mertia, Sandeep, 134
Merton, Robert K., 109n1
metropole, 8–9, 56, 60, 62n2, 64, 226, 272n7
MfN (*Museum für Naturkunde*, Natural History Museum), Berlin, 196, 201, 201n3
microparasites, 176, 178
Mikisew Cree First Nation, 235–37, 240–41
Milbank Fund, 107
militant feminists, 85, 87, 90, 91, 93
military, 37, 38, 39, 54, 122n10, 153, 172
Miller, Eddington (Sir), 163, 166n14
Mills, Charles Wright, 109n1
Milwaukee Public Museum, 223
miombo (*Brachystegia*) genus of, 30
Miombo woodlands, 29
"miracle of the Han" (한강의 기적), 114, 115
"miracle of the Rhine" (*Wirtschaftwunder*), 115
miscegenation, 212–15, 217, 218
missionaries, 5, 20–22, 24, 176, 225–26, 258, 271

Mitchell, Jean, 24
Mitchell, Timothy, 11, 34, 161
mobility, 230, 233
modernism, 7, 9, 67–69, 72–73, 73n4, 126
modernity, 1, 168, 206–8, 210, 220, 237, 270
Modi, Narenda, 136
MoMA (Museum of Modern Art), 124, 125, 128
monetary economy, 22–23, 25
Monier-Williams, Monier, 5–6
monoculture, plants, 119, 195
moral failure, hunger as, 99
Morogoro Women-Focused Afforestation Project, 33
Morton, Newton, 213, 216
MPIWG (Max Planck Institute for the History of Science), Berlin, ix, 2
Msuya, Cleopa, 34
"multi-scalarity," 64
Muñoz Marín, Luis, 106
Murphy, Michelle, 179
Murray, Hubert (1861–1940), 260–62
Museum für Naturkunde (MfN, Natural History Museum), Berlin, 196, 201, 201n3
Museum of Modern Art (MoMA), 124, 125, 128
Muslim Sunni groups, 154, 155
Mussolini, Benito, 78, 79
Mysore Modern (Nair), 206

Nair, Janaki, 206
name-worshipping rituals, Orthodox, 50, 51–52
Nash, Vaughn, 96
Nasrallah, Hassan, 156
Natalini, Adolfo, 125–26, 131n1
Nathan, Robert, 115
Nathan Report (1952), 115
national identity, 207, 212, 215–17
National Institute of Construction and Cement, 76, 80
nationalism, economic, 76, 78, 82
Nationalist Party (Partido Nacionalista, PN), 106
National Museum of Australia, 254, 261
The National System of Political Economy (List), 78
Natural History Museum (*Museum für Naturkunde*, MfN), Berlin, 196, 201, 201n3
nature, ix, 12, 29, 68–69, 193

Nazis, 79, 87–88, 91, 101, 271
Neel, James, 213
Nehru, Jawaharlal, 134, 208
"New Hebridean," 23
New Hebrides (Vanuatu): census, 20–26, 268; Ni-Vanuatu, 20, 21, 25, 26n1
New Hebrides Condominium (1906–1980), 26n2
New Poor Law (1834), 97, 98
"New Village Movement" (*Saemaul Undong*) campaign, 120–21
New York, 128, 129, 219n15
NGOs, 33, 133, 136, 137, 138, 139, 143n18
Nichols, Maryland, 176
Nihon Chisso, 122n7
Nippon Kōei, 65–67, 73n3, 74n5
Nitchitsu (Nitrogenous Fertilizer Corporation), 65–66
NITI Aayog, 136
Nitrogenous Fertilizer Corporation (Nitchitsu), 65–66
Ni-Vanuatu, 20, 21, 25, 26n1. *See also* New Hebrides
Nkrumah, Kwame, 172
Noguchi, Shitagau, 122n7
non-human entities, 2, 13–14, 23, 25, 86, 271
North Korea, 66, 113–14, 116–18, 122n6, 122n14
North Vietnam, 23–24
Nowell, William, 199
Nya Kiloh (*Chromolaena odorata*, Siam weed), 245–49, 251–52, 252nn5–6
nyarong (spirits), 171
Nyerere, Julius, 31

OAS (Organization of American States), 212, 213
oil, 31–33, 81–82, 193, 235–36, 240
Olbrechts, Fran, 146
Oldenziel, Ruth, 24
The Old Man of Mikeno, 220–21, 221, 227–28
Oliver Twist (Dickens), 97
onchocerciasis (river blindness), 177, 180
OPEC, 31
OpenEducation AI hackathon, 134
open pit mining, 235
Operation Bootstrap, 106
Opus Dei, 79
orangutans, Bornean, 175n21; colonial experimentation and, 169, 170, 171, 268; decolonization and, 168–70, 172–74; hunger and, 171; modernity for, 168

Organization of American States (OAS), 212, 213
ownership model, of landscape, 152, 156, 158

Pacific Bechtel, 122n6
Pacific Islanders, 20–21, 23–24
Pacific Islands, 20, 24–25
PAHs (polycyclic aromatic hydrocarbons), 236–37
Pakistan, 56
Palestine/Israel: beehive thefts, 230–33, 234n8; with mobility, 230, 233
Der Palmenmann, 199
Papua (British) New Guinea. *See* zoomorphic wickerwork figure, Australian Administered British New Guinea
Papuan Official Collection, Australia with, 254, 257, 261–62
Parable of the Sower (Butler), 203n22
parasites, in Liberia: biomedical research and, 178–80; Firestone Company and, 176–83, 267; Firestone Company as, 176–77, 181–83; tropical diseases and, 176–81, 183, 184n10
Park Chung-Hee, 65–67, 72–73, 119, 122n9, 123n23, 187
parlement of place, Elema peoples with, 255–57, 260, 261
Parliamentary Research Service, 138
Partido Nacionalista (PN, Nationalist Party), 106
Partido Popular Democrático (PPD, Popular Democratic Party), 106, 108
Patton, Adell, 183n7
PDQ ("Pretty Damn Quick," Short Range) Committee, 38–41, 43, 44
Peace-Athabasca Delta, 235–36, 241
Pende arts, 146, 150
Pende people, 144, 145, 146, 148–50, 151n13
Pentagon, 37. *See also* COBOL
people-place, ontologies of, 255
Pérez del Pulgar, José Agustín, 79, 80
Perrine, Henry, 198
Perso-Arabic alphabet, 5
Der Pflanzer (The Planter) (journal), 199
Phillips, Charles A., 38, 39
Philosophy of Structures (Torroja), 80
pigs, 22, 25–26, 78, 108, 257, 259, 262
Pinson, Daniel, 13
Pit Arens, 82
PL480 ("Food for Peace" program), 117–18

planning, ix–xi, 3–4, 12, 42, 126, 134–35, 196, 269; colonialism and, 1–2, 7, 8–11, 14; family, 11, 106–7; guide for users, 15–16; South Korea and, 64, 66–70, 72–73; state, 9, 13, 77–78, 98, 208, 250, 252, 266; theory, 13–14, 270. *See also* colonial planning; planning moment; postcolonial planning

Planning Commission, India, 134–36

Planning Matter (Beauregard), 14–15

planning moment, 2, 25, 85–86, 93, 262; enculturating and differentiating, 267–69, 272n7; infrastructuralizing, 266–67, 272n5; spatializing, 269–70; temporalizing, 270–72; warp threads and, 6–15, 265

Planning Theory (journal), 2

plantations, 176, 182, 198, 208, 249, 251, 253n24; Firestone Company, 177–81, 183; Vietnamese laborers, 23–24

The Planter (*Der Pflanzer*) (journal), 199

plants, 10, 119, 195, 198, 202n16, 245; botany, 178, 196–97, 200–201; seedlings, 32–34, 177. *See also* forests; seeds; weeds

plasmoquine, 179, 180

PN (Partido Nacionalista, Nationalist Party), 106

"Points of Greatest Controversy" (Sammet), 40

Poli, Alessandro, 131n1

police, 85, 89–92, 231

political economy, ix, 82, 98, 102, 135, 139, 160; autarkic, 76, 78–79, 81; Malthus and, 99, 100, 102n6

politics, 173, 267; decolonization, 8, 60; geopolitics, 24, 205, 207–8, 225

politique esthétique, 146–50

polycyclic aromatic hydrocarbons (PAHs), 236–37

Popular Democratic Party (Partido Popular Democrático, PPD), 106, 108

populations, 99, 105–6, 109, 212–18. *See also* census

Porter, Libby, 10

Porter, Theodore, 8

Portuguese settlers, in Brazil, 215

postcolonial computing studies, 46, 50

postcolonialism, x, 2, 11, 25, 32, 149, 160, 173; capitalism and, 143n18; colonial spaces and, 206, 270; from five-year plans to managed speculation, 134–36; governance, 172; museums, 150; speculation and hackathons, 141; worldwide, 56–57, 64–67, 85, 89, 93, 161, 165, 205–10

postcolonial planning, 14, 136; colonial and, 4, 265–66, 269; concepts in, 16; enculturating and differentiating, 268–69; infrastructuralizing and, 266; knowledge structure and, 1, 9

poverty, 24, 31, 101, 105, 207, 217, 247; alleviation, 30, 154, 209, 251; labor and, 97–98, 100; violence and, 108; women and, 107

power, 2, 7, 64–65, 130, 207, 221, 273n14; architecture and, 124, 126, 128; grids and, 125, 128, 129; hegemonic structures of, 13; hydropower investigation reports, 66–71; imperial, 224–26; inequality and, 252; territory and state, 209

PPD (Partido Popular Democrático, Popular Democratic Party), 106, 108

"The Practical Reasons of Weeds in Indonesia" (Dove), 252n5

Praet, Lina, 147

Prahalad, C. K., 135

prefabrication, of national materials, 76, 80

Presbyterian missions, 21

"Pretty Damn Quick" (PDQ, Short Range) Committee, 38–41, 43, 44

primitiveness, 237, 254, 255

"Principles of Planology" (De Casseres), 3

private property, 155, 157–58, 182, 250

privatization, of forests, 33

problem solving, computing and, 47–49

production, xii, 22, 117–18, 121, 139, 199; charcoal, 30, 32–35; honey, 230, 231, 233. *See also* knowledge production; reproduction

protectionism, 77, 78

Public Health Service, US, 213

Puerto Rico, 106, 108, 110n17. *See also* fertility survey workforce, in Puerto Rico

Qaqqaq, Mumilaaq, 243n15

Quran, 155

race, 9, 20, 173, 206, 268; inter-racial relationships, 204–5, 210; miscegenation and, 212–15, 217, 218

racism, 49, 89, 98, 147, 180, 218, 238

rainforest, 177, 183

Rana, Aziz, 48

Ratio Club, 50, 52

RCA, 41–42
real estate, 49, 153–56
Reasoner Brothers Royal Plant Nursery, Florida, 197–98
The Red Zora and Her Gang (Held), 94n5
Red Zora (Rote Zora), 85, 87–93, 94n5
Rees, J. D., 97
refugees, 87, 98, 209, 258
Reitsch, Hannah, 172
Remington Rand, 38, 43
repatriation, 23–24, 201
reproduction, 22, 105, 107, 199–200, 238
"Republic of Korea Hydropower Investigation Report" (Korea Electric Power Company) (1962), 66, 68
residential schools, 237–38, 243n15
resistance, 11, 162–63, 178–79, 198, 251–52
Rethinking Capitalist Development (Sanyal), 143n18
Revolutionary Cells (RZ, Revolutionäre Zellen), 87
Reynolds, Lloyd, 106
Rhee, Syngman, 65, 66
rice, 101, 113, 117–19, 245, 247, 250–51, 252n9
Rice, Justus B., 179, 180
river blindness (onchocerciasis), 177, 180
roads, grids, 207
Rockefeller Foundation, 107, 179
ROK. *See* South Korea
Roman alphabet, 4–6
Roosevelt, Franklin, 105
Roosevelt, Theodore, 224
Rostow, Walt, 122n8
Rote Zora (Red Zora), 85, 87–93, 94n5
Roy, Srirupa, 208
Royal Botanic Gardens, Kew, 197, 199
Royal Museum for Central Africa, 146
Royce, Josiah, 50, 51
rubber, 79, 176–78, 181–82, 251, 267
rural economy, 102n3, 114
"rush construction," 189
Russia, 50, 135, 157, 160
RZ (Revolutionary Cells, Revolutionäre Zellen), 87

Saavedra, Angelina, 107
Saemaul Undong ("New Village Movement") campaign, 120–21
Said, Edward, 269
Salazar, António de Oliveira, 78
Sammet, Jean, 37, 39, 40, 42

Sandercock, Leonie, 13
Sanyal, Kalyan, 135, 143n18
Saraiva, Tiago, 78
Sarawak Museum of Natural History and Ethnology, 168, 170, 172–73, 175n6
Sarkar, Sreela, 53
Satō Tokihiko, 65
schistosomiasis, 178, 180
Schumpeter, Josef, 140
science, ix, 1–2, 25, 50–51, 94n4, 237–38; agricultural, 118–19; with artifacts collected, 222–23; computer, 37, 47; of planning, 3, 196; social sciences, 8, 104–6, 109, 109n1; STS, 4, 14–15, 179; technology and, x, xii, xiii, 14–15, 30, 79
Science (journal), 186
Science and Technology Studies (STS), 4, 14–15, 179
Scott, James, 7, 9, 42, 73n4, 273n8
SDF (Social Democratic Federation), 97
Second Anglo-Afghan War (1878–1880), 97, 102n3
Second Republic of Korea (1960–1961), 65
second-wave feminists, 22
"Secret Returns to Secretary of State on (1) Political Feeling (Local). (2) Communism (Local). 1948–53.," 24
seedlings, 32–34, 177
seeds, 1, 88, 137, 203n22, 246, 265, 271
seeds, German East Africa: colonialism, 195–201, 202n14, 202n16; exchange, 196, 197; Mai Collection, 196, 201; sisal agave, 195–99, 201
Seeing Like a State (Scott), 42
Seeuws, Nestor, 144, 146, 148, 149, 150
Sellow, Friedrich, 202n4
Semina Selecta (Linnaean Gardens), 196
Seminole Wars, 198
Sennett, Richard, 128
Sepilok Orangutan Rehabilitation Center, 173
Serres, Michel, 86, 94n4
sertanejo (hinterlands inhabitants), 217
settler-colonialism, 10, 52, 128, 233, 236–40, 269
settler economy, 231
settlers, 21, 23–25, 198, 215, 230–32
A Seven-Year History of Dredging (Korea Dredging Corporation), 187, 188
sex tourism, 87
sexual desire, 205

sexuality, 105, 108
sex workers, 153
Short Range (PDQ) Committee, 38–41, 43, 44
Siam, 253n14. *See also* Thailand
Siam weed (*Nya Kiloh*, *Chromolaena odorata*), 245–49, 251–52, 252nn5–6
Sibley, James, 177
Sierra Leone, 177
Silicon Valley, 47, 130, 134
silk, xii, 204
Simpson, C. L., 181
sisal agave (*Agave sisalana*), 195–99, 201
sisal hemp, 197
Sivasundaram, Sujit, 25
slavery, 10, 178, 195, 198, 214–16
Sloan, Hans, 224
"smart cities" planning, 134
Smith, Adam, 99
Smith, Howard F., 179
Smith, Wally, 141n1
Smithsonian Institution, 224
Social Democratic Federation (SDF), 97
social reproduction, 22
social sciences, 8, 104–6, 109, 109n1
sociomateriality, 14, 46, 48, 86–94, 94n4
Soja, Edward, 273n14
Solidere, 153–57
Song (960–1279) China, x–xi
Sonjasdotter, Åsa, 196, 199, 200
Southborough (Lord), 63n4
"Southern" forms of planning, 14
Southern Rhodesia/Zimbabwe, 161
South Korea (ROK), 71, 73n3, 88–89, 115, 121n4, 122n10, 122n13; agriculture and, 64, 66, 113, 116, 118–19; colonialism and, 64–66, 114–18; comprehensive national land planning policy, 64, 66, 69–70, 72–73; economy, 72, 121, 187–89; farmers in, 113–14, 117, 119; industrialization, 65, 66, 67, 70, 74n21, 117; North Korea and, 66, 113–14, 116–18, 122n6, 122n14; Park Chung-Hee and, 65–67, 72–73, 119, 122n9, 123n23, 187; postcolonialism and, 64–67; trade and, 65, 113, 120. *See also* dams, in South Korea; dredging, in South Korea; fertilizer, South Korea and
South Pacific Commission, 24
South Vietnam, 67, 120
sovereignty, 31, 81, 179, 181, 209, 248
Soviet Union, 51–52, 73n3, 116, 202n4, 207

Soyanggang River, 67–68, 72, 74n21
space, 269, 273n14
Spain, 178, 197, 267. *See also* dodecahedral silo, in Spain
Spanish Civil War (1936–1939), 78–79
spatializing, 266, 269–70
spatial planning, with grids, 11, 42
spatiotemporality, 15, 86, 91, 93
Sperling, Jan Bodo, 204
Sperry Rand, 39, 41
spirits (*nyarong*), 171
"Städtebau," 3
Stalin, Joseph, 79
standardization, 4, 38
starvation, 97, 101, 171, 237
state forests, 248–49, 251
state planners, 76, 81, 138, 187
state planning, 9, 13, 77–78, 98, 208, 250, 252, 266
Steam Assisted Gravity Drainage, 235
steel, 38, 120, 270. *See also* Hindustan Steel Plant, India
sterilizations, forced, 238
Stockholm Conference (1972), 30
Stoler, Ann Laura, 11
"story telling," 12
Strobl, Ingrid, 90–92, 95n21
Strong, Richard P., 177–78, 181
Strother, Zoë, 149–50
structural violence, 87
STS (Society for Science), 4, 14–15, 179
student movement (1968), 85, 87, 93
Sturm, Sean, 15
Stycos, Joseph, 107–8, 109
subsistence, 22–26, 29, 199
Sudan, 167n16; cotton complex and, 162–63; decolonization and, 160–61; Five-Year Plan, 163–64; national economy and, 161–65, 268
Sudan African National Union, 163
Sudan Political Service, 166n14
sugar, cotton yarn, wheat flour ("three whites"), 113, 116, 117, 122n13
suicide, 101, 243n15
Summer Schools on the Scientific Method at Harvard (1913-1915), 50, 51
Superstudio, 125, 131n1; *The Continuous Monument*, 127, 128; grids and, 124–31; *Supersurface*, 124, 125, 127–31
Supersurface (film), 124, 125, 127–31
Sup'ung Dam, 65

surnames, Brazil: adoption, 213–14, 217; miscegenation and, 212–15, 217, 218; national identity and, 215–17; OAS and, 212, 213; population genetics and, 212–18; racial ideology and, 268
Svalbard Global Seed Vault, 195, 200
Sweden, 78
swidden cultivation, 246–47, 249, 251, 252nn8–9, 253n24
Sylvania, 37
Syncrude, 235
Syria, 98, 157, 158

tailings ponds, 235–36
Taiwan, 66–67, 73n3, 116, 119, 121n4
"take-off," economic, 114, 122n8
Tanzania, 197. *See also* Dar es Salaam, Tanzania
taxidermy, 220–23, 225, 228
taxonomer, United States of America: Akeley, Carl, and, 220–28, 221; AMNH and, 224–27; colonialism and, 221–22, 227–28; life and categorization in modern museums, 223–27
taxonomy, ix, 13, 169, 197
Taylorism, 80
Technical Institute for Construction and Cement, 80
technology, 1–2, 31, 34, 206–7, 232, 270; demos in information, 133–34, 138–39, 141n1; science and, x, xii, xiii, 14–15, 30, 79
technoscience, 14–15
Temple, Richard (Sir), 100–101
temporalizing, 266, 270–72
terra nullius, 10, 205, 208
terrorism, 15, 85, 87–92, 95n21
Thailand (Siam), 120, 248, 253n14
Third World, 30–32, 88, 104, 105, 172, 213
"three-track" methodology, knowledge, 238–39, 241–42
"three whites" (sugar, cotton yarn, wheat flour), 113, 116, 117, 122n13
Thriving Chosen, 117
Tilak, Bal Gangadhar (1856–1920), 57
time, 94n4, 195, 271; spatiotemporality and, 15, 86, 91, 93; temporalizing, 266, 270–72
time fuse, 85–89, 91, 93, 95n21. *See also* EMES Sonochron, FRG and
Timmermans, Paul, 148–49
Tk'emlúps te Secwépemc First Nation, 237
Tongil rice, 118–19

Toraldo di Francia, Cristiano, 131n1
Torroja, Eduardo, 80, 81
"total planability," 88
tourism, 34, 87
town planning, 3, 126, 134
Town Planning Review, 3
toxicologists, 236, 237, 239
trade, 22, 30, 65, 78, 113, 120, 254; COVID and, 77; Elema peoples and, 261
Traditional Ecological Knowledge, 236–41
trafficking, in women, 87, 89
Treaty 8 Territory. *See* water samples, in Treaty 8 Territory
trees, 30, 198, 253n24; planting, 32, 34; rubber, 177, 182, 251
tropical diseases, 176–81, 183, 184n10
Truman, Harry S., 79
truth, 1, 9, 52, 240
Truth and Reconciliation Commission, 237
Tsing, Anna, 247
Tumin, Melvin, 106

UN (United Nations), 115, 122n6, 122n14
United Nations Korea Construction Agency (UNKRA), 115
United States (US), 65, 78, 105, 119, 140, 178–79, 183, 184n10, 219n15; Congress, 110n7; with economic aid, 113, 117–18, 120; fertility research in, 106–7; grids, 124–31, 125; imperialism, 85; LE to dollar, 167n16; military, 37, 38, 39, 122n10; with non-US researchers, 109; with nuclear energy, 82; Public Health Service, 213; in South Korea, 115, 122n10; Vietnam and, 73n3. *See also* COBOL; taxonomy, Chicago
United States Military Government in Korea (USAMGIK), 122n10
universalist knowledge, 4, 12
universal plain English, 40–42
University of Puerto Rico (UPR), 105–9
UNKRA (United Nations Korea Construction Agency), 115
Unlearning the Colonial Cultures of Planning (Porter, L.), 10
The Untouchables (Ambedkar), 62n3
US. *See* United States
USAMGIK (United States Military Government in Korea), 122n10
user's guide, planning, 15–16
US Steel, 38

INDEX 329

Vanuatu. *See* New Hebrides
Verly, Robert, 147–50
Vidler, Anthony, 126–27
Vienna Circle, 50, 52
Vietnam, 23–24, 67, 73n3, 120
Vietnam War (1965–1973), 113, 119, 120, 188
violence, xiii, 87, 108, 193; beekeepers and, 231; colonial, 169, 195, 205, 237, 238; with erasure of prior existence, 208, 209
Visual Instruction Committee, Colonial Office, 102n2

Waad, 153–54, 157
wage labor, 21, 23–25, 100, 198, 251
Wallace, Alfred Russel, 170–72, 175n7
Wallerstein, Immanuel, 104, 109, 109n1
Wang Anshi (1021–1086), xi
waqfs (Islamic charitable endowments), 154–56, 158
warp threads, 6–15, 265
"Watchman and Wesleyan Advertiser" (newspaper), 5
water resources planning, 67–68, 73. *See also* dams; dredging, in South Korea
water samples, in Treaty 8 Territory, 269, 270; cancer rates, 237; Indigenous communities and, 235–42; oil industry and, 235, 236, 240; Peace-Athabasca Delta, 235–36, 241; "three-track" methodology, 238–39, 241–42
"weediness," 247
weeds, 15, 186, 203n22, 271
weeds, Laos: LFAP, 248–51; maps and, 245, 246, 247–51; Siam, 245–49, 251–52, 252nn5–6; swidden cultivation and, 246–47, 249, 251, 253n24
West Germany. *See* Federal Republic of Germany
Wharton, Clifford, 184n10
wheat flour, cotton yarn, sugar ("three whites"), 113, 116, 117, 122n13

White Fathers, 226
Whitman, Loring, 178
WHO (World Health Organization), 213
Wiener, Norbert, 51
wildlife sanctuary, 227
Wilhelm II (Emperor), 201n3
Williams, Francis Edgar, 258–59
Winichakul, Thongchai, 248
Wirtschaftwunder ("miracle of the Rhine"), 115
Wittgenstein, Ludwig, 50, 52
Wolfe, Tom, 126
women, 32–33, 62n8, 87–89, 107–8, 139, 204–5, 210
wood, 31–33, 35n5
woodlands, Miombo, 29
workshops, 4, 47, 136–37, 144–50, 267
World Bank, 134, 135
World Economic Forum (Davos) (2011), 136
World Health Organization (WHO), 213
World Systems Theory, 104
World War I, 50, 78, 167n16, 199, 200
World War II, 23, 51, 65, 79, 160, 172, 274n17
writing, expository, 15, 16
Writings and Speeches (Ambedkar), 63n13

yams, 22, 25, 26, 257
Yaxley, John Francis, 20, 21, 24
yellow fever, 176–79, 184n10

Zaire, 144, 145. *See also* Belgian Congo
Zaramo people, 29–30
Zhu Xi (1130–1200), x–xi
zoomorphic wickerwork figure, Australian Administered British New Guinea: Australian plan, 256, 260–62; colonialism and, 255–62; Elema plan, 257–60; Elema society, 255–62; historical and political moments through, 262; Papuan Official Collection, 254, 257, 261–62; parlement of place, 255–57, 260, 261